ACID DEPOSITION
Atmospheric Processes in Eastern North America
A Review of Current Scientific Understanding

Committee on Atmospheric Transport and
Chemical Transformation in Acid Precipitation

Environmental Studies Board

Commission on Physical Sciences,
Mathematics, and Resources

National Research Council

NATIONAL ACADEMY PRESS
Washington, D.C.   1983

The National Research Council was established by the National Academy of Sciences in 1916 to associate the broad community of science and technology with the Academy's purposes of furthering knowledge and of advising the federal government. The Council operates in accordance with general policies determined by the Academy under the authority of its congressional charter of 1863, which establishes the Academy as a private, nonprofit, self-governing membership corporation. The Council has become the principal operating agency of both the National Academy of Sciences and the National Academy of Engineering in the conduct of their services to the government, the public, and the scientific and engineering communities. It is administered jointly by both Academies and the Institute of Medicine. The National Academy of Engineering and the Institute of Medicine were established in 1964 and 1970, respectively, under the charter of the National Academy of Sciences.

Library of Congress Catalog Card Number 83-61851

International Standard Book Number 0-309-03389-6

Available from

NATIONAL ACADEMY PRESS
2101 Constitution Avenue, N.W.
Washington, D.C. 20418

Printed in the United States of America

NATIONAL RESEARCH COUNCIL

2101 CONSTITUTION AVENUE   WASHINGTON, D. C. 20418

This report was undertaken by the National Academies, funded in part by a consortium of foundations*, because of the importance that we attach to illuminating the critical issues concerning acid rain. The report summarizes current scientific understanding about what is known and what is not known about relations between emissions of acid forming precursor gases and acid deposition. It should be of considerable interest to policy makers involved in regulatory decision making and in setting research priorities.

Those charged with private and public policy decisions will be particularly interested in the study Committee's evaluation of currently available scientific evidence as it affects the issue of linearity or non-linearity in relations between emissions and deposition. The public policy implications of this relationship are of major economic and political significance.

The Committee devotes a great deal of attention to discussion of areas of uncertainty that are often the result of incomplete data. But despite the uncertainties, and even though there is as yet no complete understanding of the multiplicity of complex chemical reactions and transport mechanisms, the Committee concludes that there is no evidence that the relationship between emissions and deposition in northeastern North America is substantially non-linear when averaged over a period of a year and over dimensions of the order of a million square kilometers. It is the Committee's judgment that if the emissions of sulfur dioxide from all sources in this region were reduced by the same fraction, the result would be a corresponding fractional reduction in deposition. The data reviewed by the Committee support this conclusion only in northeastern North America. The judgment is based on several pieces of evidence.

----
*Carnegie Corporation of New York
 Charles E. Culpeper Foundation, Inc.
 The William and Flora Hewlett Foundation
 The John D. and Catherine T. MacArthur Foundation
 The Andrew W. Mellon Foundation
 The Rockefeller Foundation
 The Alfred P. Sloan Foundation

One is the well documented observation that the molar ratios of sulfates to nitrates deposited at many observation points in this region are nearly constant, and are similar to the molar ratios of the integrated emissions. The Committee believes that the simplest and most direct interpretation of this observation is the one they have given. A second observation that supports the Committee's judgment comes from a 13-year run of observations at a single location in North America that shows the linear relationship between deposition at this one site and overall emissions. Although these observations were made at a single geographic point, they represent the result of well designed and carefully taken measurements. Finally, the judgment draws upon the results of recent laboratory studies that are discussed in the report.

The Academies' process of selecting a diversified, expert study Committee and subjecting the Committee's report to a substantive review was carefully followed in the preparation of this document. In fact, because of the importance of this particular question, the review procedure took a longer time than usual and involved a larger number of critical readers than is customary.

I believe that we could not have found a more expert panel or have subjected this report to a more searching review. I hope that this report will help inform the public policy decision making process and will also permit the design of research programs that will result in more complete scientific data leading to more definitive conclusions and recommendations in the future.

Sincerely,

Frank Press
Chairman

Committee on Atmospheric Transport and Chemical Transformation in Acid Precipitation

JACK CALVERT, National Center for Atmospheric Research, Chairman
JAMES N. GALLOWAY, University of Virginia
JEREMY M. HALES, Battelle Pacific Northwest Laboratories
GEORGE M. HIDY, Environmental Research & Technology, Inc.
JAY JACOBSON, Boyce Thompson Institute
ALLAN LAZRUS, National Center for Atmospheric Research
JOHN MILLER, National Oceanic and Atmospheric Administration
VOLKER MOHNEN, State University of New York, Albany

MYRON F. UMAN, Staff Officer

Environmental Studies Board

Commission on Physical Sciences,
Mathematics, and Resources

Preface

In 1981 the National Research Council issued a report titled Atmosphere-Biosphere Interactions: Toward a Better Understanding of the Consequences of Fossil Fuel Combustion. The report focused on the effects on living systems of atmospheric pollutants associated with energy production. One chapter in the report described in detail the state of knowledge concerning the effects of acid precipitation on the biosphere. Based on its survey of field data in sensitive areas and the results of experiments, the report concluded that with respect to sensitive freshwater ecosystems, "It is desirable to have precipitation with pH values no lower than 4.6 to 4.7 throughout such areas, the value at which rates of degradation are detectable by current survey methods. . . . In the most seriously affected areas (average precipitation pH of 4.1 to 4.2), this would mean a reduction of 50 percent in deposited hydrogen ions." However, a change in pH, which is measured on a logarithmic scale, from 4.1 to 4.6 corresponds to a threefold decrease in deposited hydrogen ions rather than a twofold decrease.

The conclusion was misinterpreted in the press and by others (see, for example, Science 214:38 October 2, 1981, and page A26 of The Washington Post for October 16, 1981) as a recommendation for a 50 percent reduction in emissions of the pollutant gases, sulfur dioxide and the oxides of nitrogen, that are precursors to acid precipitation. While the report concluded that the desired reduction in the deposition of hydrogen ions implies a reduction in emissions, it did not indicate how much additional control of emissions would be required to meet the goal.

As a consequence, the Chairman of the National Research Council asked the Environmental Studies Board to review the current state of knowledge about atmospheric processes that link emissions to deposition with the purpose of describing to the extent possible the consequences of the goal of reduced deposition for emissions. By how much would emissions have to be reduced to reduce the deposition of hydrogen ions by 50 percent in sensitive areas? Conversely, by how much would depositions be reduced if there were specific reductions in emissions?

Our committee was organized in January 1982 under the auspices of the board to address this question. The work of the commitee was supported by funds provided by the National Academy of Sciences.

We conducted the study by reviewing the current literature and consulting with a number of colleagues who are knowledgeable in one aspect or another of atmospheric science related to acid precipitation. To assure an up-to-date assessment, we did not restrict our review to the literature that had already been published but included where appropriate materials that had been accepted for publication in the peer-reviewed literature although not yet actually published. Although initial drafts of various sections of our report were prepared by individuals, the document was extensively reviewed and revised by the entire committee. The report represents the collective views of the members. The appendixes were prepared by individuals to provide more detailed technical discussions than seemed warranted in the report.

We are grateful to our colleagues for their enthusiastic cooperation when we asked for their data, analyses, and views. In particular, we want to express our thanks to Perry Sampson, who modified his computer model of deposition chemistry at our request. The results are described in Chapter 3. We are also grateful to Paul Altschuller and Rick Linthurst for allowing us to incorporate material into Appendix C that originally was prepared at their request for a Critical Assessment Document on Acidic Deposition (being prepared by them under contract to the U.S. Environmental Protection Agency) and to Bruce Hicks and Jake Hales, who wrote that material. Noor Gillania and D.E. Patterson also graciously permitted John Miller to prepare Appendix B based on work they originally prepared for the Critical Assessment Document. William Stockwell carried out numerous simulations of the complex chemistry of the atmosphere, and Brian Heikes assisted with our analysis of observational data. Ginger Caldwell helped in our review of statistical methods of analysis.

We also want to thank the members of the NRC staff who assisted in our work. Myron F. Uman, Janis Friedman, Kate Nesbet, and Janet Stoll provided staff support. Editorial assistance was provided by Robert C. Rooney, Jacqueline Boraks, Christine McShane, and Roseanne Price. The manuscript was processed under the capable and patient leadership of Estelle Miller, and the graphics were prepared by the Design and Production Department of the National Academy Press under the direction of James Gormley.

Finally, it is a pleasure to acknowledge the enthusiasm, dedication, and cooperation of my colleagues on the committee. They assumed responsibility for a difficult analysis in a highly charged political atmosphere, and they carried out that responsibility with distinction.

Jack Calvert, Chairman
Committee on Atmospheric
Transport and Chemical
Transformation in Acid
Precipitation

Contents

ACID DEPOSITION
Atmospheric Processes in Eastern North America
*A Review of Current
Scientific Understanding*

Summary

During the past 25 years in Europe and the past 10 years
in North America, scientific evidence has accumulated
suggesting that air pollution resulting from emissions of
oxides of sulfur and nitrogen may have significant adverse
effects on ecosystems even when the pollutants or their
reaction products are deposited from the air in locations
remote from the major sources of the pollution. Some
constituents of air pollution are acids or become acidic
when they reach the Earth's surface and interact with
water, soil, or plant life. Several studies have documented the potentially harmful effects of the deposition
of acids on ecosystems, which are of particular concern
in areas with low geochemical capacities for neutralizing
the acidic inputs (such as parts of the northeastern
quadrant of North America, the Appalachian Mountains, and
some of the mountainous areas of western North America).
Although the pollutants may be deposited in dry form or
in rain, snow, or fog, the deposition phenomenon is often
called acid rain or acid precipitation. In this report
we use the term acid deposition to encompass both wet and
dry processes.

The question of what, if anything, to do about acid
deposition is a complex one, involving generation and
interpretation of scientific evidence, assessment of
risks, costs, and benefits, and political considerations,
both domestic and international. This report deals with
a small, but important, part of the analysis that currently is being conducted to answer the question--the
scientific evidence concerning the relationships between
emissions of acid-forming precursor gases and deposition
of potentially harmful pollutants. Our purpose is to
assess the current state of scientific information that
can be marshaled to describe those relationships in the

1

hope that our assessment will be valuable to decision makers in government and in the private sector. We focus on conditions in portions of eastern North America, for which more information is available than for other areas of the continent.

The central issues of concern in this report are the adequacy of current scientific understanding about the relationships between emissions and deposition, the extent to which the relationships are strongly nonlinear, and the extent to which distant sources contribute to deposition in ecologically sensitive, remote areas. We have reviewed the available scientific evidence that pertains to the issues of nonlinearity in the relationships between emissions and deposition and long-range transport. In the report we describe the current state of understanding about atmospheric processes (Chapter 2 and Appendixes A, B, and C), review the development of theoretical models (Chapter 3), and analyze the available observational evidence for source-receptor relationships (Chapter 4). Much remains to be learned about the detailed mechanisms involved and their relative importance for the relationships between emissions and deposition. As scientists, our training leads us to be concerned about the current limits of our understanding of the relevant processes and the uncertainties associated with assessing cause and effect. Much of our report, therefore, has been devoted to exploring the areas of uncertainty in understanding of the phenomena. Continuing research on acid deposition is needed to resolve or reduce the uncertainties and thereby to provide information useful in making more informed public-policy decisions regarding acid deposition (Chapter 5).

Our findings and conclusions are summarized below.

STATUS OF SCIENTIFIC KNOWLEDGE

Current scientific understanding of the relationships between emissions of precursor gases, such as sulfur dioxide (SO_2) and the oxides of nitrogen (NO_x), and deposition of acids or acid-forming substances, such as sulfuric acid (H_2SO_4), nitric acid (HNO_3), the anions sulfate ($SO_4^=$) and nitrate (NO_3^-), and the cation ammonium (NH_4^+), is based on theoretical considerations, the results of modeling exercises, and analysis of observational data.

Data

Data are limited that can be used to characterize air
quality, meteorological conditions, and emissions from
which relationships between patterns of emissions and air
quality in rural areas might be discerned in North
America. Most of the historical data on air quality
describe urban conditions. The most reliable information
on rural conditions comes mainly from a single study in
the northeastern United States, the Sulfate Regional
Experiment (SURE), performed in 1977-1978. A long-term
record (18 years) of reasonably reliable data on deposition chemistry is available at only one site in North
America. Reliable data on regional precipitation
chemistry have been collected only over the past 4 or 5
years through monitoring networks set up in the United
States and Canada. There are no regional data from
observations of dry deposition.

Available data on precipitation chemistry and on
annual average ambient concentrations of SO_2, NO_x,
sulfate, ammonium, and nitric acid indicate elevated
levels of pollutants in the air and acidic substances in
precipitation over much of eastern North America. Ambient
concentrations are much higher than can be accounted for
by emissions from natural sources on a regional scale.
The geographical distributions of SO_2 and sulfate
differ somewhat: SO_2 concentrations are more localized
in the regions around major concentrations of sources,
and ambient sulfate aerosols appear to be more widely
distributed. The distributions of sulfate in precipitation are similar to those of sulfate in the ambient air.
Currently the molar concentrations of nitrate and sulfate
in precipitation are roughly comparable over much of the
eastern United States.

Ambient concentrations of air pollutants are highly
variable over time, whereas rates of emissions of the
precursor gases SO_2 and NO_x are less variable. Differences in temporal behavior are due in large measure to
the variability of meteorological conditions and, for
secondary pollutants such as sulfate and ozone, the
chemical reactivity of the atmosphere and the amount of
solar radiation. Concentrations of sulfate in both the
air and precipitation tend to reach their maximum values
in summer in the northeastern United States; seasonal
variations are less evident in the Midwest and Southeast.
In many areas, ambient SO_2 and NO_x concentrations are
highest in the winter, although no measurements of these

parameters in ecologically sensitive areas have been reported. Nitrate in precipitation shows much less seasonal variation in the northeastern United States. Both ambient sulfate concentrations and sulfate in precipitation have been shown to be statistically related to aerometric parameters. For example, variations in ambient sulfate concentrations (measured at ground level) are related to variations in SO_2 and ozone concentrations, relative humidity, winds, and ventilation, whereas sulfate in precipitation is related to winds, the types and rates of precipitation, and ambient concentrations of sulfur oxides. Ambient nitrate data are not amenable to similar analyses because of uncertainties in the analytical chemical methods.

Not all sulfates and nitrates in the air or in precipitation contribute to the acidity of the air or precipitation. Acidity in solution is a function of the concentration of hydrogen ions. Some sulfate and nitrate in the air and in precipitation may be associated with cations other than hydrogen, such as ions of calcium or ammonium. Thus the acidity of deposition is the result of influences of the variety of cations and anions that may be present and in general cannot be identified with one or two anions. However, once deposited, sulfate and nitrate associated with cations other than hydrogen, such as ammonium, may still result in acidification of eco- systems as a result of biological and chemical inter- actions in soils and water.

Meteorological Processes

One of the greatest difficulties in establishing relationships between sources of pollution and conditions in ecologically sensitive areas is that of accounting for the influences of atmospheric processes on the behavior of pollutants. These processes include the large-scale transport of air masses, atmospheric mixing near the Earth's surface, physical and chemical reactions among pollutants and naturally present species, deposition of gases and suspended particles, and cloud processes leading to precipitation. Transport, mixing, physical and chemical reactions, and cloud processes are respon- sible directly or indirectly for the distribution and rate of deposition of pollutants to the ground. Our empirical and theoretical understanding of the processes is strong in some aspects and weak in others.

Meteorological processes control the transport and dispersion of pollutants from their sources. Regions of northeastern North America that are considered to be sensitive to acid deposition are subjected to highly variable meteorological conditions, with the result that, in addition to local sources, a number of geographically widespread source regions are likely to contribute to the total deposition of acid-forming chemicals. Nonetheless, empirical analyses suggest that many of the precipitating air masses—and therefore most of the pollution-related ions dissolved in precipitation—reaching several sensitive, remote areas of the northeastern United States and southeastern Canada have their origins in upwind regions to the south and southwest. Because of the high variability in synoptic-scale meteorological phenomena affecting sensitive areas, however, all sources in eastern North America must be considered as contributing in one degree or another to the phenomenon of acid deposition. Evidence exists for long-range transport of pollutants leading to acid deposition, but the relative contributions of specific source regions to specific receptor sites currently remain unknown.

Models

Methods are available for estimating the effects of emissions of SO_2 and NO_x on regional distributions of ambient concentrations of these gases. The methods include statistical analysis of observational data and theoretical calculations using deterministic models of the chemical and physical processes involved. The methods employed to date, however, do not produce reliable estimates of spatial and temporal distributions of acidity, partly because of incomplete inventories of basic compounds in the air that neutralize some of the acidity.

The actual limitations of predictive models for calculating the effects of sources on sulfate deposition are not well defined because of deficiencies in knowledge of atmospheric processes and the lack of coherent regional data on air quality and deposition. No studies of the validity or limitations of models for nitrate concentrations have been performed. No model for wet deposition of acidity has been developed that takes account of sources and distributions of all the important ions in precipitation.

The deterministic models that are available employ by

necessity approximations to the atmospheric processes that are known or hypothesized to be important to acid deposition. All deterministic models are limited in their usefulness by the sparseness of meteorological data compared with that needed as input to project reliably the movement and mixing of air masses.

ISSUES

Applicability of Models to Decisions on Control Strategy

Results from current air-quality models applied to regional-scale processes have provided guidance on the significance of dynamic processes influencing sulfur deposition. The results of the models are qualitatively consistent with observations, thus demonstrating important temporal and spatial scales of the source-receptor relationships. Qualitatively, the models have pointed to the importance of certain geographical groupings of (SO_2) sources and the potential influence of the sources on certain receptor areas. However, current models have not provided results that enable us to have confidence in their ability to translate SO_2 emissions from specific sources or localized groupings of sources to influences on specific sensitive receptors. Little has been done in modeling to translate NO_x emissions into nitrate deposition or to link sulfate and nitrate to acid (H^+) deposition. A predictive capability that includes accounting for important cations is considered an essential requirement where long-range transport processes are involved.

Because of the simplifying assumptions made in order to develop practical, economical regional-scale air-quality models and because data are not available to validate or verify the models, workers in the field generally have only limited confidence in current results. The models and their results are useful research tools; but given the state of knowledge of the physics and chemistry of the atmosphere in the context of long-range transport in air pollution, we advise caution in using deterministic models to project changes in patterns of deposition on the basis of changes in patterns of emissions of precursor gases.

For practical purposes, deterministic models have been and will continue to be used for research on atmospheric transport and transformation chemistry, and this use may lead to improvements in our understanding of deposition of acidifying substances in eastern North America. Before models can be relied on for development of refined strategies for dealing with acid deposition that take into account the specific locations of sources and receptors, however, we will need a greatly improved base of meteorological data and more precise treatment in the models of both chemical and meteorological aspects such as vertical distributions of meteorological variables, including clouds. In the near term, we believe that collection and analysis of field data are likely to lead to improved understanding more quickly than refinement of deterministic models. In fact, such data are needed to improve the models themselves. Confidence in control strategies will be strengthened to the extent that they are founded on scientifically sound, verified models.

Laboratory evidence suggests that an alternative model of the chemical processes involved in acid deposition may be postulated that represents the gas-phase reactions leading to the oxidation of SO_2 more correctly than the model of Rodhe, Crutzen, and Vanderpol, which has been widely used for this purpose. In keeping with results of laboratory experiments, the alternative model employs a series of reactions that results in oxidation of SO_2 without net consumption of a major oxidant, the hydroxy radical. When these gas-phase reactions are incorporated into the model, the previously reported nonlinearity in the relationship between changes in ambient concentrations of SO_2 and changes in ambient concentrations of sulfate aerosol is greatly reduced (see Chapter 3).

Nonlinearity

There is, admittedly, much to be learned about the relationships between emissions and deposition. However, on the basis of analysis of currently available data in eastern North America and within the limits of uncertainty associated with errors in the data and in estimating emissions, we conclude that there is no evidence for a strong nonlinearity in the relationships between long-term average emissions and deposition. This conclusion is based on analysis of available data on historical trends (mainly at the Hubbard Brook Experimental Forest in New Hampshire), the ratios of pollutants in emissions and deposition, and comparison of the percentages of emissions of SO_2 and NO_x that are deposited as sulfate and nitrate in precipitation (see Chapter 4); it is also supported by theoretical calculations taking account of the latest results of realistic laboratory studies (see Chapter 3).

The only available direct evidence of strong non-linearity between average emissions and average deposition of sulfur compounds is found in the extensive data taken in Europe over the past 25 years. Analysis of the European data has focused on historical trends in bulk sulfate deposition and on the spatial distribution of the ratio of sulfate to nitrate in monthly bulk samples. The analysis has also employed the original Rodhe-Crutzen-Vanderpol model. The observed trends in Europe are somewhat uncertain because of changes in sampling and analytical techniques and analytical laboratories throughout the period (see Chapter 4).

Reasonably reliable historical data indicating trends in North America are available only from the Hubbard Brook site (see Chapter 4). These data, from 13 years of weekly bulk samples, show no evidence of strong nonlinearity.

Analysis of the spatial distributions of the molar ratio of SO_2 to NO_x in emissions and the molar ratio of sulfate to nitrate in precipitation provides additional though indirect evidence that there is no strong nonlinearity in the relationships between long-term average emissions and deposition in eastern North America (see Chapter 4). The molar ratio of sulfate to nitrate in precipitation does not vary substantially over a large region in eastern North America; and for annual average data, it is similar to the average molar ratio of SO_2 to NO_x in emissions. Analysis of the data from the midwestern and northeastern United States also indicates that the percentage of emitted SO_2 that is deposited as sulfate in precipitation is approximately equal to the percentage of emitted NO_x deposited as wet nitrate in that region. Since the conversion of NO_2 to HNO_3 and its subsequent incorporation into cloud water are believed to be relatively rapid and efficient processes, the data suggest that the combined gaseous and aqueous conversions of SO_2 to sulfate are similarly efficient. It is therefore improbable that the oxidation of SO_2 is sufficiently hindered by a lack of oxidant to cause a disproportionately small reduction in sulfate concentrations in precipitation as a result of a given reduction in SO_2 emissions.

The North American data therefore suggest that (a) whatever atmospheric processes are taking place, pollutants are being thoroughly mixed over a region with linear dimensions up to 1,000 km, and (b) the formation of sulfate is neither enhanced nor retarded relative to the formation of nitrate. The conclusion is clouded by three types of uncertainty: the limited amount and uncertain quality of the data, the natural variability of atmospheric processes, and the lack of firm understanding of the physical and chemical processes involved.

If improved measurements indicate that the relationships between emissions and deposition in Europe and eastern North America are different, the differences between the two regions in meteorology, or latitude, or other factors, such as the spatial distribution of sources, may be responsible.

Influences of Local and Distant Sources

Theoretical and observational evidence exists for the long-range transport of air pollutants leading to acid deposition (see Chapters 3 and 4). However, the relative importance for deposition at specific sites of long-range transport from distant sources as compared with more direct influences of local sources cannot be determined from currently available data (see Chapter 4) or reliably estimated using currently available models (see Chapter 3).

Trends in the historical data at the Hubbard Brook Experimental Forest appear to reflect general trends in emissions (see Chapter 4). Available meteorological analyses of trajectories of precipitating systems at three locations in the Northeast (Whiteface Mountain and Ithaca in New York and south central Ontario) indicate that much of the acidity in precipitation--as well as much of the precipitation--comes from air masses arriving from the South and Southwest.

Based on the analysis of spatial distributions of the annual average molar ratios of pollutants in emissions and deposition, it appears that the atmospheric processes in eastern North America lead to a thorough mixing of pollutants, making it difficult to distinguish between effects of distant and local sources (see Chapter 4).

IMPLICATIONS FOR EMISSION-CONTROL STRATEGIES

The implications of our findings and conclusions for choosing among possible emission-control strategies, should they be deemed necessary, are limited. We do not believe it is practical at this time to rely upon currently available models to distinguish among alternative strategies. In the absence of other methods, analysis of observational data provides guidance for assessing the consequences of changing SO_2 emissions for wet deposition of sulfate.

If we assume that all other factors, including meteorology, remain unchanged, the annual average concentration of sulfate in precipitation at a given site should be reduced in proportion to a reduction in SO_2 and sulfate transported to that site from a source or region of sources. If ambient concentrations of NO_x, nonmethane hydrocarbons, and basic substances (such as ammonia and calcium carbonate) remain unchanged, a reduction in sulfate deposition will result in at least as great a reduction in the deposition of hydrogen ion.

It can be stated as a rule of thumb that the farther a source is from a given receptor site, the smaller its influence on that site will be per unit mass emitted. Analysis of air-mass trajectories and modeling may provide insight into the relative contributions of subregional groupings of sources to sulfate deposition in ecologically sensitive areas. Interpretation of this information, however, is subjective, and it will entail considerable judgment in assigning zones of influence of sources, even for long-term averages. This subjectivity has been a source of major differences in expert opinion, and it will continue to be until scientific knowledge improves considerably.

On the basis of analysis of the spatial distributions of the molar ratios of pollutants in emissions and deposition and assuming that all other emissions and conditions remain unchanged, we would expect that if the molar ratio in emissions in eastern North America were changed by changing SO_2 emissions, a similar change would occur in the ratio of sulfate to nitrate in wet deposition. If, as described in Chapter 4, dry deposition is linearly proportional to emissions, then the average annual ratio in total deposition in the region should also respond to changes in the emission ratio. Because the analysis is based on spatial distributions, its applicability is limited to circumstances in which the spatial distribution of emissions is not changed. Because we cannot rely on current models or analyses of air-mass trajectories, we cannot objectively predict the consequences for deposition in ecologically sensitive areas of changing the spatial pattern of emissions in eastern North America, such as by reducing emissions in one area by a larger percentage than in other areas.

RESEARCH NEEDS

We believe that extensive laboratory, field, and modeling studies should be continued if we are to establish the physical and chemical mechanisms governing acid deposition (see Chapter 5). It appears to us, however, that useful information about the delivery of acids to ecologically sensitive areas by transport and transformation processes can be determined more quickly by direct empirical observation in the field than by other means. Although the results of such field studies may not yield complete detailed descriptions of the interactions of all the processes involved, the studies are likely to provide basic phenomenological evidence with sufficient reliability to form a basis for improving the near-term strategy for dealing with the problem of acid deposition in eastern North America. Indeed, the data are essential to enhance theoretical understanding and to develop improved deposition models. In the long term, however, the ultimate strategy for dealing with acid deposition will depend on the application of realistic, validated models.

▌ Introduction

During the past 25 years in Europe and the past 10 years in North America, scientific evidence has accumulated suggesting that air pollution resulting from emissions of hydrocarbons and oxides of sulfur and nitrogen may have significant adverse effects on ecosystems even when the pollutants or their reaction products are deposited from the air in locations remote from the major sources of the pollution (National Research Council 1981, Environment '82 Committee 1982). Some constituents of air pollution are acids or become acidic when they reach the Earth's surface and interact with water, soil, or plant life. Several studies have documented the potentially harmful effects of the deposition of acids on ecosystems (NRC 1981, National Research Council of Canada 1981, Overrein et al. 1980, Drablos and Tollen 1980). Although the pollutants may be deposited in dry form or in rain, snow, or fog, the deposition phenomenon is often called acid rain or acid precipitation. In this report we use the term acid deposition to encompass both wet and dry processes.

DEPOSITION ACIDITY

An acid is a chemical substance that, in water, provides an excess of hydrogen ions (H^+) to the solution.[1] In solutions, the electrical charges of positively charged ions (cations) balance those of negatively charged ions (anions). In acid precipitation, excess hydrogen ions are usually balanced by sulfate ($SO_4^=$), nitrate (NO_3^-), and to a lesser extent chloride (Cl^-) ions. There may, in general, be other cations in addition to H^+ present in precipitation. Organic acids, for example, are

12

found in all areas, but they are important as donors of
hydrogen ions only in remote areas where concentrations of
sulfate and nitrate are lower (Galloway et al. 1982). Acid
deposition in dry form consists of gases such as sulfur
dioxide (SO_2), nitrogen oxides (NO_x), and nitric acid
vapor (HNO_3) as well as particles containing sulfates,
nitrates, and chlorides.

Acids occur naturally in the atmosphere because, for
example, of the dissolution of carbon dioxide (CO_2) in
water or the oxidation of naturally occurring compounds of
sulfur and nitrogen. The "natural" acidity of rainwater,
measured as pH,[2] is often assumed to be pH 5.6, which is
an idealized value calculated for pure water in equilibrium
with atmospheric concentrations of CO_2. However, the
presence of other naturally occurring species, such as
SO_2, ammonia, organic compounds, and windblown dust, can
lead to "natural" values of pH between 4.9 and 6.5
(Charlson and Rodhe 1982, Galloway et al. 1982).

The presence of compounds of sulfur and nitrogen of
anthropogenic origin tends to increase the acidity (lower
the pH) of precipitation. More than half the acidity of
precipitation averaged over the globe may be due to natural
sources, but anthropogenic sources may dominate in some
regions. For example, in eastern North America (i.e., east
of the Mississippi River) 90 to 95 percent of precipitation
acidity may be the result of human activities, although
natural sources may also be important at times in specific
locations (U.S./Canada Work Group #2 1982).

Figure 1.1 shows the mean value of pH in precipitation
weighted by the amount of precipitation in the United
States and Canada in 1980. There are no known natural
causes that can account for either the distribution or the
value of acidity in eastern North America. The region of
highest acidity does, however, correspond to the regions of
heavy industrialization and urbanization along the Ohio
River Valley and the Eastern Seaboard, where anthropogenic
emissions of sulfur dioxide (Figure 1.2), nitrogen oxides
(Figure 1.3), and hydrocarbons are high. Figures 1.4 and
1.5 indicate the spatial distributions of sulfate and
nitrate, respectively, weighted by the amount of precipitation that was deposited in North America in 1980. The
data were obtained from several monitoring networks in the
United States and Canada.

Trends in acid deposition in North America have been
difficult to discern, and data with which to assess them
are sparse. Comparisons of historical data (for example,
Cogbill and Likens 1974) have been questioned because of

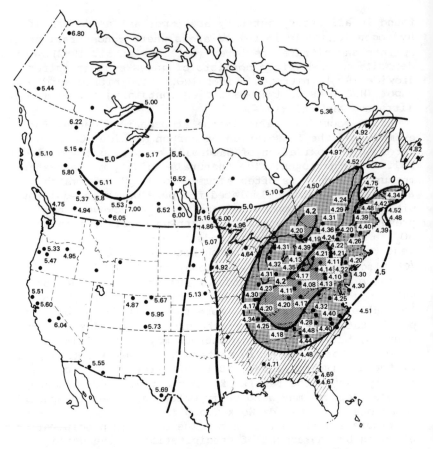

FIGURE 1.1 Annual mean value of pH in precipitation weighted by the amount of precipitation in the United States and Canada for 1980. SOURCE: U.S./Canada Work Group #2 (1982).

difficulties associated with comparing data obtained by means of different experimental methods of uncertain comparability at different sites at different times and because of difficulties in taking into account the influence of neutralizing substances on the data (Hanson and Hidy 1982, Stensland and Semonin 1982). A long-term (18-year) record of reasonably reliable data on deposition chemistry is available at only one site in North America (see Chapter 4).

The relationship between emissions and deposition in North America is complicated by changes that are not reflected in data on aggregate emissions. For example, because of concern about urban air pollution in the 1960s, there has been a tendency since then to build large new facilities away from urban centers and to use tall stacks to eject emissions at higher altitudes, hence promoting dispersal and dilution of the pollutants. Pollution control equipment installed during this period also changed the chemical and physical characteristics of the emissions, substantially reducing direct emissions of sulfates and neutralizing substances in fly ash. Thus, while total emissions of SO_2 in the United States increased between 1960 and 1970, urban concentrations of SO_2 decreased (Altshuller 1980). Almost all available data on air quality reflect urban conditions. Only recently have extensive networks of monitors been established in rural areas.

FIGURE 1.2 Representative values of SO_2 emissions in the United States and Canada in 1980 (thousands of metric tonnes). SOURCE: U.S./Canada Work Group #3B (1982).

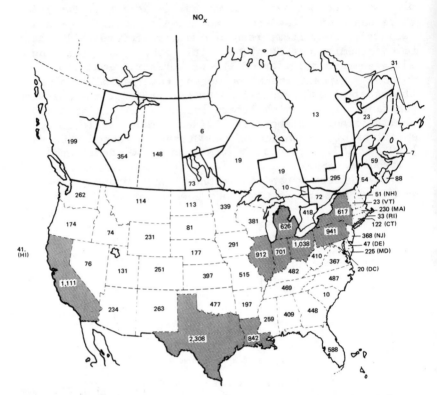

FIGURE 1.3 Representative values of NO_x emmissions in the United States and Canada in 1980 (thousands of metric tonnes). SOURCE: U.S./Canada Work Group #3B (1982).

ENVIRONMENTAL EFFECTS

Atmospheric deposition involves three components: emissions, deposition, and effects on receptors. Certain aspects of the effects of atmospheric deposition are of particular importance for the development of effective policies for emission control. They concern the significance for receptors of (1) physical and chemical states of deposited materials and (2) rates and reversibility of acidification.

In discussing these issues, it is helpful to distinguish between primary, secondary, and tertiary receptors according to their proximity to the initially deposited material. Primary receptors experience direct contact with atmospheric pollutants. Examples are the surfaces of

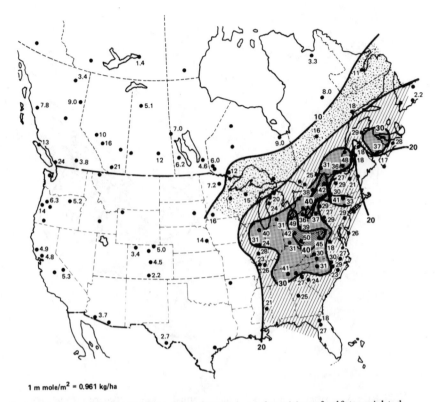

1 m mole/m² = 0.961 kg/ha

FIGURE 1.4   Spatial distribution of mean annual wet deposition of sulfate weighted by the amount of precipitation in North America in 1980 (mmoles/m²). SOURCE: U.S./Canada Work Group #2 (1982).

structures and materials, the outer foliage of vegetative canopies, and the surfaces of soils that are not protected by vegetative canopies. Secondary receptors are subject to wet and dry deposition indirectly and only after the pollutants have been in contact with other materials. Examples include the inner foliage of vegetative canopies, soil underneath vegetation, and subsurface layers of exposed soils. Tertiary receptors are even further removed from the point of initial contact with deposition. Examples are subsoil, underlying rock formations, streams and lakes that receive most of their water from runoff from the watershed, and lake and stream sediments. The rates of transfer and mixing of materials are affected by the proximity of the receptors to the point of initial deposition and by their mass and other physical and chemical properties.

18

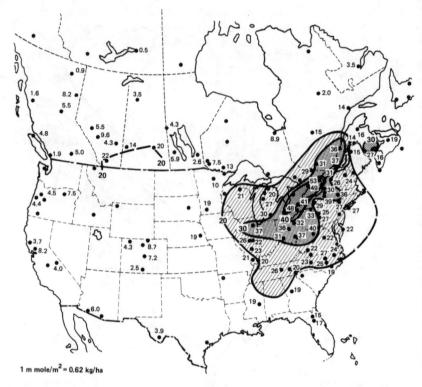

1 m mole/m² = 0.62 kg/ha

FIGURE 1.5 Spatial distribution of mean annual wet deposition of nitrate weighted
by the amount of precipitation in North America in 1980 (mmoles/m²). SOURCE:
U.S./Canada Work Group #2 (1982).

Physical and Chemical States
of Deposited Materials

The effects of atmospheric deposition on primary receptors
depend on the physical state (solid, liquid, or gaseous)
and the chemical state (e.g., sulfur or nitrogen species)
of the deposited materials (NRC 1981). The physical form
of deposited material determines its availability for
reaction, whereas its chemical form determines its reactivity. Tertiary receptors are less responsive to the
physical and chemical form of atmospheric deposition than
primary and secondary receptors because of dilution. The
effects of acid and acidifying ions (hydrogen, sulfate,
nitrate, and ammonium) are dependent in part on the
accompanying rates of deposition of neutralizing cations

(calcium). Hydrogen ions are harmful to the extent that the receptors cannot prevent or compensate for changes in acidity or the consequences of acidification.

Many chemical compounds of both sulfur and nitrogen are naturally present in soils and are involved in chemical and biological transformations in soils and vegetation. Sulfur and nitrogen are essential nutrients required for plant growth. There are many differences in the properties and biological action of the compounds of the two elements, and there are differences in the types of transformations they undergo in the environment. Biological processes (e.g., metabolic action, decomposition) have a great influence on nitrogen transformations, while both geological processes (e.g., weathering) and microbial transformations strongly affect the sulfur cycle. There is a larger pool of endogenous nitrogen than of sulfur in organisms, and larger amounts of nitrogen than sulfur are required for plant growth. The two nutrients are closely related, so that addition of one element to an ecosystem allows greater biological utilization of the other (Turner and Lambert 1980). The optimum molar ratio of sulfur to nitrogen in terrestrial ecosystems is approximately 0.03.

Nitrogen usually is efficiently metabolized in undisturbed ecosystems (Likens et al. 1977), while sulfur frequently is not retained by forest soils (Abrahamsen 1980). For aquatic ecosystems, therefore, sulfur is more important for acidification than nitrogen. Alkaline as well as acidic cations accompany the movement of sulfate from soils to aquatic systems; consequently, acidification of soil is more likely to occur from excessive sulfate deposition than from excessive nitrate deposition. The spring flush of acids into aquatic systems may, however, be closely associated with the accumulation of nitrate and sulfate in snowpack (Galloway and Dillon 1982, McLean 1981).

One of the important effects of acidification is the potential mobilization of elements from soils due to increased solubility and subsequent uptake by vegetation or movement to aquatic systems. Aluminum is present in bound form in many soils, and it can be dissolved and become available for accumulation by organisms to which it can be toxic. Dissolution of aluminum or other metals depends on the amount of water passing over a surface; solubility generally is enhanced in an acidic solution. Thus heavy rainfall exceeding surface evaporation--even with low acid content--can mobilize ions over time. This mobilization from the soil may be enhanced when acid-forming materials also are deposited from the atmosphere and washed away.

However, insufficient understanding of the interactions between soils and groundwater currently precludes estimation of the rate of mobilization of metals.

Major factors determining effects of atmospheric pollutants on forests are (1) the chemical nature and loading of deposited elements, (2) the ion exchange characteristics of soils, (3) the residence times and hydrological pathways of water through the watershed, (4) the nature and extent of existing vegetation, and (5) the geochemical activity of bedrock and soils (Evans et al. 1981, Zinke 1980). All forest ecosystems are not expected to respond to acid deposition in the same way. Effects are likely to be site-specific and dependent on the relative contributions of external and internal sources of acidity.

Major factors determining effects of atmospheric pollutants on lakes and streams are (1) the total loadings of particular compounds, (2) hydrological pathways through the terrestrial systems upstream of the water body, (3) the ion-exchange characteristics of soils in terrestrial systems upstream, (4) the residence times of water in the terrestrial systems, and (5) the geochemical reactivity of the bedrock and soils of the terrestrial systems.

Reversibility and Irreversibility

Ecosystems are repeatedly stressed by natural disasters, extreme climatic and meteorological events, and human influences such as changes in land use and pollution. Responses may be reversible or irreversible, depending on the stress, the receptor, and the time span of interest. For example, a river may carve a new channel after a flood, an effect that may be considered irreversible except by human intervention. Over a period of hundreds of years, however, the channel may fill with silt, so even this effect can be "reversed." A lake may become turbid with sediment and organic matter after a heavy rain, an effect that usually is reversed rather rapidly by natural processes. So the consequences of extreme events often are reversed by natural processes over time; as a result, considerations of the reversibility or irreversibility of effects of acid deposition should take account of the time span of interest.

The most common effect of stress on an ecosystem, such as may be caused by exposure to pollutants, is retrogression to conditions typical of an earlier stage of ecological succession. Reduction in species diversity and simplification of ecosystem structure are typical responses to pollutants (Whittaker 1975). These changes often are accompanied by alterations in productivity. When exposures to pollutants are reduced or eliminated, natural processes may return systems either to their previous pathways of succession or to different successional pathways.

The extent to which effects of acid deposition are reversible depends on the receptor and the type of effect. Although there have been no clear demonstrations of effects on terrestrial systems to date, it is reasonable to believe that adverse effects on primary receptors are more readily reversed than those on tertiary receptors. For example, the yield of one annual crop might be reduced by the contact of acidic deposition with foliage or flowers, but a subsequent crop may be less severely affected if less acid is deposited. Damage to trees and perennial plants, particularly those that retain foliage for several years (most conifers), may be less easily reversed because of the long period required for regeneration and recovery of most woody plants. When both the aquatic and the terrestrial ecosystems are acidified in an area in which rates of mineralization and decomposition of organic matter are low, reversibility is unlikely.

Acid deposition results in net accumulations of certain elements and net losses of others in ecosystems over long time scales; the identity of the elements in each category and their rates of change vary with the ecosystem and rates of deposition. The effects of slow but persistent changes may not be apparent for many generations. Signs of these changes may be observed, but the time scales for occurrence of irreversible changes are difficult to predict, because the processes that produce and consume hydrogen ions and the reactions that affect the accumulation and loss of elements are complex and poorly understood. Extensive regions of North America (Figure 1.6) and northern Europe have little geochemical acid-neutralizing capacity. Perhaps only in retrospect will we know with certainty that systems have changed, and the reversibility of these effects by natural processes might require far more time than the period initially required to cause the changes due to anthropogenic acidification.

OTHER RELATED REGIONAL AIR POLLUTION PHENOMENA

In addition to the atmospheric processes affecting acid deposition, there are other regional air pollution

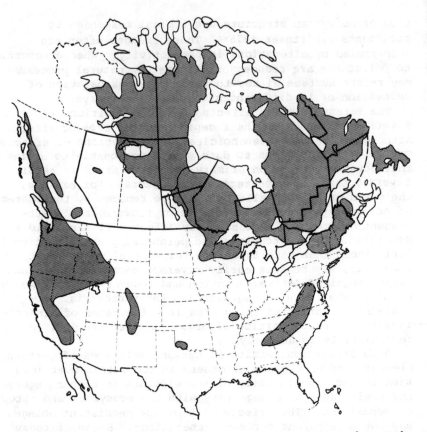

FIGURE 1.6 Regions of North America with low geochemical capacity for neutralizing acid deposition. SOURCE: Galloway and Cowling (1978).

phenomena of consequence for environmental quality that are related to acid deposition in that they are the result of similar chemical and physical processes acting on the same pollutants.

One is the occurrence of elevated concentrations of ozone (O_3) in polluted air masses extending over several hundred to a thousand kilometers (Vukovich et al. 1977, Wolff et al. 1977). Episodes of elevated ozone occur in summer under conditions that also lead to high atmospheric concentrations of sulfate aerosol, which is eventually removed by precipitation. The events are believed to be associated with increased concentrations of precursor gases, such as nitrogen oxides and hydrocarbons, that undergo reactions to form oxidants under conditions of large-scale atmospheric stagnation.

A second related phenomenon is the impairment of
visibility by optically dense haze extending over large
geographic areas (Trijonis and Shapland 1979). The
phenomenon, which has been recognized in the eastern United
States for some time, occurs during episodes of high
relative humidity. Degradation of visibility has also been
observed in the West (Macias et al. 1981, Trijonis 1979)
and in the Arctic (Rahn and McCaffrey 1980). Although the
optical characteristics of the atmosphere are linked to
natural climatic factors, such as relative humidity, Husar
and Patterson (1980) found an apparent association in the
historical record of changes in visibility with changes in
the combustion of fossil fuels. There seems little doubt
that sulfate aerosols and other fine particles play a
significant role in regional haze (NRC 1980). Haze in the
Arctic in winter has been attributed to long-range transport of air masses polluted with sulfates and particulate
carbon from sources in northern Europe (Rahn and McCaffrey
1980).

PURPOSE OF THE STUDY

The question of what, if anything, to do about acid
deposition is a complex one, involving generation and
interpretation of scientific evidence, assessment of risks,
costs, and benefits, and both domestic and international
political considerations. This report deals with a small
but important part of the analysis that is currently being
conducted to answer the question--the scientific evidence
concerning the relationships between emissions of precursor
gases and deposition of potentially harmful pollutants.
Our purpose is to assess the current state of scientific
information that can be marshaled to describe those
relationships in the hope that our assessment will be
useful to decisionmakers in government and in the private
sector.

The impetus for our work has been proposals for the
adoption of policies to control emissions of sulfur dioxide
and nitrogen oxides (beyond current limitations on
emissions from new facilities) as a means of reducing acid
deposition and hence alleviating reported and anticipated
damage from that deposition. The operators of sources of
the pollution (mostly electric utilities, industrial
boilers, and motor vehicles) reasonably wish to ensure that
the costs they--and their customers--would bear as a result
of control policies are commensurate with any benefits to

be obtained. A critical link in the evaluation of benefits is the estimation of the reduction in the deposition of acids that would accompany a reduction in emissions. A recent report by the National Research Council (NRC 1981) concluded that current rates of deposition of hydrogen ions should be reduced by about 50 percent (corresponding to an increase in pH of 0.3 unit) if sensitive regions of eastern North America are to be protected from adverse effects of the deposition. If this goal were adopted, by how much would emissions have to be reduced? Conversely, by how much would deposition rates be reduced if there were specific reductions in emission rates?

Our committee was organized to assess current scientific understanding about atmospheric processes that might be applied to answering these questions. Our objective was to determine what conclusions can be drawn from the state of knowledge late in 1982 about the relationships between emissions and deposition. Essential facts we faced in our work are that the subject under study is complex, the scientific evidence is evolving, and uncertainties in current understanding remain. Nevertheless, our goal required that we take account of both the theoretical understanding and observational evidence that are available today and make our best scientific judgment about their meaning.

If national policy on acid deposition is to be made on the basis of the scientific information currently available, that policy could take several forms, including maintaining the status quo. Other policies might incorporate uniform reductions (rollback) in emissions, might be designed to achieve the maximum possible environmental benefit, or might be carefully engineered to bring risks, costs, and benefits into optimal balance. The different options require scientific and technical information in different degrees of detail. In addition, they all, to one degree or another, must account for uncertainties in understanding. Decisions on almost all issues of public policy--including military affairs, the economy, and social welfare no less than environmental issues--are routinely made in light of uncertainties in knowledge. Provided uncertainties are taken into account, sufficient information is available for deciding what, if anything, to do about acid deposition.

Recognizing that uncertainties in scientific understanding about acid deposition currently exist and that uncertainties are likely to exist to some degree into the future, we believe that, whatever the near-term decision on acid-deposition policy, research and development should proceed with the goal of supporting more advanced and sophisticated future policies. Laboratory and field research on atmospheric processes will be extremely important in this effort. In the meantime, it seems prudent to adopt policies that are flexible and adaptable to the changing base of scientific understanding. As research continues, it can be hoped that our ability to design carefully constructed optimal strategies will continually improve.

ORGANIZATION OF THE REPORT

This report describes the state of knowledge as of the end of 1982 regarding the atmospheric processes relating emissions of precursor gases and acid deposition. It does not include a detailed examination of the effects of acidic or acidifying substances on ecosystems once deposited. For such a treatment, see NRC (1981).

Chapter 2 is a general review of the current theoretical understanding of the major atmospheric processes involved in acid deposition: transport and dispersion, chemical transformation, and deposition. More complete reviews are contained in the appendixes. Chapter 3 is a general review of the theoretical models currently used or proposed for assessing the relationships between sources and receptors. Chapter 4 reviews and analyzes field data on deposition in order to develop a phenomenological understanding of acid deposition in North America. Needed research is described in Chapter 5. The focus of the report is on conditions in portions of eastern North America, for which more information is available than for regions elsewhere on the continent.

NOTES

1. More precisely, acidity in aqueous solutions is a function of the concentration of the hydrated hydrogen ion (H_3O^+), which is also called the hydronium ion. For convenience, we adopt the conventional notation, referring to H_3O^+ as H^+. In solutions, the product of the molar concentration of H^+ with that of the hydroxide ion (OH^-) is approximately constant (about 10^{-14} at 25°C). As acid is added to water, the concentration of H^+ increases and that of OH^- decreases so that the product remains

constant. By an excess concentration of H^+, we mean
that the concentration of H^+ is greater than that of
OH^-.
2. The acidity or alkalinity of a solution is
measured on the pH scale. A solution that is neutral
(neither acidic or alkaline) has pH 7.0. Decreasing pH
indicates increasing acidity. The pH scale is logarithmic
(pH equals the negative logarithm to the base 10 of the
hydrogen ion concentration), so a solution of pH 4.0 is
10 times more acidic than one of pH 5.0.

REFERENCES

Abrahamsen, F. 1980. Acid precipitation, plant nutrients
and forest growth. Pp. 58-63 in Proceedings of the
International Conference on Ecological Impact of Acid
Precipitation, D. Drablos and A. Tollan, eds. Oslo:
SNSF Project. Norwegian Council for Scientific and
Industrial Research.
Altshuller, A.P. 1980. Seasonal and episodic trends in
sulfate concentrations (1963-1978) in the eastern
United States. Environ. Sci. Technol. 14:1337-1348.
Charlson, R.J., and H. Rodhe. 1982. Factors controlling
the acidity of natural rainwater, Nature 295:683-685.
Cogbill, C.V., and G.E. Likens. 1974. Acid precipitation
in the northeastern United States. Water Resources
Res. 10:1133-1137.
Cronan, C.S. 1980. Consequences of sulfuric acid inputs
to a forest soil. Pp. 336-343 in Atmospheric Sulfur
Deposition: Environment Impact and Health Effects,
D.S. Shriner, C.R. Richard, and S.E. Lindberg, eds.
Ann Arbor, Mich.: Ann Arbor Science Publishers.
Drablos, D., and A. Tollan, eds. 1980. Proceedings of the
International Conference on Ecological Impact of Acid
Precipitation, Sandefjord, Norway, March 11-14, 1980.
Oslo: SNSF Project. Norwegian Council for Scientific
and Industrial Research.
Environment '82 Committee. 1982. Acidification Today and
Tomorrow. Translated by S. Harper. Stockholm: Swedish
Ministry of Agriculture.
Evans, L.S., G.R. Hendry, D.J. Stensland, D.W. Johnson,
and A.J. Francis. 1980. Acid precipitation:
considerations for an air quality standard. Water,
Air, Soil Pollut. 16:469-509.
Galloway, J.N., and E.B. Cowling. 1978. The effects of
precipitation on aquatic and terrestrial ecosystems:

a proposed precipitation chemistry network. J. Air Poll. Control Assoc. 28:229-235.

Galloway, J.N., and P.J. Dillon. 1982. Effects of acidic deposition: the importance of nitrogen. Stockholm Conference on Acidification of the Environment. Stockholm: Swedish Ministry of Agriculture.

Galloway, J.N., G.E. Likens, W.C. Keene, and J.M. Miller. 1982. The composition of precipitation in remote areas of the world. J. Geophys. Res. 11:8771-8786.

Hansen, D.A., and G.M. Hidy. 1982. Review of questions regarding rain acidity data. Atmos. Environ. 15:1597-1604.

Husar, R.B., and D.E. Patterson. 1980. Regional scale air pollution: source and effects. Ann. N.Y. Acad. Sci. 338:399-417.

Last, F.T., G.E. Likens, B. Ulrich, and L. Walloe. 1980. Acid precipitation--progress and problems. Pp. 10-12 in Proceedings of the International Conference on Ecological Impact of Acid Precipitation, D. Drablos and A. Tollan, eds. Oslo: SNSF Project. Norwegian Council for Scientific Research.

Likens, G.E., F.H. Bormann, R.S. Pierce, J.S. Eaton, and N.M. Johnson. 1977. Biogeochemistry of a Forested Ecosystem. New York: Springer-Verlag.

Macias, E., J. Zwicker, and W.W. White. 1981. Regional haze case studies in the southwestern U.S. II. Source contributions. Atmos. Environ. 15:1987-1999.

McLean, R.A.N. 1981. The relative contribution of sulfuric and nitric acids in acid rain to the acidification of the ecosystem: implications for control strategies. J. Air Pollut. Control Assoc. 31:1184-1187.

National Research Council. 1980. Controlling Airborne Particles. Washington, D.C.: National Academy Press.

National Research Council. 1981. Atmosphere-Biosphere Interactions: Toward a Better Understanding of the Ecological Consequences of Fossil Fuel Combustion. Washington, D.C.: National Academy Press.

National Research Council of Canada. 1981. Acidification in the Canadian aquatic environment: scientific criteria for assessing the effects of acid deposition on aquatic ecosystems. NRCC No. 18475. Ottawa: National Research Council of Canada.

Overrein, L.H., N.M. Seip, and A. Tollan. 1980. Acid Precipitation--Effects on Forest and Fish. Final report of the SNSF Project 1972-1980. Oslo: Norwegian Council for Scientific and Industrial Research.

Rahn, K., and R. McCaffrey. 1980. On the origin and transport of the winter arctic aerosol. Ann. N.Y. Acad. Sci. 338:486-503.

Stensland, G.J., and R.G. Semonin. 1982. Another interpretation of the pH trend in the United States. Bull. Am. Meteorol. Soc. 63:1277-1284.

Trijonis, J. 1979. Visibility in the Southwest--an explanation of the historical data base. Atmos. Environ. 13:833-844.

Trijonis, J., and R. Shapland. 1979. Existing Visibility in the U.S. Report EPA 450/5-79-010. Research Triangle Park, N.C.: U.S. Environmental Protection Agency.

Turner, J., and M.J. Lambert. 1980. Sulfur nutrition of forests. Pp. 321-333 in Atmospheric Sulfur Deposition: Environment Impact and Health Effects, D.S. Shriner, C.R. Richard, and S.E. Lindberg, eds. Ann Arbor, Mich.: Ann Arbor Science Publishers.

U.S./Canada Work Group #2. 1982. Atmospheric Science and Analysis. Final Report. H.L. Ferguson and L. Machta, cochairmen. Washington, D.C.: U.S. Environmental Protection Agency.

U.S./Canada Work Group #3B. 1982. Emissions, Costs and Engineering Assessment. Final Report. M.E. Rivers and K.W. Riegel, cochairmen. Washington, D.C.: U.S. Environmental Protection Agency.

Ulrich, B. 1980. Production and consumption of hydrogen ions in the ecosphere. Pp. 225-282 in Effects of Acid Precipitation on Terrestrial Ecosystems, T.C. Hutchinson and M. Havas, eds. New York: Plenum Press.

Vukovich, F.M., W. Bach, B. Crissman, and W. King. 1977. On the relationship between high ozone in the rural surface layer and high pressure systems. Atmos. Environ. 11:967-984.

Whittaker, R.H. 1975. Communities and Ecosystems. 2nd ed. New York: Macmillan Publishing Company.

Wolff, G., P.J. Lioy, G. Wright, R. Meyers, and R. Cederwall. 1977. An investigation of long range transport of ozone across the midwestern and eastern U.S. Atmos. Environ. 11:709-802.

Zinke, P.J. 1980. Influence of chronic air pollution on mineral cycling in forests. Pp. 88-99 in Proceedings of the Symposium on Effects of Air Pollutants on Mediterranean and Temperate Forest Ecosystems. General Technical Report PSW-43. Pacific Southwest Forest and Range Experiment Station, U.S. Forest Service. Washington, D.C.: U.S. Department of Agriculture.

2 Atmospheric Processes

Physical and chemical processes in the atmosphere determine the fates of emissions of precursor gases and hence the exposures of primary receptors to pollutants. In this chapter we review current understanding of these atmospheric processes in light of the need to characterize relationships between emissions and deposition. For convenience, we consider these atmospheric processes as occurring in a sequence of clearly defined steps (Figure 2.1). The separate processes are as follows:

   I. Transport and mixing
  II. Chemical reactions in the homogeneous gas phase (dry reaction)
 III. Dry deposition
  IV. Attachment
   V. Chemical reactions in the homogeneous aqueous phase (wet reaction)
  VI. Wet deposition

Heterogeneous chemical processes may occur between gases and liquids adsorbed on solid surfaces, although these are generally considered to be less important in the development of acid deposition than the homogeneous processes. We therefore do not consider heterogeneous processes in this report.

Each of the separate processes takes a certain amount of time; the sum of the processing times along any particular pathway is the source-receptor transport time for the pollutant along that pathway. The processing times are extremely variable, depending strongly on meteorological processes, ambient conditions, and the presence and concentrations of various chemical species. Many of the steps are reversible, so that itinerant

29

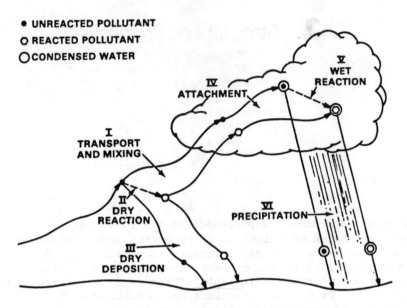

FIGURE 2.1 Atmospheric pathways leading to acid deposition.

pollutant molecules may undergo repeated cycling with
corresponding lengthening of the effective processing
times.

TRANSPORT AND MIXING

The transport of pollution by normal atmospheric advection and mixing is a vitally important influence on
deposition phenomena, and it works both directly and
indirectly. Transport phenomena directly determine where
the pollution goes before it is deposited and therefore
affect the atmospheric residence time of pollutant
materials. Dry deposition, for example, is often limited
by the speed at which the atmosphere can vertically
transport pollution to the proximity of the surface.

Transport can indirectly affect pollutant deposition
in a number of ways. Transport processes, for example,
bring pollution into contact with storm systems, where
precipitation scavenging occurs. Transport also can
introduce pollutants into environments more (or less)
conducive to transformation chemistry. This complex of
interactions links transport with the other processes
shown in Figure 2.1.

It has been the usual practice to divide atmospheric transport processes into two categories. The first, usually termed advection, pertains to the net motion of a parcel of air as it drifts with the mean wind. The second category, diffusion, pertains to the intermixing of the parcel with its surroundings. Historically the distinction between atmospheric advection and diffusion has not been totally clear. Quite often, for example, atmospheric transport models incorporate diffusionlike terms to account for time-averaging of meandering plumes, when in fact the physical processes described have little to do with actual intermixing of materials. Similar treatments often arise in transport models using grids to approximate the desired solutions numerically. Advection processes occurring on scales smaller than the grid spacing escape resolution by the system and thus are often lumped in terms of pseudo-diffusion processes (see Appendix B).

Such approximations are often inescapable. They do, however, contribute significantly to the uncertainty in our ability to model atmospheric pollution, and they obscure the meaning of diffusion in such processes. It is therefore important to remember that advection and mixing are indeed distinct transport phenomena that can lead to different behavior of parcels of polluted air.

The distances associated with pollution transport obviously depend strongly on how long the pollutant resides in the atmosphere and thus is available for action by the advection-diffusion process. In this context it is important to note that atmospheric residence times for typical power plant pollutants (sulfur compounds, for example) are rather uniformly distributed; some pollutant molecules are deposited from the atmosphere relatively quickly and thus at locations near the source, whereas others are deposited more slowly and thus much farther away. On the basis of the best current estimates, it is not unusual for the transport distance of a given pollutant molecule to be of the order of hundreds or even a thousand kilometers. It also is not unusual, however, for a molecule to be deposited close to the source. From this one can conclude that while long-range transport processes certainly are important, shorter-range phenomena are occurring as well.

Another factor that must be taken into account in assessing transport is the height at which pollutants are released into the atmosphere. One approach to local air-quality problems has been to increase the height of stacks in accordance with the notion that the higher the point of release the less the pollutants would affect the surrounding area. This approach to pollution control was applied to large new plants where taller stacks were constructed. At older plants, relatively short stacks were replaced with taller ones. A number of factors, such as meteorology and terrain, influence how the height of an individual stack affects dispersal of a given pollutant, so it is difficult to evaluate the effectiveness of tall stacks for dispersal in general.

Recently Koerber (1982) studied a set of 62 coal-fired power plants in the Ohio River Valley. He developed a measure of the potential for long-range transport that involved physical stack height, plume rise, and mixing height. Figure 2.2 shows the temporal trend in Koerber's parameter between 1950 and 1980. The implication of this and other work is that stack heights must be taken into account when assessing source-receptor relationships involving long-range transport.

Because of cumulative uncertainties, the trajectories and times associated with long-range transport are much more difficult to estimate than their shorter-range counterparts. Early, very crude attempts to simulate long-range phenomena simply employed local wind roses and straight trajectories from the sources in question. The obvious deficiencies associated with this approach prompted further efforts to develop curved-trajectory simulations, which were driven by conceptualized, time-evolving wind fields.

The curved-trajectory approaches, while representing a major advancement over their straight-line predecessors, suffered from two major disadvantages. The first of these was that the data from which the wind fields were derived were usually extremely sparse in both time and space--a problem that becomes particularly severe under complex meteorological conditions involving fronts and storm systems. Although a variety of sophisticated interpolation techniques has been advanced subsequently to offset this problem, the poor coverage of meteorological data in both space and time remains particularly troublesome.

The second major problem associated with these types of trajectory approaches is caused by mass motions of air vertically and vertical wind shear, i.e., the dependence of wind speed and direction on altitude. Early trajectory simulations, based on constant-altitude wind fields, soon were replaced by layer-averaged or constant-pressure

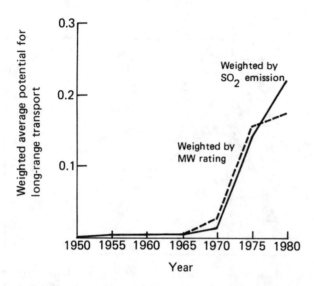

FIGURE 2.2   Trend in long-range transport potential for 62 sources in the Ohio River Valley. SOURCE: Koerber (1982).

surface versions to overcome this disadvantage partially. On the basis of thermodynamic arguments, it is
expected that vertical motions of air parcels should
adhere rather closely to constant-entropy surfaces in the
atmosphere, and from this a few "isentropic" trajectory
simulations have evolved as well. However, vertical
motions caused by the heat released or absorbed during
cloud formation are not taken into account by either
method.

Although curved-trajectory simulations can produce
rather reliable results in simple meteorological situations, they are fraught with uncertainty when conditions
become complex, such as near frontal systems. Some idea
of this uncertainty may be gained from Figure 2.3, which
shows the results of two different calculations of a
trajectory under the same conditions in the vicinity of a
frontal storm. One calculation (solid curves) uses the
assumption of isentropic transport, while the other
(dashed curves) employs isobaric transport. After 24
hours, the calculated positions of the two hypothetical
air parcels are several hundred kilometers apart (also
see Chapter 3 on the treatment of transport and mixing in
models as well as Appendix B).

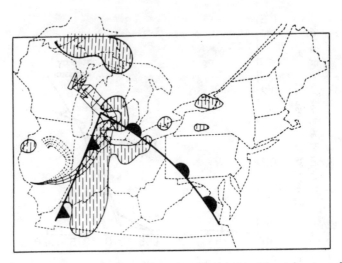

FIGURE 2.3 Calculated plume trajectories in the vicinity of frontal systems for 24 hours after release. Solid plumes were calculated on the assumption of isentropic vertical motion; dashed plumes were calculated on the assumption of isobaric motion. Shading indicates area of precipitation. SOURCE: Adapted from Davis and Wendell (1977).

The difficulties associated with wind shears and vertical motions could be largely overcome if the vertical motions were indeed known. Several weather prediction models produce such information, albeit prognostic in nature, rather coarse in scale, and dedicated more to upper regions of the troposphere than to the planetary boundary layer. In addition, a few mesoscale dynamic models currently exist that can supply initial estimates, at least, of information of this type. The application of such techniques for pollution trajectory simulation has been comparatively limited owing to the complexity of the modeling process and the expense of the simulations. A summary of several of the trajectory simulation techniques discussed here appears in Chapter 3 (Table 3.2). They have been applied to form composite regional pollution models.

A major factor contributing to the uncertainty in long-range trajectory simulations stems directly from our current inability to measure long-range transport. Several balloon studies have been attempted, but they have been less than satisfactory because of technical difficulties, the balloons' supposed inability to track vertical motions exactly, and statistical problems associated with tagging a stochastic system with too few units. Chemical tracers have not been particularly successful to date owing to detection difficulties over large distance scales. Tracer techniques are evolving rapidly, however, and it is not unreasonable to expect some highly significant results to emerge from experiments with tracers in the next 5 years. For example, during the summer of 1983 six releases of tracers are to be made from Ohio and Ontario under the Cross Appalachian Tracer Experiment (CAPTEX). A wide arc of measurement sites will be set up over 600 km downwind of the releases. This experiment will be the first step in a long-range tracer program.

An important feature of the long-range transport of air pollutants is that the plumes from individual sources may become so dilute and so thoroughly mixed far downwind of major source areas that the attribution of specific parcels of polluted air to specific sources is imprac-tical. In these cases, the sources contribute pollutants to air masses that may be considered to be entrained in synoptic-scale meteorological systems. The classic example of mixing occurs in large stagnant air masses that occur most frequently in summer in the eastern United States (see Chapter 4). The motion of air masses on the synoptic scale may be important for understanding acid deposition in areas remote from major source regions. The average flows across North America are shown in Figure 2.4, which illustrates that the region in which acid deposition is currently thought to be an environ-mental problem is also a region of intense interaction between tropical marine and arctic air masses.

CHEMICAL TRANSFORMATION

During transport through the atmosphere, SO_2, NO_x, hydrocarbons, and their oxidation products participate in complex chemical reactions that transform the primary pollutants into sulfates and nitrates. The transformation processes are important because, as we discuss later, deposition of the primary pollutants and that of their transformation products are governed by different processes.

There are many chemical pathways through which SO_2 and NO_x in the atmosphere can be transformed (oxidized) into sulfate and nitrate compounds, including homogeneous

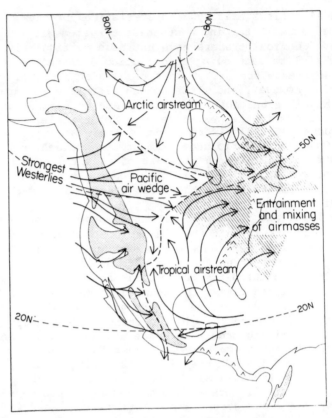

FIGURE 2.4 Surface flows across North America, illustrating the area of complex entrainment and mixing of air masses in the eastern portion of the continent. SOURCE: Bryson and Hare (1974).

processes that take place in the gas phase and in liquid droplets or heterogeneous processes that take place on the surfaces of particles or droplets. Figure 2.5 indicates the pathways by which SO_2 and NO_x are transformed into gaseous- and aqueous-phase acids. Field studies indicate that the relative importance of gas- and liquid-phase reactions depends on meteorological conditions, such as the presence of clouds, relative humidity, intensity of solar radiation, and the presence and concentrations of other pollutants.

A comprehensive review of homogeneous gas- and solution-phase atmospheric chemistry associated with acid deposition is presented in Appendix A. The appendix

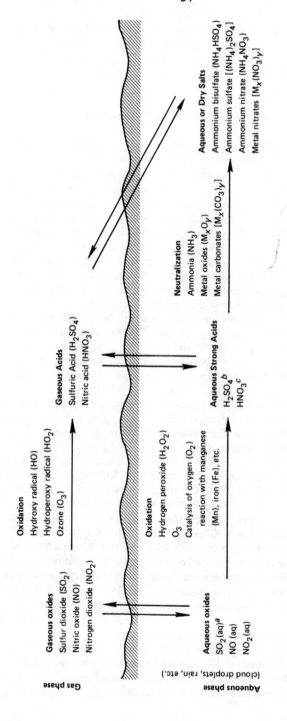

FIGURE 2.5 Pathways for the formation of atmospheric sulfate and nitrate.

includes detailed descriptions of alternative oxidation
pathways and analyses of reaction rates. General
descriptions and conclusions, drawn from this material,
are presented below.

Homogeneous Gas-Phase Reactions

SO_2 and NO_x have been observed in the atmosphere to be
oxidized through homogeneous gas-phase reactions at rates
of a few and 20 to 30 percent/h, respectively (Step II in
Figure 2.1). The observed rates cannot be explained by
direct oxidation by atmospheric oxygen, reactions that
occur too slowly for typical concentrations of pollutants
and, in the case of SO_2, in the absence of catalysts.
Similarly, although there are direct pathways to the
formation of sulfuric and nitric acids beginning with
absorption of solar radiation by SO_2 and NO_2, respectively, these processes also appear to be unimportant
under typical conditions in the troposphere.

According to current understanding, most of the
gas-phase chemistry in the lower atmosphere that results
in oxidation of SO_2, nitric oxide (NO), and nitrogen
dioxide (NO_2) entails reactions with a variety of
highly reactive intermediates--excited molecules, atoms,
and free radicals (neutral fragments of stable
molecules)--that are generated in reactions initiated by
the absorption of solar radiation by trace gases. The
most important of the intermediates for gas phase
oxidation appears to be the hydroxy radical, HO.

The hydroxy radical can be formed in the troposphere
by a number of reactions. A common process begins with
dissociation of NO_2 by absorption of sunlight, which
forms a highly reactive oxygen atom that combines quickly
with a diatomic oxygen molecule to form the triatomic
oxygen molecule, ozone (O_3). Ozone may be photodis-
sociated, yielding an electronically excited diatomic
molecule of oxygen and an electronically excited oxygen
atom, $O(^1D)$, which reacts readily with a water molecule
to form HO. The hydroxy radical, unlike many radicals
that are fragments of complex molecules containing
carbon, does not react readily with molecular oxygen; HO
survives in the atmosphere to react with most impurity
gases, such as hydrocarbons, aldehydes, NO, NO_2, SO_2,
and carbon monoxide (CO). Reactions between HO and
several impurity gases produce additional classes of
reactive transient species, which, in turn, react with

atmospheric constituents to form additional reactive species. For example, reactions of HO with CO and hydrocarbons produce peroxy radicals; peroxy radicals react rapidly with NO to form NO_2 and alkoxy, acyloxy, and other HO radicals.

The net result of all of these interactions is a large number of chemical pathways for oxidation of SO_2 and NO_x to sulfuric acid (H_2SO_4) and nitric acid (HNO_3), respectively, many of which depend initially on the formation of HO. A sequence of these reactions can be constructed in which a single HO radical may oxidize CO, hydrocarbon, or aldehyde, followed by oxidation of NO to NO_2 accompanied by production of additional HO radicals. Repeated cycling of the sequence results in continued oxidation of NO to NO_2 and relatively constant concentrations of HO.

There are a number of gaseous-phase chemical reactions between SO_2 and reactive transient species that may lead to formation of H_2SO_4; these reactions, along with currently accepted values for the reaction rates, are listed in Appendix A. While many of the rate constants are known only with an uncertainty of 50 percent, it appears as if the most important reaction is that between SO_2 and HO, yielding $HOSO_2$. Evidence is good that this reaction ultimately leads to the generation of sulfuric acid, and a number of pathways for this subsequent reaction have been explored. Which of these pathways is most important is still unknown, but it is likely that the oxidation of SO_2 by HO is a chain-propagating reaction.

The principal agents for oxidizing NO to NO_2 are ozone and peroxy radicals, whereas NO_2 is oxidized to HNO_3 by a well-characterized reaction with HO (Appendix A).

According to current understanding, then, the rates at which sulfuric and nitric acids are formed in homogeneous gas-phase reactions depend on ambient concentrations of the hydroxy radical. Direct measurement of HO in the atmosphere is difficult, but both theoretical and experimental estimates are available from which to estimate rates of conversion from SO_2 and NO_2 to H_2SO_4 and HNO_3, respectively. Using the rate constants listed in Appendix A, we find that for high concentrations of HO--characteristic of polluted summer sunny skies--SO_2 will be converted to H_2SO_4 by reaction with HO at a daily averaged rate of about 0.7 percent/h (16.4 percent per 24-h period), whereas NO_x

is converted at an average rate of about 6.2 percent/h (100 percent per 16-h period). At HO concentrations typical of winter sunny weather in a polluted atmosphere, the average rates are roughly 0.12 and 1.1 percent/h (3 and 25 percent per 24-h period), respectively. These rates, of course, depend on HO concentration and therefore fall rapidly after sunset since HO is formed largely by photochemical processes.

While contributions to oxidation of SO_2 from other reactions may be important in some circumstances, the rates reported in Appendix A are consistent with those observed in urban plumes in the absence of clouds. Observed conversion rates for NO_2 in cloud-free conditions are also consistent with the estimates presented in Appendix A.

Homogeneous Aqueous-Phase Reactions

Oxidation of SO_2 is rapid in water, often of the order of 100 percent/h. Rates of aqueous-phase oxidation of SO_2 are typically much higher than those of gas-phase oxidation. The lifetimes of individual clouds, however, are short, so that the long-term average oxidation rate in cloudy air may be similar to that in the gas phase. When SO_2 dissolves in water, several species are formed: the hydrate $SO_2 \cdot H_2O$ and the ions HSO_3^- (bisulfite), $SO_3^=$ (sulfite), and H^+ (hydrogen ion). As the concentration of H^+ increases, a solution becomes more acidic, corresponding to lower values of pH. The concentration of total dissolved sulfur, designated S(IV), in water in equilibrium with gaseous SO_2 at a specified partial pressure is inversely related to the concentration of H^+ in the solution. That is, SO_2 is less soluble in more acidic (lower pH) solutions. As indicated in Appendix A, equilibrium between gas-phase SO_2 and total dissolved sulfur is established quickly, so S(IV) in cloud droplets or liquid aerosol particles is a function of pH and the ambient gaseous concentration of SO_2.

Current evidence suggests that two agents, hydrogen peroxide (H_2O_2) and ozone (O_3), may be primarily responsible for oxidizing S(IV) to H_2SO_4 in atmospheric water for typical concentrations of pollutants (step V in Figure 2.1). Other possible oxidation pathways exist, including homogeneous reactions involving the hydroxy radical or aqueous-phase NO_3 and catalytic reactions involving soot or ions of manganese and iron. Current

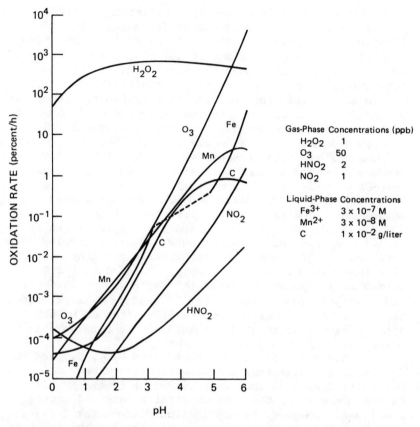

FIGURE 2.6 Theoretical rates of liquid-phase oxidation of SO_2 assuming 5 ppb of SO_2, 1 ml/m³ of water in air, and concentrations of impurities as shown. SOURCE: Martin (1983).

theoretical understanding of the oxidation rates of dissolved SO_2 by the various proposed mechanisms for typical impurity concentrations is shown in Figure A.13 of Appendix A and reproduced here as Figure 2.6. The figure shows that oxidation by H_2O_2 predominates under the specified condition except for high values of pH (low acidity), in which case O_3 may be an important reactant as well. Because the pH of aerosol droplets and cloud water is generally measured to be below 5, H_2O_2 is currently regarded as the most important oxidizing agent in the aqueous-phase chemistry of the formation of sulfuric acid.

As demonstrated in Figure 2.6, oxidation of S(IV) by H_2O_2 is an effective process that is relatively independent of pH. The other mechanisms that have been studied are strong functions of pH; they are likely to be active early in the process of acidification of droplets but rapidly become ineffective as acidification proceeds. The relative positions on plots such as Figure 2.6 of the curves for the less important mechanisms depend somewhat on the assumed concentrations of the impurities.

Both H_2O_2 and to some degree O_3 have their origins in homogeneous gas-phase reactions. The chemistry of ozone production was described earlier. The major homogeneous sources of H_2O_2 in the troposphere are reactions involving the hydroperoxy radical (HO_2). In the polluted atmosphere there is strong competition for reaction with HO_2 among NO, NO_2, aldehydes, and other species, so the efficiency with which H_2O_2 is generated is a complex function of the concentrations of these and other impurities (see Appendix A). In theory, the amount of H_2O_2 formed in gas-phase reactions and taken up in cloud water and precipitation is sufficient to oxidize a large fraction of S(IV). Ozone is also readily taken up by atmospheric water, although its solubility is considerably lower than that of H_2O_2.

If homogeneous air masses containing H_2O_2, O_3, and SO_2 encounter aqueous aerosol droplets, cloud water, or precipitation, solution-phase oxidation of the SO_2 is favored because of the high conversion rates of these reactions. However, the optimal conditions for formation of H_2O_2 and O_3 in the troposphere differ, so the relative importance of the two aqueous-phase oxidants will depend on conditions in the gas phase that determine the relative rates of production of the two oxidants. For example, as detailed in Appendix A, conditions of low NO_x concentrations and high levels of hydrocarbons and aldehydes favor formation of H_2O_2, whereas high concentrations of NO_2, hydrocarbons, and aldehydes favor O_3 production.

While H_2O_2 is thought to play an important role in oxidizing aqueous phase SO_2 in the atmosphere (Figure 2.6), it has only recently become possible to measure concentrations of H_2O_2 reliably in ambient air or in cloud water or precipitation, because of deficiencies in experimental techniques.

In comparison with current understanding of various pathways for the formation of H_2SO_4 in atmospheric water, little is known about the solution-phase chemistry that results in formation of nitric acid in aerosols, cloud water, or precipitation. Both theoretical and experimental evidence, described in Appendix A, suggest that dinitrogen pentoxide (N_2O_5) formed in gas-phase reactions between ozone and NO_2 may be efficiently scavenged by water droplets to form nitric acid directly. Sufficient data are not yet available on which to base evaluations of the importance of this or other mechanisms to the formation of HNO_3 observed in the atmosphere.

Most clouds evaporate before precipitation can develop. Therefore cloud processes can affect the chemical nature of sulfur and nitrogen compounds in the absence of precipitation and can contribute to their redistribution in the troposphere.

Relative Roles of Gaseous- and Aqueous-Phase Chemistry

In recent years the role of aqueous-phase chemistry in the development of acid deposition has received increased attention. The results of field and laboratory studies suggest that although rates of oxidation of SO_2 in the gas phase are relatively slow, the relative importance of gas-phase and solution-phase oxidation varies, depending on a variety of meteorological conditions, such as the extent of cloud cover, relative humidity, presence and concentrations of various pollutants, intensity of solar radiation, and amount of precipitation. Although solution-phase conversion rates can be considerably higher than those in the gas phase, air masses over the eastern United States are likely to be relatively free of clouds and precipitation a large fraction of the time, so both gaseous- and aqueous-phase processes must in general be regarded as contributing to acid formation. It is also clear from the discussion in Appendix A that formation of sulfuric and nitric acids in liquid aerosols, cloud droplets, and precipitation depends on gas-phase reactions to supply the necessary reactants.

The clearest evidence that gas-phase reactants contribute to solution-phase formation of acids was obtained in the Acid Precipitation Experiment (APEX) described briefly in Appendix A. In this experiment, the constituents of dry air were measured prior to the time the air mass ascended to produce a large area of precipitation characteristic of a warm front. Measurements were also made of samples of cloud water and of precipitation at the base of the cloud. The results indicated that both nitric and sulfuric acids were formed rapidly in the cloud, although the oxidizing agent remained unidentified because of weaknesses in the analytical methods.

Data from another experiment are now available showing an appreciable rate of conversion of SO_2 to H_2SO_4 at night in clouds over coastal waters, indicating an oxidation process other than reaction with the hydroxy radical, which is present in significant concentrations only in daytime.

The importance in atmospheric chemistry of aqueous-phase processes taking place in clouds is illustrated theoretically in Figure 2.7, which gives the results of calculations that combine homogeneous gas-phase chemistry with the current picture of aqueous reactions (Environmental Research & Technology, Inc., and MEP, Inc. 1982). The figure shows the progress of oxidation in clear air (beginning with NO and NO_2 concentrations of 10 ppb, concentration of reactive hydrocarbon vapors of 200 ppb, SO_2 concentration of 5 ppb, and $SO_4^=$ concentration of 2 $\mu g/m^3$) and the effects of introducing a cloud with 1 g/m^3 of liquid water at 1400 h. In theory the insertion of cloud water causes dramatic decreases in atmospheric concentrations of H_2O_2, HNO_3, SO_2, and $SO_4^=$. The behavior of NO, NO_2, O_3 and peroxyacetylnitrate (PAN) was not strongly influenced by the presence of cloud water. The example demonstrates that clouds have the potential to dominate chemical interactions involving water-soluble or water-scavengable constituents. Field experiments are required to determine if this dramatic effect actually occurs in the atmosphere.

DEPOSITION

Dry Deposition

The term dry deposition is used to denote a variety of processes by which pollutant gases and aerosol particles reach the Earth's surface, including the surfaces of both living and inanimate objects on the ground (vegetation and buildings, for example). The processes depend on concentrations of the pollutants and small-scale meteorological effects near the surface as well as on the characteristics of the receiving surface.

Superficially, dry deposition seems to be almost trivially simple in comparison with other aspects of the relationships between emissions and deposition. Dry

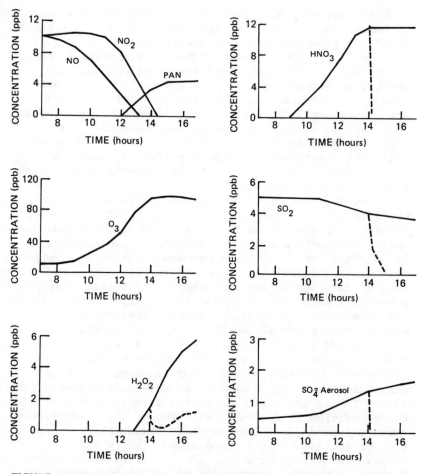

FIGURE 2.7 Theoretical calculations of gas and aerosol concentrations as a function of time for gas-phase reactions only (solid line) and with the introduction of cloud water (dashed line) at 1400 hours. SOURCE: Environmental Research & Technology, Inc., and MEP, Inc. (1982).

deposition takes place at the Earth's surface and thus is inactive in the volume of the atmosphere in which chemical transformation and processes leading to wet deposition occur.

In fact, however, dry deposition is incompletely understood. Uncertainties in dry deposition may be an important source of error in today's regional modeling efforts.

There are several reasons for the current uncertainties in understanding dry deposition (Appendix C). The first is that dry-deposition rates are extremely difficult to measure. Although a number of possible techniques exist (Hicks et al. 1981) and considerable effort has been devoted to developing appropriate methods for measuring fluxes to surfaces, the base of high-quality data is still distressingly small. Furthermore, the more reliable data that do exist tend to have been obtained under experimentally convenient conditions (for example, high pollutant concentrations, uniform terrain) and thus reflect only a small subset of the potentially important environmental conditions.

A second reason for uncertainty in dry-deposition rates is a consequence of the complexity of the physical processes in the atmosphere. As indicated in Figure 2.8, several mechanisms convey pollutants to the surface, and it is often not clear which processes dominate under which conditions. Especially important in this regard are the near-surface mechanisms for aerosol particles, such as inertial impact, phoresis, and electrical effects. Uncertainties in this area are currently substantial, especially for deposition to surfaces of vegetation.

The third reason arises from uncertainties in the characteristics of the substrate on which materials are deposited. Contrary to the superficial view that dry deposition is purely a surface phenomenon, phenomena both at and in the substrate can play a role in determining the deposition flux. It is well known, for example, that stomatal openings on leaf surfaces influence the deposition of gases such as SO_2 and ozone. Soils and building materials have been shown to "saturate" with depositing gases. Re-emission of sulfur compounds from plant surfaces has been detected. All of these results render the concept of a simple boundary condition approach to dry deposition somewhat questionable; the corresponding uncertainties are again large.

These difficulties combine to give a number of widely varying estimates for the temporal and spatial scales of dry deposition of specific pollutants. As a rule of thumb, for sulfur and nitrogen compounds at least, dry deposition is taken on the average to be about as effective as wet deposition in pollutant removal. About one third of sulfur emissions is transported out of the continent. Thus roughly one third of northeastern emissions is assumed to be dry-deposited on the North American continent (see the section in Chapter 3 on material balance).

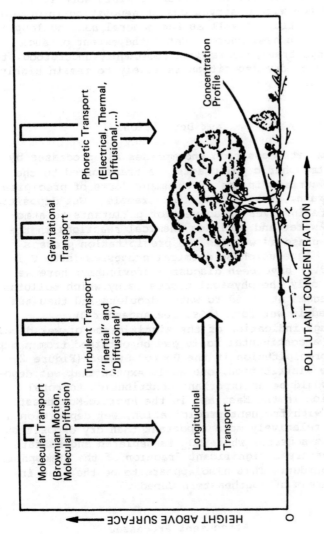

FIGURE 2.8 Pathways for dry deposition.

There are also studies, however, that obtain a scale length for dry deposition in excess of 10^4 km for some species (Slinn 1983), strongly suggesting interaction with global circulation patterns. This work is in concordance with observations of deposition in Greenland and the Arctic, as well as the general haze buildup in the northern hemisphere. Until the extent of such long-range transport is more thoroughly understood, the modeling of dry deposition is likely to remain highly uncertain.

Wet Deposition

The term wet deposition encompasses all processes by which atmospheric pollutants are transported to the Earth's surface in one of the many forms of precipitation: rain, snow, or fog, for example. Wet deposition therefore involves attachment of pollutants to atmospheric water and includes chemical reactions in the aqueous phase as well as the precipitation process itself. Aqueous phase chemical processes (step V in Figure 2.1) have been discussed previously; here we address only the physical processes by which pollutants first become attached to water droplets and then are deposited in wet form (also see Appendix C).

A rough indication of the significance of wet deposition on a continental scale can be obtained from a map of annual precipitation in the United States (Figure 2.9). From the distribution, one would expect that wet deposition would be an important contribution to total deposition in the East and in the Pacific Northwest. In regions with frequent precipitation, wet deposition also becomes relatively more important than dry deposition far away from sources, where SO_2 is depleted and sulfate particles are a significant fraction of the atmospheric sulfur burden. This also appears to be the case in remote areas of southeastern Canada.

Attachment Processes

The physical processes by which pollutants become attached to droplets and other falling hydrometeors such as ice crystals (step IV in Figure 2.1) have been the subject of extensive research, and a number of technical

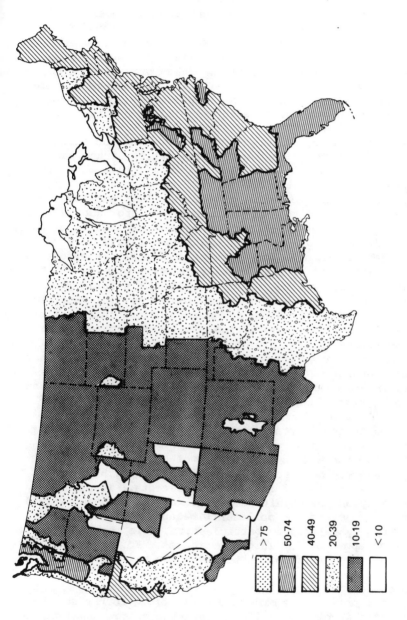

FIGURE 2.9 Average annual precipitation in the United States. SOURCE: GCA Corporation (1981).

reviews of current knowledge in this area are available (see, for example, Slinn 1983).

The most important attachment process under most in-cloud conditions is undoubtedly nucleation. Nucleation is a kinetic process in which water molecules condense from the vapor phase onto a suitable surface. Dust and pollutant aerosol particles provide such surfaces in the air. The result is a cloud of droplets (or ice crystals) containing the pollutant. The droplets may grow by the same process (condensation) or may lose water by evaporation.

The tendency of water vapor to condense on aerosol particles depends on the characteristics of the particles and the degree of saturation of the air with water vapor. As a consequence the aerosol and associated cloud particles compete for water molecules. Some particles will capture water with high efficiency and grow substantially in size. Others will acquire only small amounts of water, whereas still others will remain essentially "dry" elements. In addition, some particles may be effective for nucleation of ice crystals, whereas others will be active only for the formation of liquid water. The nucleating capability of a particular aerosol particle is determined by its size, its morphological characteristics, and its chemical composition. Acid-forming particles, by their very nature, are chemically competitive for water vapor and thus tend to participate actively as condensation nuclei for liquid water. This attribute enhances their propensity to become scavenged early in storms and has a significant effect on the nature of the acid-precipitation formation process.

Diffusional attachment, as its name implies, results from diffusion of the pollutant molecule or particle through the air to the surface of a water droplet. The process may be effective in the case of both suspended cloud elements and falling hydrometeors. It depends chiefly on the magnitude of the molecular (or Brownian) diffusivity of the pollutant; because diffusivity is inversely related to particle size, this mechanism is less important for larger particles. For practical purposes, diffusional attachment can be ignored for particles with radii of more than a few tenths of a micrometer.

The motion of a molecule or particle to the surface of a water droplet by diffusion depends on the gradient in the concentration of the pollutant in the vicinity of the surface. Thus, if the cloud or precipitation droplet can accommodate the influx of pollutant readily (for example, the pollutant is highly soluble in water), it will effectively depopulate the adjacent air, thus making a steep concentration gradient and encouraging further diffusion to the droplet. If for some reason (such as particle "bounce off" or low gas solubility) the droplet cannot accommodate the pollutant, further diffusion to the droplet will be discouraged. If the cloud or precipitation droplet supplies the pollutant to the local air through an outgassing mechanism, the concentration gradient will be reversed and diffusion will carry the pollutant away from the droplet. In general, diffusional attachment processes are sufficiently well understood to allow their mathematical description with reasonable accuracy.

Inertial attachment arises by virtue of the facts that pollution particles and scavenging droplets are constantly in motion and that both have finite volume and mass. The most important example of inertial attachment is the impaction of aerosols by falling hydrometeors. In this case, the hydrometeor falls under the influence of gravity, sweeping out a volume in space. Collisions occur between the falling hydrometeors and some aerosol particles, resulting in attachment.

The effectiveness of impaction depends on the size of both the aerosol particle and the hydrometeor; mathematical formulas exist to estimate the magnitudes of these processes. Impaction generally becomes unimportant for aerosols less than a few micrometers in size. In this context it is interesting to note that a two-stage capture mechanism can exist, in which a small aerosol first grows through nucleation to form a larger droplet that is then captured by inertial attachment. This two-stage process, called accretion, is an essential factor in the generation of precipitation in clouds and has been postulated as an important mechanism in scavenging pollutants below clouds.

A second example of inertial attachment is turbulent collision. In this case, the particles and scavenging elements, subjected to a turbulent field, collide because of dissimilar dynamic responses to velocity fluctuations in the local air. This scavenging mechanism is thought to be of secondary importance and has received comparatively little attention in the literature, although some recent theoretical analyses have suggested it to be significant for droplets and particles of specific sizes.

52

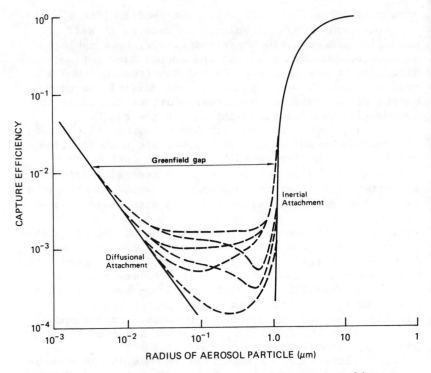

FIGURE 2.10   Theoretical scavenging efficiency of a falling raindrop of diameter
0.31 mm as a function of aerosol particle size. SOURCE: Adapted from Pruppacher
and Klett (1978).

Although the diffusional and inertial attachment
processes are efficient for capturing very fine and very
coarse particles, respectively, neither mechanism is
effective for particles in the range of 0.1 to 5 μm.
The resulting minimum in capture efficiency as a function
of particle size, shown schematically in Figure 2.10, is
known as the Greenfield gap.

Depending on circumstances, there are several
additional attachment mechanisms (including accretion via
the two-stage nucleation-impaction mechanism mentioned
earlier) that can operate in the Greenfield gap. The
processes include turbulent deposition, electrical
attraction, and phoretic effects (see Appendix C for
details). As indicated by the dashed lines in Figure
2.10, these mechanisms can significantly relieve the
Greenfield effect under appropriate circumstances
(Appendix C).

From this discussion, it should be evident that the aggregate of possible attachment processes comprises a complex system that is difficult to characterize mathematically. This complexity, combined with the processes of formation and delivery of precipitation that occur both consecutively and concurrently, provides a major source of uncertainty in current models of regional pollution transport and deposition.

REFERENCES

Bryson, R.A., and F.K. Hare. 1974. Climates of North America. World Survey of Climatology, Vol. 11. New York: Elsevier Scientific Publishing Company.

Davis, W.E., and L.L. Wendell. 1977. Some Effects of Isentropic Vertical Motion Simulation in a Regional Scale Quasi-Lagrangian Air Quality Model. BNWL-2100 PT 3. Richland, Wash.: Battelle Pacific Northwest Laboratories.

Environmental Research & Technology, Inc., and MEP, Inc. 1982. Models for Long Range and Mesoscale Transport and Deposition of Atmospheric Pollutants. Phase I: Modeling System Design. Report SYMAP-101. Toronto, Ontario: Ontario Ministry of Environment.

GCA Corporation. 1981. Acid Rain Information Book. Final Report. DOE/EP-0018. Prepared for the U.S. Department of Energy. Springfield, Va.: National Technical Information Service.

Hicks, B.B., M.L. Wesely, and J.L. Durham. 1981. Critique of methods to measure dry deposition; concise summary of workshop. Presented at the 1981 National Meeting of the American Chemical Society, Atlanta. Ann Arbor: Ann Arbor Scientific Publications.

Koerber, W.M. 1982. Trends in SO_2 emissions and associated release height for Ohio River Valley Power Plants. In Proceedings of the 75th Annual Meeting of the Air Pollution Control Association, paper 82-10.5, New Orleans. Pittsburgh: Air Pollution Control Association.

Martin, L.R. 1983. Kinetic studies of sulfite oxidation in aqueous solutions. In Acid Precipitation: SO_2, NO, and NO_2 Oxidation Mechanisms: Atmospheric Considerations. Ann Arbor, Mich.: Ann Arbor Scientific Publications. In press.

54

Pruppacher, H.R., and J.D. Klett. 1978. Microphysics of
Clouds and Precipitation. Boston: D. Reidel
Publishing Company.
Slinn, W.G.N. 1983. Precipitation scavenging. In
Atmospheric Sciences and Power Production. D.
Randerson, ed. Washington, D.C.: U.S. Department of
Energy. In press.

3 Theoretical Models of Regional Air Quality

Analysis of the spatial and temporal behavior of atmospheric parameters and climatological patterns depends on a thorough theoretical understanding of the physical and chemical processes involved. That understanding, in turn, depends on observations of phenomena in the field and in the laboratory. One purpose of analyzing the relationships between emissions of precursor gases and deposition of acidic or acid-forming substances is to develop means for assessing the potential effectiveness of alternative proposals for mitigating the adverse effects of acid deposition. Uncertainties in the current understanding of the relevant physical and chemical processes are reflected in uncertainties in analytical models of these relationships.

Construction of analytical models is a typical method by which scientists approach complex problems. For many years earth scientists have been developing knowledge about flows of substances in the environment (within and among the atmosphere, hydrosphere, biosphere, and lithosphere). All elements cycle naturally through the environment; sulfur and nitrogen are two prominent examples. Models have been developed--some conceptual, some empirical, some theoretical--to organize that knowledge in ways that allow predictions to be made that are subject to testing. In recent years, this analytical approach has taken on considerable practical importance, because of the need to assess the implications of anthropogenic disturbances on natural ecological processes. So it is with models of acid deposition.

In this report we are concerned with only part of the phenomenon of acid deposition: the relationships between emissions and deposition. Models of the cycling of sub-

stances in the hydrosphere, biosphere, and lithosphere
are beyond the scope of this report.

Models of the distribution of emissions through the
atmosphere and their subsequent deposition can be divided
into two classes: theoretical and empirical. Empirical
models consist of analyses of observations in the field;
Chapter 4 deals with empirical approaches used to manipu-
late data and test hypotheses. In the class of theoretical models are both deterministic calculations and
estimates of material balance (or budgets); the current
state of the art in these approaches is described below.

MATERIAL BALANCE

The method of material balance or budgeting involves
assessing the gross flows of a substance into and out of
compartments of the environment. The compartments are
defined for the purposes of analysis; they are generally
large, so that detailed behavior of constituents is not
considered. Leaving out the detail, of course, means
that the results may provide only general guidance and
understanding.

The most straightforward approach to budgets for acid
deposition is to segment processes into one or more
compartments, allowing flow between the compartments
(e.g., Charlson et al. 1978). Budgets for sulfur in the
atmosphere have been constructed for the global atmosphere (Granat 1976) and for regions of Europe and
eastern North America (e.g., Galloway and Whelpdale 1980,
Granat et al. 1976, Shinn and Lynn 1979). A summary of
two budgets for eastern North America is shown in Table
3.1; these calculations were made for each category by
somewhat different means. They present a qualitatively
similar (but quantitatively different) picture of the
sulfur oxide transport and deposition in the eastern
United States as well as export to the Atlantic Ocean.
Other than giving estimates for the average annual
deposition over large areas, these types of calculations
reveal little about the consequences of changing anthropogenic emissions of sulfur or nitrogen. They also
provide no guidance about the deposition of acid-producing
material on specific regions that are ecologically
sensitive. They do, however, provide a sense of the
scale of exports of atmospheric pollutants from one
region to another.

TABLE 3.1 Comparison of Atmospheric Sulfur Budget
Estimates for the Eastern United States[a] and Northeastern
United States[b] in teragrams (million metric tonnes) per year

	Galloway and Whelpdale (1980)[a]	Shinn and Lynn (1979)[b]
Input		
Man-made emissions	14	7.5
Natural emissions	0.5	0.6
Inflow from oceans	0.2	—
Inflow from west	0.4	—
Transboundary flow	0.7	—
	15.8	8.1
Output		
Transboundary flow	2.0	(1.1)
Wet deposition	2.5	1.5
Dry deposition	3.3	2.5
Outflow to oceans	3.9	3.0
	11.7	8.1

[a]Area east of 92° W (Mississippi River).
[b]Connecticut, Delaware, Illinois, Indiana, Kentucky, Maryland,
Massachusetts, Michigan, New Jersey, New York, Ohio, Pennsylvania,
Rhode Island, Virginia, and West Virginia.

One application of the method has been to assess the transport of pollutants across international boundaries. Because certain pollutants, particularly sulfates and nitrates, may be transported large distances from the sources of their precursor gases, air pollution is an interstate and even an international issue. Not all the sulfur and nitrogen emitted from sources in the United States comes to the ground in the United States, and not all the sulfur and nitrogen that comes to the ground in the United States is emitted from sources in the United States. The same, of course, can be said for states and regions within the United States.

It has been estimated that, of the total sulfur emitted to the atmosphere in the eastern part of the United States, about one third is transported to the western Atlantic Ocean and beyond, while roughly one sixth is exported to Canada. The remainder, about one half, falls in the United States (Galloway and Whelpdale 1980).

The fraction of the exports of atmospheric sulfur from the United States to Canada that is deposited in Canada is unknown. It has been hypothesized that the fraction of Canadian emissions of sulfur that falls in Canada is larger than the fraction of U.S. exports to Canada that falls in Canada. This supposition can be explained by the differences in the deposition processes for SO_2 and sulfates and the fact that U.S. exports of atmospheric sulfur to Canada are likely to be richer in sulfates than Canadian emissions. Nevertheless, more sulfur is deposited than emitted in eastern Canada (Galloway and Whelpdale 1980), so U.S. exports can account for substantial quantities of the sulfur deposited there.

DETERMINISTIC MODELS

Most of the effort to develop models of acid deposition during the past decade has been devoted to deterministic descriptions of the distribution of sulfur oxides in plumes. The work has grown from efforts to develop plume models for studying effects of emissions on ambient concentrations of pollutants at relatively small distances from sources. Current models used to analyze regional pollution problems such as acid deposition apply to areas of the order of 10^6 km^2 and focus on long-term (annual) average behavior, taking into account emissions, airflow, mixing, chemical transformations, and both wet and dry deposition. Generally, chemical transformations and deposition processes are treated parametrically, whereas transport is calculated using available data on wind fields, for example. The models are based on sets of continuity equations for concentrations of the species of interest; the continuity equations are coupled through terms representing the production and destruction of species in chemical reactions. The equations are solved using computers.

In effect, deterministic models represent detailed material balance calculations analogous to the compartmentalization approach mentioned earlier, but in this case the compartments in the atmosphere are much smaller, so detailed behavior must be included.

Once confidence in deterministic models has been achieved, through testing and verifying, it should be possible to use them to assess the potential consequences of alternative proposals for mitigating acid deposition, since sensitivity tests would be feasible with this type of model.

There is a variety of regional models for average deposition rates of sulfur oxides over eastern North America (e.g., U.S./Canada Work Group #2 1982). The models use different approximations to characterize atmospheric processes (Table 3.2). They have not been verified systematically because of a lack of observational data. However, testing and initial comparisons of several models for annual averages indicate that their accuracy in estimating either ambient SO_x concentrations or wet-deposition rates is inadequate for quantitative assessment of the effects of emissions from specific sources (U.S./Canada Work Group #2 1982). Initial comparisons show no preference by performance for a specific model for application to the situation in eastern North America, although from the limited number of comparisons currently available, it appears as if models that treat meteorological parameters in a gross statistical sense appear to perform as well as the more sophisticated models (U.S./Canada Work Group #2 1982).

At least three models (SURADS, RTM-II, and STEM) are capable of simulating regional sulfate pollution episodes over eastern North America (Table 3.2). These models use added sophistication in treating atmospheric processes, including incorporating multilevel winds and mixing, diurnally varying chemistry according to photochemical modeling, and variable dry-deposition rates. However, the SURADS model has not incorporated cloud processes and wet deposition in published applications. Tests of the SURADS model against the data from the Sulfate Regional Experiment yielded promising results for ambient sulfate conditions but less satisfactory results for sulfur dioxide concentrations (Mueller and Hidy 1982a). The other two models, RTM-II and STEM, incorporate cloud processes and other aspects of precipitation chemistry, but their performance in comparison with observations has not been reported.

Treatment of Transport and Mixing

Because long-range transport is at the heart of the controversies surrounding acid deposition, we review here the ways in which regional-scale models typically treat trajectory analysis.

Meteorologists have approached the transport problem in a number of ways. The simplest method is to use observed values of horizontal winds at specified altitudes to calculate by interpolation where the winds would carry a given air parcel containing the material of interest (i.e., Lagrangian or trajectory model). This type of trajectory model has been widely used and is referred to

TABLE 3.2 Characteristics of Several Regional-Scale Air Pollution Models Used to Assess Acid Deposition

Characteristic	AES[a]	ASTRAP[a]	CAPITA[a]	ENAMAP-1[a]	MEP[a]	MOE[a]
Type	Lagrangian-box	Statistical–trajectory	Monte Carlo	Puff–trajectory	Lagrangian	Statistical
Output	Monthly SO_2 and $SO_4^=$ concentrations and dry and wet S depositions	Monthly SO_2 and $SO_4^=$ concentrations and dry depositions and bulk S wet deposition	Monthly SO_2 and $SO_4^=$ concentrations and dry and wet depositions	Monthly SO_2 and $SO_4^=$ concentrations and dry and wet depositions	Monthly concentration and dry and wet depositions of sulfur	Long-term SO_2 and $SO_4^=$ concentrations; annual dry and wet sulfur depositions
Input	Annual and seasonal SO_2 emissions; daily precipitation amounts; winds and temperatures twice daily at four heights	Annual and seasonal SO_2 emissions for six-layered grid; stack parameters for major sources; 6-h precipitation amounts; wind profiles twice daily	Annual SO_2 emissions; 6-h precipitation probabilities; twice daily upper-air wind profiles and three times daily surface winds	Annual SO_2 emissions; 3-h precipitation amounts; wind profiles twice daily	Annual and seasonal SO_2 emissions; 3-h precipitation amounts; 6-h surface pressures	Point sources; area sources treated as effective point sources; statistics of durations of wet and dry periods and average precipitation rate during wet periods; statistical treatment of winds
Number of cells in the grid	52×37	User specified	52×37	46×41	User specified	User specified
Grid size (km)	127×127	Receptor point locations	127×127	70×70	Point receptors	Point receptors
Analysis of precipitation (by preprocessor)	Objective analysis of daily precipitation amounts	Hourly data are summed to produce 6-h totals across a grid of about a 76-km spacing	No actual precipitation rates used; time averages of precipitation probabilities used for each grid square	Hourly U.S. and 6-h Canadian data are summed to produce 3-h totals	Objective analysis of 3-h amounts	Climatological lengths of Eulerian and Lagrangian wet and dry periods; rate of 1 mm/h

	Monthly climatological heights	Diurnal pattern including nocturnal surface-based inversion; maximum: 1000 (w) and 1800 (s)	Day: 800 (winter); 1200 (spring/fall); 1350 (summer); Night: 300	1150 (winter); 1300 (spring/fall); 1450 (summer)	Varies diurnally over model domain	1000
Mixed layer						
Oxidation rate for SO$_2$ (percent/h)	1.0	Varies diurnally: 0.2-5.5 (summer); avg. of 2.0; 0.1-1.5 (winter); avg. of 0.5	0.6 (winter); 1.2 (summer)	1.0	Seasonal and diurnal variation: mean of 1.0	1.0
Dry deposition velocity (cm/sec)					Seasonal and diurnal variation: mean of	
SO$_2$	0.5	Vary diurnally: 0.45 (summer avg.); 0.25 (winter avg.) (SO$_2$ and SO$_4^{2-}$ similar but not identical)	0.31 (winter); 1.20 (summer)	0.38 (winter); 0.48 (summer)	0.75	0.5
SO$_4^=$	0.1		0.07 (winter); 0.15 (summer)	0.22 (winter); 0.28 (summer)	0.25	0.05

TABLE 3.2 (Continued)

Characteristic	AES[a]	ASTRAP[a]	CAPITA[a]	ENAMAP-1[a]	MEP[a]	MOE[a]
Wet removal rate (percent/h)						
SO_2	$3 \times 10^6 \, P_{24}(t)/H$	—	0.6 PP (winter); 11.7 PP (summer)	28.0 $P_1(t)$	Dependent on pH and temperature	10.8
$SO_4^=$	$85 \times 10^6 \, P_{24}(t)/H$ where H is the mixed layer (mm)	—	5.0 PP (winter); 29.0 PP (summer) where PP = probability of precipitation (%) each 6-h period	7.0 $P_1(t)$	Precipitation rate dependent: mean of 30 $P_1(t)^{0.5}$	36.0 (used in stochastic scavenging model with $\tau_d = 56$ h and $\tau_w = 7$ h (the avg. dry- and wet-period durations)
Bulk S	—	Minimum of $100(P_6(t)/10)^{1/2}$ and 80	—	—	—	—

$P_1(t)$ = liquid precipitation rate (mm/xh)

TABLE 3.2 (Continued, Part 2)

Characteristic	RCDM-3[a]	UMACID[a]	ELSTAR[b]	RTM-II[c]	STEM-I[d]	STEM-II[e]	SURADS[f]
Type	Analytical	Puff-trajectory	Lagrangian trajectory	Hybrid Lagrangian/Eulerian	Time-dependent Eulerian	Time-dependent Eulerian	Eulerian
Output	Monthly concentration and dry and wet depositions of sulfur	Estimates of source contributions to downwind concentrations and contributions of upwind sources on receptors at 6-h time steps	Hourly concentrations of photochemical products and intermediates; dry deposition; budgets for NO_x	3-h, daily, monthly concentrations of SO_2, $SO_4^=$; wet and dry sulfur deposition	Hourly and daily SO_2 and $SO_4^=$ concentrations; instantaneous and accumulated SO_2 and $SO_4^=$ deposition; daily regional sulfur budget	Hourly and daily species concentrations; instantaneous and accumulated deposition rates	Hourly and daily SO_2, $SO_4^=$ concentrations; dry deposition rates; daily SO_2 budget
Input	Emissions and centroids of emissions for area sources; spatially averaged annual and seasonal wet and dry periods; monthly, seasonal, and annual resultant winds and persistence factors	Annual SO_2 emission rates; 3-h precipitation amounts; wind profiles twice daily	Emissions of SO_x, NO_x, HC; winds extrapolated hourly in all layers from surface conditions, temperature profiles	Daily or hourly emissions; meteorological data	Surface and upper-air winds, cloud cover and ceiling, surface roughness, evaporation rates, vertical temperature and humidity profiles	Surface and upper-air winds, cloud cover and ceiling, rain-fall rate, surface roughness, evaporation rates, vertical temperature and humidity profiles	Initial SO_2 and sulfate updated hourly; hourly SO_2 emissions; estimates of NO_x and HC emissions; radar-based precipitation; 12-h synoptic wind field aloft at fine levels; hourly surface winds; temperature profile, sunlight, cloud cover
Number of cells in the grid	70×70	41×32	Variable: 100×100	52×46×2 (episodic); 26×23×2 (long term)	User specified	User specified	Variable: e.g., 20-23
Grid size (km)	80×80	80×80	5×5	40×40 (episodic); 80×80 (long term)	80×80	Regional scale	80×80
Analysis of precipitation (by preprocessor)	Spatially averaged precipitation amounts and the average durations of wet and dry periods	Hourly data summed for 3-h periods for each 80-km grid square	Not included	Objective analysis on hourly data to generate grid-by-grid rates	—	—	Cloud chemistry and scavenging algorithm in fine layers

TABLE 3.2 (Continued)

Characteristic	RCDM-3[a]	UMACID[a]	ELSTAR[b]	RTM-II[c]	STEM-I[d]	STEM-II[e]	SURADS[f]
Mixed layer	600 (winter); 1200 (summer); 1000 (annual)	Varies only with month	4 or 5 layers as function of maximum mixing height on test day	Linear interpolation of daily minimum and maximum; spatial interpolation by spline fit	Calculated, varies diurnally	Calculated, varies diurnally	Five layers to 1500 m; preprocessors calculate mass conservative winds by layer and estimate diurnally variable mixing in each layer
Oxidation rate for SO_2 (percent/h)	Chemical conversion time scale 2.4×10^5 sec (1 percent/h)	Winter: 1.4 (day); 0.1 (night); Summer 2.8 (day); 0.2 (night)	Variable; tied to photochemical model	Function of solar zenith angle and geographic location: northeastern U.S. in summer, varies from 0.1 to 2.0 with 0.78 avg.; northern Great Plains, 0.3 avg.	—	—	Constant (0.5 to 3.0) or diurnally variable as calculated
Dry deposition velocity (cm/sec)	Weighted by the percent of dry time	Varies according to time after sunrise and land use:					
SO_2	0.50	0.10-0.55 (winter); 0.10-0.82 (summer)	Varies with aerodynamic resistance	One-dimension diffusional model, depends on land use and time of day	Calculated based on surface type, evaporation, and meteorological conditions	Calculated based on surface type, evaporation, and meteorological conditions	Diurnally variable, calculated
$SO_4^=$	0.05	0.05-0.28 (winter); 0.03-0.43 (summer)	0.55 times SO_2 rate				0.1 to 1.0 times SO_2 rate

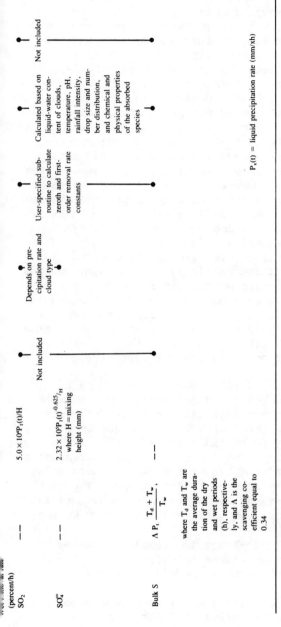

65

(percent/h)

SO$_2$ — $5.0 \times 10^6 P_t(t)/H$ Not included   Depends on precipitation rate and cloud type   User-specified subroutine to calculate zeroth and first-order removal rate constants   Calculated based on liquid-water content of clouds, temperature, pH, rainfall intensity, drop size and number distribution, and chemical and physical properties of the absorbed species   Not included

SO$_4^=$ — $2.32 \times 10^9 P_t(t)^{-0.625}/H$ where H = mixing height (mm)

Bulk S $\Lambda P_t \dfrac{T_d + T_w}{T_w}$,   —

where T_d and T_w are the average duration of the dry and wet periods (h), respectively, and Λ is the scavenging coefficient equal to 0.34

$P_x(t) =$ liquid precipitation rate (mm/xh)

[a]U.S./Canada Work Group #2 (1982).
[b]Lloyd et al. (1979).
[c]Durran et al. (1979); Liu et al. (1982); and Stewart et al. (1983).
[d]Carmichael and Peters (1983).
[e]Carmichael et al. (1983).
[f]Mueller and Hidy (1982b).

as a constant-level (isobaric) trajectory (Heffter and Ferber 1977). The vertical component of the wind is ignored in this approach, so the method does not give true trajectories in many circumstances, such as near frontal zones or over a mountain range.

A more sophisticated, but not necessarily better, approach is to assume an isentropic trajectory. In essence this method takes account of the fact that an air parcel will not in all cases follow a horizontal pressure surface but will remain on a surface of constant potential temperature. Calculations of isentropic trajectories therefore incorporate some large-scale vertical motions. However, the isentropic model still does not account for vertical motions in clouds or when isentropic surfaces intersect the terrain. The more complicated the single-trajectory models become, the less application they have in day-to-day operational use because of increased requirements for computer time and power.

Single forward trajectories are rarely used to investigate precipitation chemistry because many sources usually contribute acidic precursors along the path of flow. An alternative approach is to compare the single backward trajectory to a given precipitation event. In back-trajectory analysis a parcel is traced back in time from a given precipitation event, allowing evaluation of the contributions of possible sources along the path of the parcel.

In the absence of multiple sources along a path, single-trajectory analysis can be useful. For example, the episodic data on precipitation chemistry collected in Bermuda have been analyzed with the technique by Jickells et al. (1982). They found that when airflow was from the North American continent the acidity of the precipitation was an order of magnitude higher than when the rains were associated with flows from other directions. The North American continent evidently is the major source of acidic precursors in Bermuda, which is more than 1,000 km from the mainland.

Limitations of single-trajectory analysis in describing transport have caused meteorologists to develop multiple-trajectory Lagrangian models and models based on calculations at fixed points in space. In the latter (Eulerian) model, concentrations of atmospheric constituents at fixed points are calculated for successive intervals of time based on concentrations calculated for previous intervals. Because the fixed points in space are usually arranged in a grid pattern, the model is also known as a grid model.

Single-trajectory calculations directed backward from receptors are sometimes helpful for identifying source areas frequently contributing to deposition at selected receptor sites. Grid models are most efficiently used for weighting the contributions of many sources to several receptors.

Early work comparing back trajectories to limited event data on precipitation chemistry indicated that in the Northeast the more acidic precipitation came with flows from the Southwest (Kurtz and Schneider 1981, Miller et al. 1978). More recent studies using an improved data base show that single-trajectory analysis at a site in the Northeast suggests a regional acid precipitation phenomenon (Wilson et al. 1982).

Compared with the interior of the United States, a much larger portion of the average annual precipitation in the Northeast is associated with cyclonic disturbances (low-pressure systems). The main tracks of cyclonic storms tend to converge in the northeastern United States and southeastern Canada (see Figure 4.7, Chapter 4). Thus the areas of eastern North America regarded as being particularly sensitive to acid deposition receive precipitation from storm systems with origins at almost any latitude in the interior or the West. Additional evidence that the air shed of the sensitive area is of synoptic scale is presented in Chapter 4. To date, acid-deposition models treat transport and mixing in terms of trajectories and diffusion. Episode models in current use take account of air pollution embedded in synoptic-scale weather systems, but they require great detail in inputs on wind, temperature, and moisture fields.

Verification of any model is critical to its acceptance both as a description of scientific understanding and as a tool for analyzing policy choices. Methods are available for testing the validity of trajectory calculations. Among the first was the use of special balloons, called tetroons, that maintain a constant altitude (Pack et al. 1978). Tetroons were released and tracked on the assumption that they behave like an idealized air parcel, despite the fact that they are confined to a surface of constant density. Because of logistic and other problems, however, this method has all but been abandoned, except in short-range experiments. A more effective method of verifying trajectory calculations is to release an inert tracer, such as perfluorocarbon or an isotopically rare methane (such as CD_4), that can be measured 1,000 km downwind or more. Plans are now being made to release tracers under a number of meteorological conditions to evaluate trajectory calculations.

While techniques for computing trajectories have advanced markedly over the past decade, they still are plagued by a host of uncertainties. One difficulty that all such techniques share is the sparsity of meteorological data. Upper-air soundings are made only twice daily at widely separated stations in the United States.

More fundamentally, trajectory models incorporate simplifying assumptions, such as isobaric or isentropic behavior, that only approximate the true behavior of air masses. Consequently, in situations in which the assumptions are not valid, model calculations may be unreliable. An example is the vicinity of storm systems, in which complex flows and thermodynamic processes limit severely the validity of the single-layered approach.

Two different calculations of a trajectory in the vicinity of a frontal storm system appear in Figure 2.3, Chapter 2. One uses a constant-pressure (isobaric) surface and the other a constant-entropy (isentropic) surface (Davis and Wendell 1977). The trajectories were calculated for the same conditions and start from the same point (near St. Louis). The calculated positions of the two theoretical air parcels after 24 hours of transport are several hundred kilometers apart. Although the isentropic trajectory is expected to be closer to reality, Figure 2.3 contains no information about the location of the "real" trajectory in this case. The plot provides a graphic illustration of the level of uncertainty in current capabilities for predicting trajectories.

The weaknesses in trajectory models are of particular concern because it is precisely in storm environments--for which current models are weakest--that it is most important to trace pollutants with accuracy. Although transport theory suggests that errors in estimates of trajectories in clear air will tend to cancel if averages are taken over multiple events, large systematic biases may be expected in storm situations. Such uncertainties make attribution of specific deposition events to specific sources particularly difficult.

Treatment of Transformation Chemistry

The extent to which realistic chemistry is incorporated into current regional models varies significantly. Most models used recently (see Table 3.2) employ absolutely no chemistry but include instead only a fixed SO_2 transformation rate (usually 1 to 4 percent/h). Results from such models should not be relied on in the development of control strategies for regional air-quality problems in which chemical phenomena play a central role.

Attempts have been made by Rodhe et al. (1981) to include some chemistry (19 reactions) in their transformation-transport model. However, this work includes only a rudimentary scheme of gas-phase reactions involving only molecules of O_3, CO, CH_4, C_2H_4, CH_2O, H_2O_2, NO, NO_2, HNO_3, and H_2SO_4 with intermediates of HO, $O(^1D)$, $O(^3P)$, and HO_2. One aqueous-phase reaction was included superficially as $SO_2 + H_2O_2 + "CLOUD" \rightarrow H_2SO_4$.

It is interesting, however, that Rodhe et al. concluded (p. 139) that "When dealing with a coupled chemical system like the one we have studied one may not assume a proportional dependence of concentrations on emission rates." They suggest, as we do, that models must include the essential chemistry that controls the oxidation of SO_2, NO, NO_2, and hydrocarbons in the atmosphere. It is almost certainly true that the scheme of Rodhe et al. fails to do this properly.

Other chemical reaction schemes have been adapted for regional-scale models. Mueller and Hidy (1982b) have reported the application of a homogeneous gas-phase reaction scheme in the SURADS model. This calculation is run to generate a diurnally varying oxidation rate for sulfate that accounts for hydrocarbon vapor/nitrogen oxide emissions and their atmospheric distribution. This model currently is being extended to incorporate cloud chemistry in the chemical processing. It has been used in an integrated stepwise manner in the ELSTAR trajectory model (e.g., Lloyd et al. 1979). The STEM series of models (e.g., Carmichael et al. 1983) and the air shed model reported by Stewart et al. (1983) incorporate a significant number of gas-phase reactions and cloud processes into an Eulerian grid calculation. The performance of the chemistry in these very complex numerical simulation schemes has not yet been compared with simpler schemes or with each other.

Treatment of Dry and Wet Deposition

Dry deposition is usually treated parametrically in models rather than in terms of the fundamental physics of the processes. For example, simple treatments assume that the rate at which gaseous SO_2 or sulfate aerosol particles reach the ground is directly proportional to the atmospheric concentrations of the respective pollutants near the ground. The constant of proportionality, called the deposition velocity, is often assumed for simplicity to be independent of spatial coordinates and time. Because dry-deposition processes depend on characteristics of ground cover, surface roughness, and stability of the air layer immediately above the surface, more sophisticated treatments use variable deposition velocities according to the spatial dependence of these conditions. The deposition velocity may also be taken to vary diurnally or can be estimated from aerodynamic parameters, such as surface winds. Resistance to gas-phase transfers associated with the stomata of leaves or the assimilation capacity of underlying surfaces can also be incorporated in theoretical treatments. The dry-deposition velocities for sulfate aerosol particles are usually assumed to be somewhat less (between 10 and 50 percent) than those for gaseous SO_2. More generally, the rate of particle deposition is dependent on size, but currently available models have not yet taken this dependence into account. Dry deposition of NO_2 differs somewhat from that of SO_2 and is sometimes taken to be as little as one half that of SO_2.

There are no regionally representative measurements of dry deposition. Therefore the parameterizations adopted in models have not been tested extensively. Instead they are based on limited experiments in the field or in wind tunnels. Thus the limitations of the parameterizations have not been established quantitatively. On the basis of comparison of model outputs with measurements, parameterization of dry deposition should be known within a factor of 2, but the range of measurements is much larger than this (e.g., Sehmel 1980).

The parameterization of wet deposition is extremely difficult for episode calculations because of the variability in time and space of cloud cover, cloud depth, and precipitation. Over a long period of time, such as a year, averaged wet deposition is assumed to be proportional to the total quantity of precipitation or the precipitation rate. The parameterization of wet deposition typically is one of two types, both of which should be considered as being in rudimentary stages of development.

The first class of parameterization involves the scavenging coefficient, which is defined as the rate at which an air pollutant is incorporated into precipitation per unit volume divided by the local concentration of the pollutant. If the scavenging coefficient is known, then wet removal can be calculated in a relatively straight-forward manner on the basis of airborne pollutant concentrations. Although in principle the scavenging coefficient varies in space and time, the usual practice is to use "storm-average" values.

The second class of parameterization involves the scavenging ratio, which is the ratio of the concentration of the pollutant in precipitation at ground level to the concentration of the pollutant in air at ground level. The scavenging ratio is somewhat easier to apply in practice than the scavenging coefficient because the former does not require a vertical integration of pollutant concentration to obtain the deposition rate.

Both the scavenging coefficient and the scavenging ratio have well-established theoretical bases. Considerable theoretical work has been applied to their physical evaluation. The parameterizations suffer from the difficulty, however, that they consolidate the effects of a large number of complex physical processes. As a consequence the parameters conceal a great deal of uncertainty, particularly when aqueous-phase transformation processes are involved. The parameters actually used in regional models tend to be based more on empirical data than on analysis of actual mechanistic behavior. The proportionality coefficients are often deduced from comparison of ambient air concentrations and precipitation chemistry or from semiempirical models (e.g., Scott 1978, 1981). Recent diagnostic models of scavenging appear to be moderately successful in resolving some of these individual mechanisms, and these techniques can be expected to be applied to regional models in the future.

Considering the crudeness of parameterization of wet and dry deposition, it is surprising that the linear models perform better in estimating the long-term average wet deposition of sulfate than in estimating the ambient concentrations.

Linearity or Nonlinearity in Theoretical Models

Physical Processes

A deterministic source-receptor model generally provides
the solutions (or approximations to the solutions) of a
set of conservation equations for each species of the form

$$\frac{\partial c_i}{\partial t} = -\vec{\nabla} \cdot c_i \vec{v} - w_i + r_i.$$

The relationship, called the continuity equation, simply
expresses a material balance for pollutant species i in
terms of the time rate of change of its time-averaged
concentration (c_i), the flow and diffusion field
($\vec{\nabla} \cdot c_i \vec{v}$), its rate of removal by deposition
processes (w_i), and its net rate of production by
chemical reaction (r_i). Each of the parameters in the
equation in general varies as a function of position $\vec{x}$
and time. This equation, together with suitable boundary
and initial conditions, is considered to be a complete
mathematical description of any constituent in the air.
   A profusion of methods has been applied for solving
equations of this type, and a correspondingly large
number of "models" for solutions exists. The characteristics of a number of models are given in Table 3.2. A
comprehensive discussion of the methods of solution used
in these models is beyond the scope of this report; it is
important, however, to recognize that, whatever the class
to which a particular model belongs, the model still may
be described in terms of the continuity equation or its
derivatives and may be viewed as a solution to these
equations with appropriate initial and boundary
conditions.
   Recently, in connection with proposals for mitigating
acid deposition, the question has arisen as to whether
deposition rates are linear functions of emissions. It
is therefore appropriate to discriminate between linear
and nonlinear models and to indicate their significant
differences.
   The continuity equation and its boundary conditions
can be mathematically linear or nonlinear, depending on
whether nonlinear operators act on the dependent variable
c_i, the concentration of species i. A linear operator
L is one that satisfies the relationship

$$L(\alpha c_i + \beta c_j) = \alpha L(c_i) + \beta L(c_j),$$

where α and β are arbitrary constants. Thus, for example, if the net rate of production (r_i) of species i through chemical reactions is proportional to the first power of c_i, such as

$$r_i = \kappa_1 c_i,$$

r_i is linearly related to c_i. If, however, $r_i = \kappa_2 c_i^2$, the relationship is nonlinear.

While detailed discussion of the continuity equation and its solutions is beyond our current purpose, we note that solutions to linear and nonlinear forms of the continuity equation have characteristics that are of practical importance in formulating control strategies. An important feature of linear systems of this type is that the results satisfy the principle of superposition. From the characteristics of linear operations we can find the combined consequences of, say, two sources simply by summing the contributions of each calculated without regard for the other. This practice is central to Lagrangian trajectory modeling and is not applicable if nonlinear processes predominate.

Let us suppose, for example, that a simple form provides the following relationship between the magnitude of the pollution source S_{ij} at location $\bar{x}_j$ and a resulting ambient concentration at a specified location $\bar{x}$:

$$c_i(\bar{x},t) = k_{ij} S_{ij}(\bar{x}_j,t).$$

The coefficient k_{ij} depends on the separation of the source and receptor locations $(\bar{x} - \bar{x}_j)$ and is assumed to be constant. The concentration c_i is proportional to the first power of S_{ij}, which is to say that c_i is linearly related to S_{ij}. For the linear form of this type, a change in S_{ij} would result in a proportionate change in c_i. For example, a 50 percent reduction in S_{ij} would result in a 50 percent reduction in c_i.

Simple linearity does not, however, guarantee proportionate (one-for-one) reductions. Suppose, for example, that there are a number of sources (N, for example) influencing c_i and that a substantial background concentration B_i (due to natural sources) is present. In this more general case, c_i has the form

$$c_i(\bar{x},t) = \sum_{j=1}^{N} k_{ij} S_{ij} + B_i(\bar{x},t),$$

where the symbol $\sum$ indicates a sum of the contributions from each of the N sources. This result is still linear in the jth source strength, S_{ij}, but if B_i and the contributions from the other $N - 1$ sources are not very small compared with $k_{ij}S_{ij}$, a decrease in S_{ij} would produce a less than proportionate change in c_i.

Although noted here in a rather simplistic sense, the linear relationship is the underlying basis for the so-called linear-rollback strategy of emission reduction.

Because of the complexity of atmospheric processes, it is unreasonable to expect that a simple linear equation with constant coefficients would provide an accurate description of source-receptor relationships. A more useful expression has the form

$$c_{ij}(\bar{x},t) = \sum_{j=1}^{N} p_{ij}S_{ij} + B_i(\bar{x},t),$$

where the parameter p depends on atmospheric variables, such as wind speed and rainfall rate, and changes in time and space appropriately to reflect the dependence of the atmospheric variables on time and space. Most regional source-receptor models in current use yield results of this type. Although p is functional in form, this expression is still a linear equation and necessarily stems from a linear form of the continuity equation. The implications regarding superposition and rollback discussed in the context of the simpler linear result apply here as well, provided that the functional dependence of p does not inadvertently include S.

It is in fact more realistic to expect that the function p depends explicitly or implicitly on S, for example, through the concentrations of other atmospheric species. In this case, we can write the solution to the continuity equation in the same form but with a new factor of proportionality q that depends explicitly on S:

$$c_{ij}(\bar{x},t) = \sum_{j=1}^{N} q_{ij}(S_{ij})S_{ij} + B_i(\bar{x},t),$$

where again the variables in the argument of q are functions of time and space. This result is nonlinear since S_{ij} appears both functionally in the argument of q_{ij} and as a first-order multiplier. The dependent variable c_i no longer depends on the first power of S_{ij} (the independent variable).

Because a number of atmospheric processes are
nonlinear, the most accurate and complete model of the
atmosphere must be nonlinear. There are computational
advantages to linear mathematical systems, however, and a
corresponding tendency to approximate physical processes
in a manner such that linearity is obtained. Such "lin-
earization" is a common practice in science and engineering; its success depends on the degree of deviation from
linearity of the phenomenon being studied as well as on
the intended application. Linearized models may do well
in simulating observed regional deposition patterns, for
example, whereas their capability to predict responses to
specific changes in emissions may be comparatively poor.

Chemical Processes

Most researchers who have analyzed regional air-quality
data have assumed a linear mechanism for transforming
SO_2 into sulfate with a rate between 0.1 and 4.0
percent/h (Table 3.2). More complex reaction schemes
generally have not been used to rationalize results from
observations. By their nature, linear models predict a
form of proportional response to emission reductions. In
some cases this has previously been demonstrated to be an
extremely poor assumption (e.g., in the control of photochemical oxidants in urban areas). Linearity may also be
a poor assumption for circumstances involving acid precipitation. There is substantial support, however, for
the argument that currently we simply do not understand
the atmospheric interactions sufficiently well to supply
the mathematical detail required by nonlinear concepts.
    Some attempts have recently been made to incorporate
chemical nonlinearities into models of acid precipitation.
Most of these schemes concentrate on the production of
acid sulfate as an aerosol rather than on aqueous-phase
processes in clouds. As described earlier in the section
on the treatment of transformation chemistry, Rodhe et
al. (1981) employed a highly simplified but seemingly
realistic chemical scheme for SO_2 and NO_x oxidation
and found some very nonlinear effects. For example,
their results indicated that a 10 percent increase in
NO_x emissions would lead to a 5 percent reduction in
sulfate concentration downwind from the source region,
and a 10 percent decrease in SO_2 emission would result
in only a 3 to 4 percent decrease in sulfate production.

Nonlinearity in the SO_2 transformation of the Rodhe
et al. model was also observed in subsequent studies by
Sampson (1982). He employed a somewhat improved hydrocarbon reaction scheme, but in other respects the
mechanism was identical to that employed by Rodhe and his
co-workers. Some of Sampson's results are reproduced in
Figure 3.1. The solid lines in the figure give Sampson's
results for the percentage change in ambient sulfate
concentration after 24, 48, and 96 hours of transport
from the source region as a function of changes in SO_2
emissions. The results of the model suggest that a
relatively small reduction in sulfate levels (roughly 15
percent) may result for long transport times (96 hours)
from a 50 percent reduction in SO_2 emissions.

The results of Rodhe et al. and Sampson should be
treated with caution. The so-called Rodhe-Crutzen-
Vanderpol model used in both studies employed specific
sequences of chemical reactions and assumed uniform
additions of polluted background air throughout the
period of transport and transformation. Different
choices of oxidation pathways and changes in the strong
background source may alter the results significantly.

For example, the dashed lines of Figure 3.1 are the
result of running Sampson's computer program without
continuous dilution of the product mixture with
background air containing sulfate (P.J. Sampson,
University of Michigan, personal communication, 1982).
The shift toward the linear curve (from the solid to the
dashed curves in Figure 3.1) is the result of eliminating
the trivial source of nonlinearity arising from the
background source, term B in the equation, c = kS + B,
considered earlier. The dashed lines of Figure 3.2 are
the result of both deleting the background source of
sulfate and selecting an alternative pathway for the
homogeneous gas phase oxidation of SO_2. Note that the
alternative assumptions give a result that is essentially
linear (with proportionality constants less than unity).

The original Rodhe-Crutzen-Vanderpol model employed
reaction (3.1) for oxidation of SO_2,

$$HO + SO_2 \rightarrow H_2SO_4, \tag{3.1}$$

whereas the modification that produced the dashed curves
of Figure 3.2 used

$$HO + SO_2 \ (+ \ O_2, \ H_2O) \rightarrow H_2SO_4 + HO_2. \tag{3.2}$$

Reaction (3.1) is a single, simplified reaction in which an attempt is made to condense the chemistry that occurs in and following the primary hydroxy radical attack on SO_2:

$$HO + SO_2(+M) \rightarrow HOSO_2(+M) \qquad (3.3)$$

[Equation (A.56) in Appendix A]. See Appendix A for a more complete discussion of the reaction.

The use of reaction (3.1) is equivalent to assuming that the addition of the hydroxy radical to SO_2 terminates the chain reactions of the HO radical, and by some undefined process the initial product of reaction (3.3) leads to H_2SO_4 without regenerating a chain-carrying species. The assumption of reaction (3.1) perturbs the atmospheric reaction cycles involving HO_2 and HO radicals, which result in the oxidation of hydrocarbons, aldehydes, CO, SO_2, NO, NO_2, and other impurity species. For example, the oxidation of CO occurs in reactions (3.4) through (3.6) by way of HO-radical attack on CO:

$$HO + CO \rightarrow H + CO_2, \qquad (3.4)$$
$$H + O_2(+M) \rightarrow HO_2(+M), \qquad (3.5)$$
$$HO_2 + NO \rightarrow HO + NO_2. \qquad (3.6)$$

Note that although an HO radical is lost in reaction (3.4), another is regenerated in reaction (3.6). Similar cycles occur involving CH_2O and the hydrocarbons, for example. Now if a reaction such as (3.1) occurs, an HO radical is removed; no further regeneration of the HO radical occurs.

In writing reaction (3.2), we assume in accordance with experience in other atmospheric reaction cycles that a chain-carrying radical (HO_2) is developed following the occurrence of reaction (3.3). For example, reaction (3.2) summarizes the net result of the sequence (3.3), (3.7), and (3.8):

$$HO + SO_2(+M) \rightarrow HOSO_2(+M), \qquad (3.3)$$
$$HOSO_2 + O_2 \rightarrow HO_2 + SO_3, \qquad (3.7)$$
$$SO_3 + H_2O \rightarrow H_2SO_4. \qquad (3.8)$$

Presumably, reaction (3.7) would often be followed by regeneration of the HO radical through reaction (3.6), at least in NO-rich polluted atmospheres.

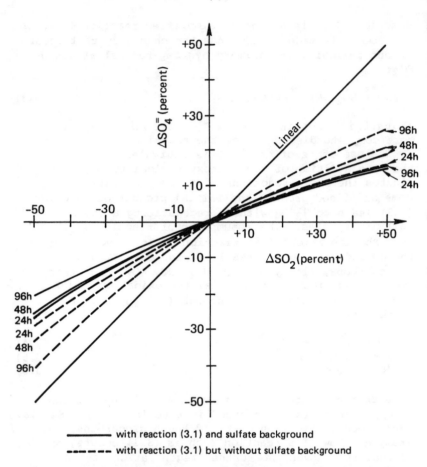

——— with reaction (3.1) and sulfate background

— — — with reaction (3.1) but without sulfate background

FIGURE 3.1   Effect of the assumption of background sulfate on the Rodhe-Crutzen-Vanderpol model for chemical transformation. SOURCE: Sampson (1982) and P.J. Sampson, University of Michigan, personal communication (1982).

The participation of reaction (3.1) results in a direct nonlinear feedback into the SO_2 oxidation mechanism, while reaction (3.2) does not seriously perturb the concentration of the hydroxy radical. The best available experimental evidence today supports the contention that the HO level in reacting mixtures of hydrocarbons, NO_x, and SO_2 is relatively insensitive to SO_2 concentrations and that the sequence (3.3), (3.7), (3.8), or some similar chain-propagating reactions, is important (Stockwell and Calvert 1983). In the experiment, Stockwell and Calvert varied the amount of

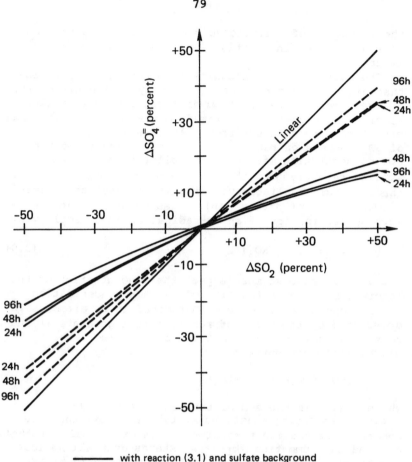

FIGURE 3.2 Effect of the assumptions of background sulfate and chain termination on the Rodhe-Crutzen-Vanderpol model for chemical transformation. SOURCE: Sampson (1982) and P.J. Sampson, University of Michigan, personal communication (1982).

SO_2 in dilute, irradiated mixtures of CO, HONO, and NO_x in air (at 1 atm), monitored the concentration of HO radicals by measuring the rate of formation of CO_2, and observed the ultimate formation of H_2SO_4 aerosol as identified by its infrared spectrum. Within the limits of experimental error, the concentration of HO radical was found to be insensitive to the concentration of SO_2 even when as much as one half of the HO radicals in the system reacted with SO_2 leading ultimately to the formation of sulfuric acid aerosol. Chain termination as implied in reaction (3.1) was found not to be important.

The main point to recognize from this discussion is that either an apparent near linearity or a nonlinearity in the model may result from different, rather subtle, simplifying assumptions related to the choice of chemical mechanism. We conclude from these results that deviations of SO_2 conversion rates from linearity with respect to SO_2 concentration may be much smaller than has been implied recently from the results of simulations employing the seemingly realistic yet simplified reaction schemes.

Generation of nitric acid in gas-phase reactions does involve termination of an HO-radical chain directly via

$$HO + NO_2 (+M) \rightarrow HNO_3 (+M), \tag{3.9}$$

and in this case we must expect the concentration of the reactant HO to be a function of the NO_2 concentration.

The concentration of the HO radical in an air mass is determined by the rates of reaction that generate it and those that destroy it. That is, at any time t the steady-state concentration of HO is given by

$$[HO] = \Sigma (R_i)_t / \Sigma k_i [A_i]_t,$$

where $\Sigma (R_i)_t$ is the sum of the rates of all HO-radical generating reactions at time t, k_i is the rate constant for the ith removal reaction of HO with reactant A_i, and the summation $\Sigma k_i [A_i]_t$ extends over all HO-loss reactions. It should be noted that reaction (3.9) is only one of several $HO-HO_2$-radical chain termination reactions that occur in the troposphere. Thus in theory the effect of small changes in the concentration of NO_2 on the concentration of HO is not expected to be dramatic. For example, computer simulations of the chemistry of the polluted atmosphere (see the mechanisms of Calvert and Stockwell 1983) show that only about 10 percent of the $HO-HO_2$-radical termination occurs through the $HO-NO_2$ reaction (3.9) for a tropospheric air mass typical of an urban, polluted area with an ambient concentration of NO_x of 100 ppb at sunrise. Air masses containing one tenth and one one-hundredth of this concentration of NO_x at sunrise, but the same levels of other pollutants as before, give about 0.1 and 0.01 percent of the total $HO-HO_2$-radical chain termination through reaction (3.9). The time dependence of the concentrations of reactants that form HO or react to destroy it are complex functions of the initial pollutant concentrations, so that the quantitative effect of the concentration of HO on NO_x initial concentration can be obtained only through detailed calculations. However, the net effect of lowering the initial NO_x concentration by a factor of 10 (from 100 to 10 ppb) while keeping all other impurities at the same fixed level of the highly polluted air mass is to lower the maximum HO concentration from 1.6×10^{-7} to 0.96×10^{-7} ppm, only a factor of about 1.7. Clearly the dependence of HO concentration is not so sensitive to NO_x concentration as one might have expected at first consideration. Thus a more detailed analysis of the complex homogeneous chemistry of the troposphere predicts that the relationship between changes in ambient concentrations of SO_2 and changes in gas-phase formation of sulfate should exhibit only small deviations from linearity. The simple theoretical considerations of Oppenheimer (1983) lead to the same conclusion.

Nonlinear conversion of SO_2 to sulfate can in theory result from the liquid-phase oxidation of SO_2 (HSO_3^-) by hydrogen peroxide (H_2O_2). For certain atmospheric conditions a limited supply of H_2O_2 may exist in the atmosphere through gas-phase reactions (3.10) to (3.12):

$$2HO_2 \rightarrow H_2O_2 + O_2, \tag{3.10}$$
$$HO_2 + H_2O \rightarrow HO_2 \cdot H_2O, \tag{3.11}$$
$$HO_2 \cdot H_2O + H_2O \rightarrow H_2O_2 + H_2O + O_2. \tag{3.12}$$

The rate of hydrogen peroxide generation in reactions (3.10) and (3.12) depends on the square of the HO_2 radical concentration.

In NO-rich polluted atmospheres, however, reaction (3.6), the rate of which is proportional to the first power of the HO_2 concentration, competes favorably for HO_2 radicals. Reaction (3.6) is very fast in NO-rich atmospheres, with the result that the generation of H_2O_2 in reactions (3.10) and (3.12) is suppressed. Although the uptake of the limited H_2O_2 into cloud water and rain will take place efficiently, for these circumstances the amounts of H_2O_2 may be significantly less than those of HSO_3^- in the water. Obviously, the oxidation of only a fraction of the HSO_3^- can occur for these conditions, and the reaction becomes oxidant limited. SO_2 in cloud water cannot be oxidized faster than the oxidant is provided to the droplet.

Note that for the case of an oxidant-limited reaction, a nonlinear response in sulfuric acid deposition will result from emission reductions. Only when SO_2 emissions are reduced so that ambient concentrations of SO_2 approach the level of the oxidant present in cloud water will a decrease in the sulfuric acid formation and deposition result. For example, if the H_2O_2 available in cloud water were consistently only 40 percent of the SO_2 that is dissolved in the cloud water at a given location, and if oxidation occurred largely through the H_2O_2-HSO_3^- reaction, then a 60 percent reduction of the SO_2 would result in no reduction in the sulfuric acid in cloud water at this location, but subsequent reductions would lead to proportionally lower acid formation and deposition. It is also possible that even with sufficient oxidant in cloud water, other substances that may also be present, such as formaldehyde, may inhibit the H_2O_2-HSO_3^- reaction. Existing analytical data for H_2O_2 in clouds do not allow an unambiguous conclusion to be reached today on the possible importance of these nonlinear effects.

Limitation of oxidant for HSO_3^- or SO_2 may arise because of physical processes as well as the chemical influence described. For example, we have noted previously that it is likely that H_2O_2 vapor present in dry air dissolves in cloud droplets to provide oxidant for the conversion of SO_2. Hydrogen peroxide is a very soluble gas and may be rained out early in some storm systems, leaving a significant fraction of SO_2 vapor unreacted.

Several types of nonlinear effects may be expected from factors not immediately related to oxidant levels. For example, as described in Chapter 2 and Appendix A, there is some evidence that SO_2 is oxidized more readily in the aqueous phase than in the gas phase. Also, increased concentrations of alkaline soil dust in the air due to drought or changing wind patterns can result in the neutralization of precipitation acidity.

In the absence of extensive measurements, we judge that nonlinear effects of SO_2 emission control on acid deposition that arise from chemical conversion mechanisms are probably small for the gas-phase conversion steps, but significant nonlinearity is anticipated for certain special conditions such as an oxidant-limited H_2O_2-HSO_3^- reaction in cloud water. However, these conditions cannot be tested from the existing data base.

FINDINGS AND CONCLUSIONS

Application of current air-quality models to regional-scale processes has provided guidance on the significance of dynamic processes influencing sulfur deposition. Theoretical models have provided results that are qualitatively consistent with empirical observations, thus demonstrating important temporal and spatial scales of source-receptor relationships. Qualitatively the models have pointed to the importance of certain geographical groupings of SO_2 sources and the potential influence of the sources on certain receptor areas. However, current models have not provided results that give confidence in their ability to translate SO_2 emissions from specific sources or localized groupings of sources to specific sensitive receptors. Little has been done in models to translate NO_x emissions into nitrate deposition or to link sulfate and nitrate to acid (H+) deposition. These capabilities are considered essential for models to be used to study the consequences of alternative control strategies in circumstances in which long-range transport processes are involved.

Because of the simplifying assumptions that are made in order to develop practical, economical regional-scale models of air quality and because data are not available to validate or verify them, researchers in the field generally have only limited confidence in current results. The models and their results are useful research tools. However, because of deficiencies in the base of meteorological data required as input and because of the sensitivity of their output to simplifying assumptions regarding both the physical and chemical processes, we do not regard currently available models as sufficiently developed to be used with confidence in predicting responses of the atmospheric system to alternative control strategies.

Despite these limitations, theoretical models are and probably will continue to be used in industrial and urban planning, for which spatial scales are smaller than those of interest in acid deposition. Given the state of knowledge of the physics and chemistry of the atmosphere in the context of long-range transport of air pollution, and given the state of the art of techniques for making quantitative estimates, we advise caution in projecting changes in deposition patterns that result from changes in emissions of precursor gases.

On the basis of laboratory evidence, we conclude that an alternative to the model of Rodhe et al. (1981), which has been widely used to represent the chemical processes involved in acid deposition, more correctly employs gas-phase reactions leading to oxidation of SO_2 that results in $HO-HO_2$-radical chain propagation. Laboratory evidence suggests that chain-terminating reactions involving SO_2 probably play only a minor role in atmospheres polluted with SO_2. When the Rodhe-Crutzen-Vanderpol model is modified so that SO_2 oxidation does not terminate chains, the nonlinearity in the relationship between changes in ambient SO_2 concentrations and changes in ambient sulfate concentrations (i.e., the commonly reported result of the Rodhe model) is greatly reduced.

Laboratory and field studies as well as theory suggest that oxidation of SO_2 in cloud water is rapid and complete, provided that concentrations of oxidants (H_2O_2, O_3) are sufficient (see Chapter 2). Measurements of oxidant concentrations in cloud water, although limited, suggest that concentrations may be sufficient in eastern North America for complete oxidation of SO_2, except perhaps in winter. If this is the case, then strong deviations from linearity in the relationship between changes in annual average ambient SO_2 concentrations and changes in the net production of sulfate in clouds would not be expected.

The relationships between emissions of SO_2 and NO_x and the deposition of sulfuric and nitric acids are complex. Models to predict patterns of the deposition of hydrogen ion will have to account for neutralizing substances as well as sulfuric and nitric acids. Assuming that the ambient molar concentrations of NO_x and basic substances (such as ammonia and calcium carbonate) remain unchanged, we conclude that a reduction in sulfate deposition will result in at least as great a reduction in the deposition of hydrogen ion.

REFERENCES

Calvert, J.G., and W.R. Stockwell. 1983. Deviations from the O_3-NO-NO_2 photostationary state in tropospheric chemistry. Can. J. Chem. (in press).

Carmichael, G.R., and L.K. Peters. 1983. An Eulerian transport/chemistry removal model for SO_2 and sulfate. Atmos. Environ. (in press).

Carmichael, G.R., T. Kitada, and L.K. Peters. 1983. A second generation combined transport/chemistry model for regional transport of SO_x and NO_x compounds. In Proceedings of the 13th International Technical Meeting on Air Pollution and Its Application. Isle des Embiez, France. September 1982. North Atlantic Treaty Organization.

Charlson, R.J., D.F. Covert, P.V. Larson, and A.P. Waggoner. 1978. Chemical properties of tropospheric sulfur aerosols. Atmos. Environ. 12:39-53.

Davis, W.E., and L.L. Wendell. 1977. Some Effects of Isentropic Vertical Motion Simulation in a Regional Scale Quasi-Lagrangian Air Quality Model. BNWL-2100 PT 3. Richland, Wash.: Battelle Pacific Northwest Laboratories.

Durran, D.R., M.J. Meldgin, M.-K. Liu, T. Thoem, and D. Henderson. 1979. A study of long-range air pollution problems related to coal development in the Northern Great Plains. Atmos. Environ. 13:1021-1037.

Galloway, J.N., and D.M. Whelpdale. 1980. An atmospheric sulfur budget for eastern North America. Atmos. Environ. 14:409-417.

Granat, L. 1976. A global atmospheric sulphur budget. SCOPE report No. 7. Ecol. Bull. (Stockholm) 22:102-122.

Granat, L., H. Rodhe, and R.O. Hallberg. 1976. The global sulphur cycle. SCOPE Report No. 7. Ecol. Bull. (Stockholm) 22:89-134.

Heffter, J.L., and G.L. Ferber. 1977. Development and verification of the ARL regional-continental transport and dispersion model. Pp. 400-407 in Proceedings of the Joint Conference on Application of Air Pollution Meteorology. Boston, Mass.: American Meteorological Society.

Jickells, T., A. Knap, T. Church, J. Galloway, and J. Miller. 1982. Acid rain in Bermuda. Nature 297:55-57.

Kurtz, J., and W.A. Schneider. 1981. An analysis of acid precipitation in south-central Ontario using air parcel trajectories. Atmos. Environ. 15:1111-1116.

Liu, M.-K., D.A. Stewart, and D. Henderson. 1982. A mathematical model for the analysis of acid deposition. J. Appl. Meteorol. 21:859-873.

Lloyd, A.C., F.W. Lurmann, D.A. Godden, J.F. Hutchins, A.Q. Eschenroeder, and R.A. Nordsieck. 1979. Development of the ELSTAR Photochemical Air Quality Simulation Model and Its Evaluation Relative to the LARPP Data Base. NTIS PB-80-188-139. Springfield, Va.: National Technical Information Service.

Miller, J.M., J.N. Galloway, and G.E. Likens. 1978.
Origins of air masses producing acid precipitation at
Ithaca, New York: a preliminary report. Geophys. Res.
Lett. 5:757-760.

Mueller, P.K., and G.M. Hidy. 1982a. The Sulfate Regional
Experiment. Report of Findings. Report EA-1901. Palo
Alto, Calif.: Electric Power Research Institute.

Mueller, P.K., and G.M. Hidy. 1982b. The Sulfate Regional
Experiment: Regional Air Quality Modeling
Documentation. Report EA-1907. Palo Alto, Calif.:
Electric Power Research Institute.

Oppenheimer, M. 1983. The relationship of sulfur
emissions to sulfate in precipitation. Atmos.
Environ. 17: 451-460.

Pack, D.H., G.F. Ferber, J.L. Heffter, K. Telegadad, J.K.
Angell, W.H. Hoecker, and L. Machta. 1978. Meteorology
of long-range transport. Atmos. Environ. 12:425-444.

Rodhe, H., P. Crutzen, and A. Vanderpol. 1981. Formation
of sulfuric and nitric acid in the atmosphere during
long range transport. Tellus 33:132-141.

Sampson, P.J. 1982. On the Linearity of Sulfur Dioxide to
Sulfate Conversion in Regional Scale Models.
Washington, D.C.: Office of Technology Assessment.

Scott, B.C. 1978. Parameterization of sulfate removal by
precipitation. J. Appl. Meteorol. 17:375-1389.

Scott, B.C. 1981. Sulfate washout ratios in winter
storms. J. Appl. Meteorol. 20:619-625.

Sehmel, G. 1980. Particle and gas dry deposition: A
review. Atmos. Environ. 14:983-1012.

Shinn, J.H., and S. Lynn. 1979. Do man-made sources
affect the sulfur cycle of northeastern states?
Environ. Sci. Technol. 13:1062-1067.

Stewart, D.A., R.E. Morris, M.-K. Liu, and D. Henderson.
1983. Evaluation of an episodic regional transport
model for multi-day sulfate episode. Atmos. Environ.
(in press).

Stockwell, W.R., and J.G. Calvert. 1983. The mechanism of
the $HO-SO_2$ reaction. Atmos. Environ. (in press).

U.S./Canada Work Group #2. 1982. Atmospheric Sciences and
Analysis. Final Report. H.L. Ferguson and L. Machta,
cochairmen. Washington, D.C.: U.S. Environmental
Protection Agency.

Wilson, J.W., V.A. Mohnen, and J.A. Kadlecek. 1982. Wet
deposition variability as observed by MAP3S. Atmos.
Environ. 16:1667-1676.

4 Empirical Observations and Source-Receptor Relationships

The analysis of data taken in the field complements the development of theoretical models as a means for understanding both the phenomenon of acid deposition and the responses of the atmospheric system to alternative emission-control strategies. Field measurements not only reveal insights into the nature of the atmospheric processes involved in the deposition of acid-forming materials but also may hold the greater promise of providing direct, unequivocal evidence from which responses to mitigating strategies might be assessed. Data from which the spatial and temporal distributions of SO_2 and NO_x over large regions of North America might be derived have not been available until recently. Similarly, reliable sampling of the chemistry of precipitation in North America is a recent development. Even so, sampling of ambient pollutants in the atmosphere and sampling of those in precipitation generally have not been simultaneous. There are no direct measurements of regional dry deposition for gases or particles. [For a description of the monitoring systems in North America and data interpretation, see U.S./Canada Work Group #2 (1982).]

Although data are relatively sparse in North America, those that are available tell us much about the phenomenon. A more extensive data base exists for northern Europe, where atmospheric SO_2 and sulfate in precipitation have been monitored systematically for several decades. Much of our understanding of acid deposition has come from analysis of the European experience and the transfer to and replication of those data on this continent. That being so, differences in patterns of emissions, climatic factors, ground cover, and influences of marine air between Europe and North

America require that our understanding be tested against North American data.

In this chapter we review the existing data for North America and use the data to assess the extent to which the relationships between emissions and deposition can be inferred from observations. The limitations of available data, based on conventional sampling and measurement methods, are discussed first. Results from some pertinent field programs are surveyed to illustrate the importance of meteorological processes for the variability of ambient air pollutants and acid-forming components in precipitation. Results of statistical analyses of data on ambient concentrations of pollutants are reviewed, and data on both emissions and deposition are analyzed to infer the influence of sources on deposition.

Our survey is not a comprehensive one; it focuses on the results that bear most directly on atmospheric transport and transformation. In this context, the issues addressed are whether data are sufficient to infer (1) the extent to which a given region of sources affects receptor sites in remote locations and (2) the importance of nonlinear processes in determining the relationships between the magnitudes of emissions in source regions and the quantities of acid-forming substances deposited in receiving regions.

One of the greatest difficulties in establishing relationships between sources of pollution and conditions at receptors is accounting for the influences of atmospheric processes on the behavior of pollutants. The atmospheric processes involved include airflow, mixing, and chemical transformations. These processes are responsible, directly or indirectly, for the distribution and rate of deposition. Attempts to discern the influences of atmospheric processes on source-receptor relationships have taken different routes, including (1) descriptive accounts of observations, (2) analysis of data segregated by airflow from source areas (trajectories), (3) statistical analyses of data, and (4) inference from source tracers. In this chapter we describe the approaches used in these types of analyses and the results obtained. We also analyze existing data on emissions and deposition to discern trends and the relative behavior of sulfur and nitrogen emissions in the atmosphere.

AEROMETRIC DATA AND THEIR LIMITATIONS

Most of the historical data on ambient air concentrations describe urban conditions. A large body of monitoring data exists in the National Aerometric Data Base (NADB), but few of these observations have been analyzed or interpreted. One of the more reliable and complete data sets available that describes regional air quality in the eastern United States was taken during a single year, 1977-1978: the Sulfate Regional Experiment (SURE) (Mueller and Hidy 1983). Unfortunately, few precipitation chemistry data were collected during the period of the study. In contrast to other data sets, those obtained from the SURE experiment have been analyzed in detail. Complementary data are or will be available from the western United States for 1980-1982 as a result of several studies (Allard et al. 1981, Pitchford et al. 1981). Observations of precipitation chemistry have been made periodically in the mid-1950s, from 1959 to 1966, and from 1972 to date in the programs described, for example, by Wisniewski and Kinsman (1982). Air-quality and precipitation data have been collected in Canada since 1974 in the Canadian (CANSAP) monitoring program.

Precipitation data provide a direct measure of wet deposition. Since there are no direct measurements of dry deposition, regional patterns are estimated from ambient concentrations and deposition velocities (Appendix C). Since deposition velocities depend on the airflow near the surface, surface properties, and cover, these calculations are believed to be uncertain for quantitative evaluation.

The quality of the data describing regional-scale processes is variable. The precision and accuracy of the measurements are generally not well defined in work reported prior to the late 1970s. Recent programs, however, have made considerable progress in reporting data with supplemental information on the calibration of instruments, as well as the errors contained in the observations. A definitive discussion of aerometric data is included in Mueller and Hidy (1983). The quality of recent data on wet deposition is less formally documented, but extensive information on the comparability of data is available from the Illinois Water Survey Laboratories, the Department of Energy's Environmental Measurement Laboratories in New York, and the Environmental Protection Agency's Environmental Monitoring Systems Laboratory. A statistical analysis of data from two independent

precipitation sampling networks (SURE and MAP3S) covering
a similar region in the eastern United States indicated
that there is good agreement between data on the
concentrations of the major ions, H^+, sulfate, nitrate,
and ammonium, between the two networks (Pack 1980).
Documentation of the quality of data is essential to
ensure that comparisons will be possible with data from
future studies. Experiments and monitoring programs
should incorporate such efforts to avoid or at least
minimize controversies about data interpretation. The
integrity of the sampling and analytical methods employed
is another important consideration in measurement quality.
Despite years of effort in the development of methods,
the instrumentation used in past and current programs is
subject to serious interferences. Uncertainties in
results can be large. Interferences or ambiguities in
the methods are particularly serious under rural or
remote conditions, in which concentrations of pollutants
or deposition levels are low.

Listed in Table 4.1 are examples of sampling or measurement uncertainties that exist in currently available
methods of sampling and analysis. From the table it is
evident that few of the important chemically related
measurements can be made without ambiguity because of
uncertainties in the methods.

RELATIONSHIPS AMONG AEROMETRIC PARAMETERS

Atmospheric measurements provide a basic description of
spatial and temporal distributions of pollutants. On a
regional scale, the distribution of SO_2 shows strong
gradients near sources, but airborne sulfate shows
relatively weak gradients. Concentrations of both SO_2
and $SO_4^=$ are elevated over the industrialized or
urbanized parts of the eastern United States (Figures 4.1
and 4.2). High concentrations of sulfur oxides are found
through the Ohio River Valley into Pennsylvania and New
York to the Atlantic Coast. From this region of high
concentration, there is a strong gradient in ambient
SO_2 concentration northeastward toward the Adirondacks,
Vermont, and New Hampshire. The distribution of nitrogen
oxide concentrations is generally similar to that of
sulfur oxides, but it may show more localized maxima near
urban areas (e.g., Mueller and Hidy 1983).
Paralleling the distributions of ambient concentrations
are the distributions of sulfate and nitrate in precipi-

TABLE 4.1 Uncertainties in Conventional Sampling and Analytical Methods

Parameter	Method	Comments
Sulfur dioxide	Flame photometric	Water vapor and CO_2 interference near detection limit
Particulate sulfate	Filter substance collection and wet extraction and ion chromatography	Potential SO_2 adsorption on sample or filter substrate, especially glass-fiber material
Nitrogen dioxide	Chemiluminescent	Interference from HNO_3, N_2O_5, organic nitrates
Nitric acid	Nylon absorber/wet extraction; denuder tubes	NH_4NO_3 volatility; possible interferences
Nitrogen pentoxide	Method not available	—
Particulate nitrate	Filter collection/wet extraction and ion chromatography	Potential N_xO_y and HNO_3 vapor adsorption on filter mat, particularly glass fiber
Ammonia	Absorption on impregnated filter/water extraction and ion chromatography	Uncertain absorption efficiency
Acidity in aerosol	NH_3 denuder tubes with filter collection/wet extraction and acid-base titration	Potential interferences from ambient ammonia absorption from basic particles
Wet deposition rate	Wet collector/wash and chemical analysis	Seal on bucket can leak contaminating sample, collector may fail to open on small rainfall
Sulfate in water	Ion chromatography	Reliable technique
Nitrate in water	Ion chromatography	Reliable technique
Acidity in water	pH meter	Care must be taken on multiple, repeated electrode calibration and wash
Dry deposition of gases	Chemical absorption plate	Geometry dependent, does not simulate vegetation
Dry deposition of particles	Flux meter	Insufficient time resolution to be compatible with turbulent fluctuations

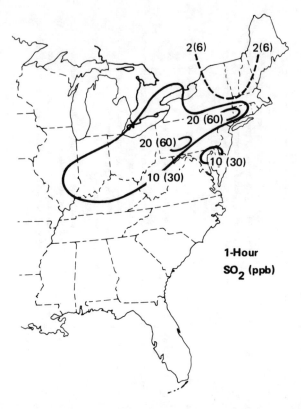

FIGURE 4.1 Composite spatial distribution of 1-hour average concentrations (ppb) of SO_2 from one representative month in each season between 1977 and 1978. This average is approximately equivalent to an annual average. Dashed isopleth is based on limited quantities of SURE data for 1977-1978 and on data taken at Whiteface Mountain, New York, after 1978. Numbers in parentheses are calculated values of dry deposition rates in kilograms of sulfur per hectare per year assuming a uniform deposition velocity of 0.8 cm/s. SOURCE: Data on concentrations from Mueller et al. (1980) and, for Whiteface Mountain, from V. Mohnen, State University of New York, Albany, personal communication (1983).

tation, as indicated in Figures 4.3 and 4.4. These distributions may be compared with that of the hydrogen ion concentration in precipitation in Figure 4.5. From these figures, the pattern of deposition of hydrogen ion in precipitation appears to correspond to regions of elevated sulfate and nitrate concentrations. This finding does not necessarily follow without accounting for all the cations (such as NH_4^+ and Ca^{++}) and anions that

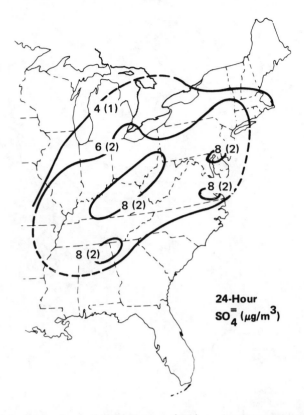

FIGURE 4.2   Composite spatial distribution of 24-hour average concentrations of sulfate ($\mu g/m^3$) from one representative month in each season between 1977 and 1978. This average is approximately equivalent to an annual average. Numbers in parentheses are calculated values of dry-deposition rates in kilograms of sulfur per hectare per year assuming a uniform deposition velocity of 0.2 cm/s. SOURCE: Data on concentrations from Mueller et al. (1980).

may be important factors affecting the acidity of precipitation.

A qualitative comparison between dry- and wet-deposition rates can be made from observation. Wet-deposition rates for sulfate are shown in Figure 4.3 for data taken in 1980. Estimated dry-deposition rates for SO_2 and particulates are shown in parentheses in Figure 4.1 and 4.2 (estimated values in kilograms of sulfur per hectare per year). The rates have been calculated using uniform deposition velocities considered typical of values reported in the literature for SO_2

94

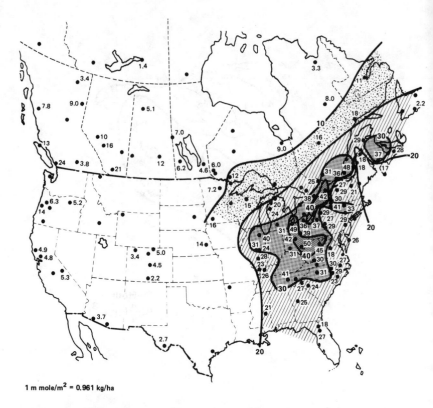

1 m mole/m² = 0.961 kg/ha

FIGURE 4.3   Spatial distribution of mean annual wet deposition of sulfate weighted
by the amount of precipitation in North America in 1980 (mmoles/m²). SOURCE:
U.S./Canada Work Group #2 (1982).

gas and for submicrometer particles (sulfate). Note that
in much of the region of high ambient concentrations, dry
deposition apparently dominates total deposition. This
result suggests that dry deposition of SO_2 exceeds wet
deposition in parts of the Ohio River Valley and the
Ohio-Pennyslvania-western New York area but becomes
progressively less important farther from the region of
major emissions, to the northeast in New England and
Canada. At large distances from sources, ambient
concentrations of sulfur oxides are low; wet deposition
will dominate dry deposition far from sources if
precipitation is significant.

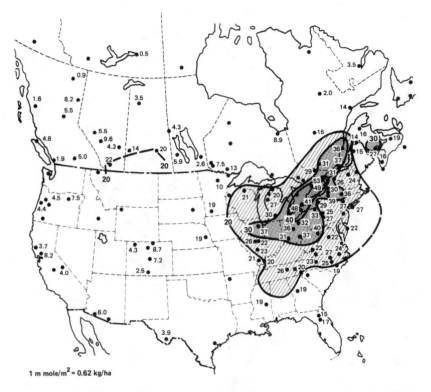

FIGURE 4.4 Spatial distribution of mean annual wet deposition of nitrate weighted by the amount of precipitation in North America in 1980 (mmoles/m^2). SOURCE: U.S./Canada Work Group #2 (1982).

THE INFLUENCE OF METEOROLOGICAL CONDITIONS

The spatial distributions of deposition for acid-forming materials are similar, and elevated concentrations appear to be associated with areas in which emissions are high. The temporal behavior of sulfur oxides and nitrogen oxides is dominated by meteorological variability. For example, analysis of the SURE data indicates that the variability in ambient concentrations of SO_2 was as much as an order of magnitude greater than the variability in SO_2 emissions over the eastern United States (Mueller and Hidy 1983). This finding suggests that advanced and sophisticated analyses are needed to separate the influences of emissions from those of aerometric parameters in such data. These analyses have

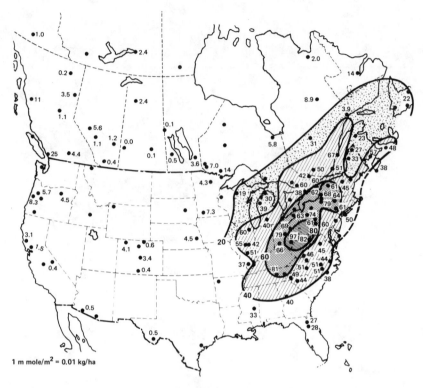

FIGURE 4.5 Spatial distribution of mean annual wet deposition of hydrogen ion weighted by the amount of precipitation in North America in 1980 (mmoles/m²). SOURCE: U.S./Canada Work Group #2 (1982).

followed two directions in the recent literature. The first is descriptive, taking into account the behavior of sulfur oxides in the atmosphere as a function of climatological or meteorological features. The second stems from the first by adapting statistical techniques to separate influences of different processes. In this section we review results of analyses of data according to meteorological conditions and studies of air-mass trajectories. In the next section we present the results of statistical methods of analysis.

Classification of Meteorological Conditions

One of the primary conclusions of many studies of field observations is the absence of a definitive relationship between SO_2 emissions and sulfate concentrations in dry air or in precipitation patterns. Sulfur emission rates are relatively constant, whereas the concentrations of sulfate aerosol and SO_2 in the air are highly variable and dominated by meteorological conditions (Electrical Power Research Institute 1981). The concentrations of sulfate particles tend to be high in summer, presumably because of more rapid photochemical oxidation of SO_2, high in the central and western region of high-pressure systems that move slowly toward the Atlantic, and high in maritime tropical air emanating from the coastal region of the Gulf of Mexico. Concentrations of SO_2 also tend to be high under stagnant air conditions of slow-moving, high-pressure, anticyclonic systems but tend to be low in the maritime tropical air. Trajectory analyses applied to specific transient anticyclonic systems have further demonstrated the tendency of such systems to favor air-stagnation situations in which high levels of pollutants accumulate (King and Vukovich 1982).

Evidence based on chemical and meteorological analyses as well as satellite photos suggests that, on occasion, pollutants originating in the Midwest and Northeast can be entrained in the clockwise flow around a transient high-pressure system and transported to the Gulf Coast region. The pollutants are believed then to be advected back through the Midwest and Northeast in the south-westerly flow of maritime tropical air (Wolff et al. 1981, 1982). The slow-moving flow of southwesterly maritime air from the southern states entrains pollutants from sources in its path and eventually moves over the northeastern United States and eastern Canada. Summer-time convective storms that occur in this air tend to be quite acidic compared with precipitation during the cooler months (MAP3S/RAINE Research Community 1982, Raynor and Hayes 1982a).

Precipitation during autumn, winter, and spring in eastern North America also tends to occur in moist southwesterly air that frequently emanates from the southern states. During these seasons, most of the precipitation occurs in the vicinity of warm fronts associated with cyclonic low-pressure storm systems. Precipitation develops because the southwesterly air travels faster than the warm front, which represents the boundary of colder and heavier air lying to the north of it. The southerly air ascends over the cold air and is cooled, producing large areas of precipitation. The situation is illustrated in Figure 4.6. The center of

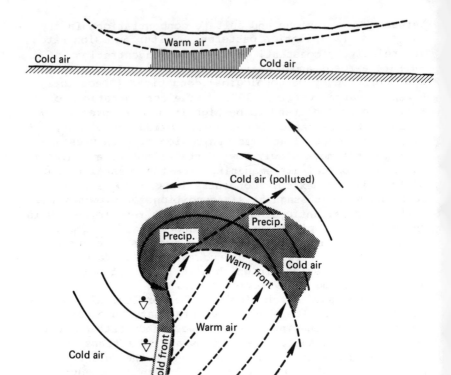

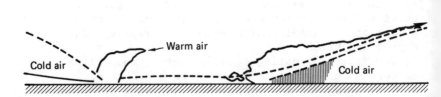

FIGURE 4.6 Idealized depiction of air masses and precipitation typically associated with warm fronts and low-pressure systems during autumn, winter, and spring in eastern North America. SOURCE: Lazrus et al. (1982).

the low-pressure system tends to travel toward the Northeast, and naturally the associated warm front moves with it. A map of preferred tracks of cyclonic weather systems (Figure 4.7) indicates the tendency for movement toward the Northeast. The bands of precipitation associated with the warm front move northeastward, depositing pollutants incorporated by the southwesterly air during its passage from southern or midwestern states. Although warm frontal precipitation is less acidic than summertime convective rain, warm fronts may deposit more acidic material in the Northeast because they deposit more precipitation (Raynor and Hayes 1982a). Figure 4.8 shows precipitation and wet deposition of chemical species as a function of the type of precipitation and type of synoptic weather system at Upton, New York.

The data of the SURE network have made it possible to quantify pollution episodes in terms of duration, sulfate concentration, extent of area, frequency of occurrence, and association with meteorological conditions. This information is summarized in Tables 4.2 and 4.3. Table 4.2 defines events (episodes) according to the number of stations with sulfate readings in specific ranges. The notations in Table 4.3 referring to meteorological type are illustrated in Figure 4.9. Warm southwesterly maritime tropical air (indicated as mT) referred to above occurs about 12 percent of the days in the region of the SURE network. The air involved in warm frontal precipitation is included in a transitional category indicated as Tr in Figure 4.9 and Table 4.3. Air masses of continental polar origin (cP) are identified with large areas of high barometric pressure and anticyclonic circulation. Continental polar cold (cPk) air refers to flow from central Canada southwest. Air-mass stagnation under the high pressure area is referred to as cP2 conditions, while cPw refers to warm airflow from south to north on the west side of the anticyclone.

Recent studies of fluctuations in SO_2 and sulfate concentrations along with other evidence indicate that sulfate aerosol advected from distant origins makes a large contribution to the particulates in the New York City area only during the summer (Lioy and Morandi 1982, Tanner and Leaderer 1982).

A correlation analysis of sulfate particle concentrations in dry air between various SURE sites indicates that significant regions of correlation generally extend 200 to 300 km from major SO_2 sources during pollution episodes and rarely beyond 500 km (Electric Power Research

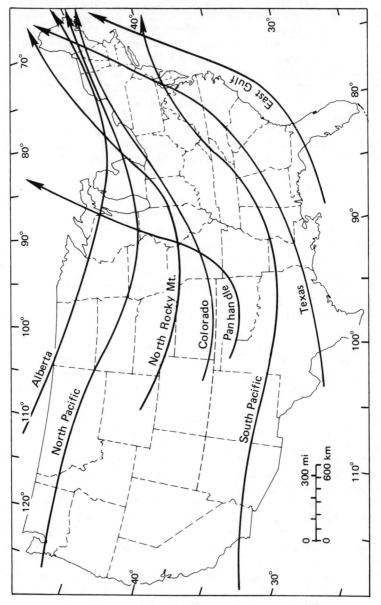

FIGURE 4.7 Main tracks of cyclonic (low-pressure) systems across the United States. SOURCE: Eichenlaub (1979).

101

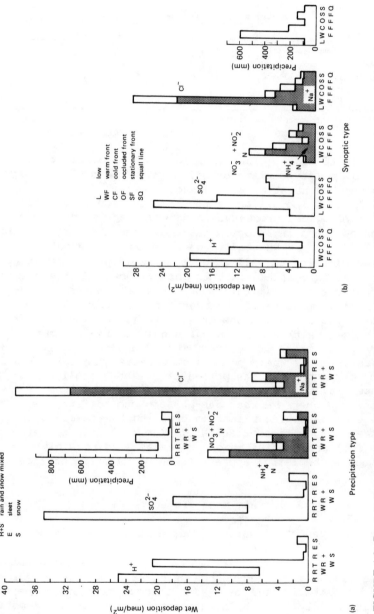

FIGURE 4.8 Precipitation and wet deposition of chemical species as a function of (a) precipitation type and (b) type of synoptic weather system at Upton, New York. SOURCE: Raynor and Hayes (1982a).

TABLE 4.2 Definitions of Pollution Events by Intensity and Coverage of Sulfate Concentration

Event Group	Percentage of Stations with Sulfate Concentrations			Days per Event—	
	>10 mg/m³	>15 mg/m³	>20 mg/m³	Range	Mean
Enlarged regional	40 to 93	25 to 72	10 to 61 ⎱	3 to 11	5
Regional	40 to 70	15 to 25	5 to 10 ⎰		
Subregional	25 to 50	5 to 15	<5	1 to 7	2
Nonregional	<35	<5	0	—	—
No event	0	0	0	—	—

SOURCE: Electric Power Research Institute (1981).

Institute 1981). The exception to this observation occurs during ducting of pollutants from southern regions to the Northeast by southwesterly air.

Another indication of the relative importance of meteorological processes in determining wet deposition can be found in European data. In eastern North America, the warmer months tend to be the period of greatest precipitation as well as of most rapid photochemistry. At certain coastal sites in Europe, however, the season of greatest precipitation does not correspond to the season of most rapid photochemistry. Figure 4.10 shows the seasonal variations in sulfate (in excess of that from sea salt) and amount of precipitation for three groups of sites in Europe. The triangles indicate coastal sites where periods of greatest precipitation do not correspond to the period of greatest sunlight intensity and hence to accelerated photochemistry. These cases demonstrate that gas-phase photochemistry may not be the dominating factor controlling the seasonal trend of wet deposition of sulfate. The influence of meteorological and precipitation processes (and possibly of aqueous chemistry) appears to be of major importance.

Air-Mass Trajectories

The influences of various source regions on given receptor sites have been calculated by following the trajectories of the air parcels bearing the pollutants. The most widely used method calculates the trajectories along isobaric surfaces. The inaccuracies of the method arise mainly from the assumption that vertical air motions can be ignored, because the necessary meteorological data on which the calculation depends are available only twice

3 of 392

103

TABLE 4.3 Annual Percentage of Event Days by Air-Mass Category

Event Group	Air-Mass Category[a]					Annual Percentage of Days in Each Event Group
	cPk	cP2	cPw	Tr	mT	
Regional	0	12	7	2	9	30
Subregional	0	4	4	4	3	15
Nonregional	2	10	2	18	0.5	32.5
No event	6	4	0.5	12	0	22.5
Annual percentage of days in each air mass	8	30	13.5	36	12.5	100

SOURCE: Electric Power Research Institute (1981).
[a]See Figure 4.9 and the text for descriptions of the air-mass categories.

daily at widely spaced locations, and from the difficulties associated with calculating trajectories in the
vicinity of precipitating systems. The periods over
which samples of precipitation are collected are usually
so long (one week or one month) that the source regions
of the air masses contributing to the sample may change
during a single collection period. This problem may even
arise in event sampling in the case of cyclonic storm
systems in which the precipitation falls from air masses
that frequently change markedly over a period of hours.

Recent studies of air-mass trajectories at Whiteface
Mountain in the Adirondacks (Wilson et al. 1982) indicate
that 62 percent of the sulfate ion and 65 percent of the
nitrate ion are deposited by precipitation associated
with air parcels emanating from the Ohio River Valley and
midwestern states (Figure 4.11). Wilson et al. also
point out that 56 percent of the total annual precipitation at Whiteface Mountain is carried by those air parcels
and that the wet deposition of acid may be related more
to this fact than to the higher concentrations in the
precipitation originating upwind to the south and
southwest. They also point out that the geographical
gradient of acid deposition (normalized to the amount of
rainfall) is not very steep. The normalized depositions
vary from 35 mg/m^2 of sulfate at the source region to
27 mg/m^2 at Ithaca, New York, and University Park,
Pennsylvania, to 18 mg/m^2 at Whiteface Mountain. They
postulate that atmospheric mixing of pollutants together
with varying patterns of precipitation tend to make the
spatial distributions of pollutant material deposited per
unit amount of precipitation relatively uniform when
considered over time scales of a year or more.

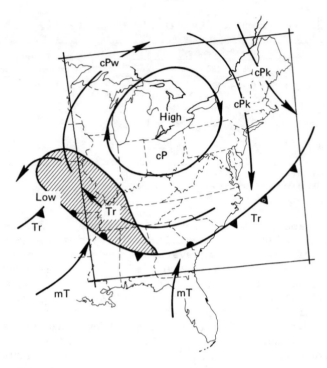

FIGURE 4.9   Categories of air masses used to classify regional air quality in Table 4.3.
SOURCE: Electric Power Research Institute (1981).

The tendency of atmospheric mixing processes and
varying precipitation patterns to attenuate the gradients
expected between source and receptor regions is illustrated by isopleth maps of annual deposition of hydrogen
ion (Figure 4.5), sulfate (Figure 4.3), and nitrate
(Figure 4.4), which place most of the Adirondacks, a
receptor region, in the same isopleth area as the Ohio
River Valley, a source region.

A similar trajectory study at Ithaca, New York
(Henderson and Weingartner 1982), yielded the percentage
deposition for trace constituents in precipitation as a
function of the trajectory of the air parcel and time of
year. From October through March the percentage depositions from the southwest quadrant were H^+, 72 percent;
$SO_4^=$, 77 percent; NO_3^-, 70 percent; and NH_4^+, 84
percent. Of the total precipitation during the period,
71 percent emanated from the southwest quadrant. Similarly, the values from April to September were H^+, 63

percent; $SO_4^=$, 68 percent; NO_3^-, 63 percent; NH_4^+,
76 percent; and precipitation, 52 percent.

It is interesting to note the following: (1) during
the cool period the percent deposition of acidic substances corresponds closely to the percent deposition of
precipitation; (2) during the warm period southwesterly
precipitation deposits a higher percentage of acid per
unit percentage of precipitation; (3) the Midwest is an
important source of the ammonium ion (derived from
gaseous ammonia), which tends to neutralize sulfuric
acid; and (4) nitric acid, derived from NO_x emissions
that are partially due to vehicular traffic (about 44
percent of NO_x emissions nationwide), is associated
with southwesterly precipitation systems. The authors
found no direct correlation between emissions and
depositions but were able to develop empirical relationships with rainfall rate and air-mass velocity that

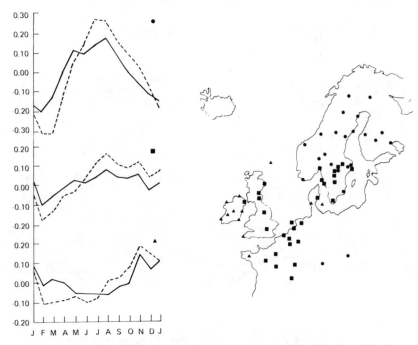

FIGURE 4.10 Seasonal variation in deposition of excess sulfate (solid curve) and
amount of precipitation (dashed curve) for three groups of stations in Europe.
SOURCE: Granat (1978).

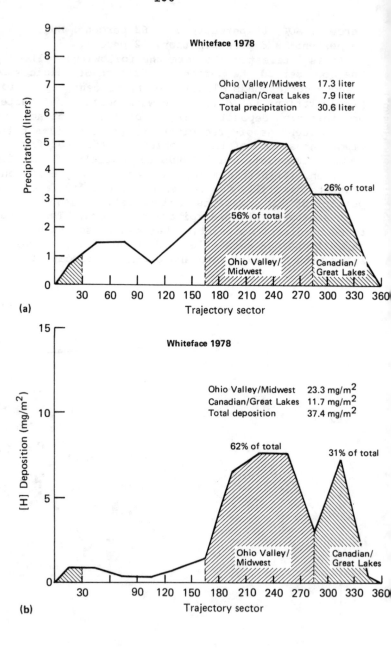

Whiteface 1978

Ohio Valley/Midwest   17.3 liter
Canadian/Great Lakes   7.9 liter
Total precipitation   30.6 liter

26% of total

56% of total

Ohio Valley/
Midwest

Canadian/
Great Lakes

Precipitation (liters)

Trajectory sector

(a)

Whiteface 1978

Ohio Valley/Midwest   23.3 mg/m^2
Canadian/Great Lakes   11.7 mg/m^2
Total deposition   37.4 mg/m^2

62% of total

31% of total

Ohio Valley/
Midwest

Canadian/
Great Lakes

[H] Deposition (mg/m^2)

Trajectory sector

(b)

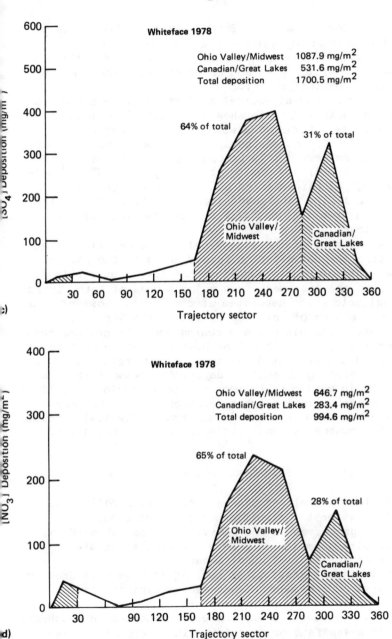

FIGURE 4.11 (a) Precipitation, (b) hydrogen ion concentration, (c) sulfate ion concentration, and (d) nitrate ion concentration in precipitation as a function of the directional sector through which the air parcel passed to reach Whiteface Mountain, New York, in 1978. SOURCE: Wilson et al. (1982).

accounted for 92 percent of the variability of sulfate deposition with SO_2 emissions along the air-parcel trajectory. Similarly, 93 percent of the variability of nitrate deposition with NO_2 emissions was accounted for in these empirical relationships.

A third recent air-parcel trajectory study conducted in south central Ontario showed similar directional expectancies. During the warm period the depositions associated with the southerly and southwesterly air-mass trajectories were H^+, 75 percent; $SO_4^=$, 70 percent; and NO_3^-, 57 percent; and during the cold period, H^+, 85 percent; $SO_4^=$, 87 percent; and NO_3^-, 86 percent (Kurtz and Schneider 1981). The percentage of total precipitation emanating from the southerly and southwestern sectors was not indicated.

The evidence of the air-parcel trajectory studies at specific receptor sites in remote areas of the northeastern United States suggests that the industrialized region to the southwest is a major source of both nitric and sulfuric acids deposited at the sites. However, the high percentage of acid deposition in the Northeast associated with air parcels coming from the southwestern sector may be a result of the high percentage of total precipitation that is delivered by these parcels. Moreover, observed gradients of deposition demonstrate that atmospheric mixing processes and precipitation variability obscure the source-receptor relationship to the extent that simple transport models may not realistically relate specific source regions with specific receptor sites.

STATISTICAL METHODS OF ANALYSIS

Statistical techniques can help to develop additional perspective on the significance of meteorology and chemical processes on the behavior of ambient sulfate. The resulting empirical models are used to relate emissions and ambient conditions.

Conventional statistical techniques have been used to establish correspondences between explanatory variables such as ambient SO_2 or NO_x concentrations, other aerometric variables, and the end products of atmospheric reactions (sulfate or nitrate). In interpreting the data, it is assumed that ambient concentrations of SO_2 or NO_x are proportional to emissions. Linear regression analyses have been performed to relate covariations in the end products to covariations in the explanatory

variables. The results (estimates of regression
coefficients) were sometimes inconsistent from location
to location and often indicated negative correlations
among observables that should in theory be positively
correlated. Although misleading estimates may result for
a number of reasons, one of the more important problems
in aerometric data is collinearity among the explanatory
variables. Belsley et al. (1980) discussed this problem,
which results when one or more columns of the matrix of
the explanatory (or independent) variables is an approximate linear combination of the other columns. If there
are strong correlations among the explanatory variables,
there will usually be approximate linear relations among
the columns. The result is that the standard errors for
the estimated regression coefficients may be drastically
increased and thereby may even cause estimates of true
coefficients with small magnitudes to have the wrong sign.

To avoid these problems, the common features of large
bodies of aerometric data have been examined using alternative multivariate methods. Two such methods are
regression on principal components, which eliminates
problems with collinearity in explanatory variables, and
empirical orthogonal function analysis (principal-
component analysis) in which intersite covariations
between end-product variables are examined. These
methods also offer an efficient means of screening large
amounts of data for internal consistency in the
relationships between the explanatory variables.

Regression on Principal Components

Principal-component analysis (Mardia et al. 1979) applies
an orthogonal transformation to the explanatory
variables, which results in a new set of uncorrelated
variables that are linear combinations of the original
variables. The transformation matrix is the matrix of
eigenvectors of either the correlation or the covariance
matrix of the explanatory variables. The first principal
component is the linear combination of variables that
explains or accounts for the maximum variability in the
original variables. The second principal component is
that linear combination, uncorrelated with the first
principal component, that explains the maximum amount of
variability not already accounted for by the first
component. The third, fourth, and other principal
components are defined similarly. There are as many

principal components as there are linearly independent
explanatory variables; however, the first few principal
components usually account for most of the variability in
the explanatory variables. The new variables or principal components have been interpreted in terms of
physically significant factors such as chemical or
transport processes (e.g., Henry and Hidy 1979).

When the principal components are found using the
correlation matrix of the explanatory variables rather
than the covariance matrix of the explanatory variables,
the coefficients in the regression of end products on the
new variables are "standard partial regression coefficients" (Snedecor and Cochran 1967). They provide a
measure of the comparative "strength" of association
between end-product variability and the new principal-
component explanatory variables. The reason for this
interpretation is that principal components are uncorrelated, so that a coefficient pertaining to one component is linearly unrelated to coefficients pertaining
to the other components. In addition each component has
unit variance so that the component is scale-free.

Considerable progress has been made in interpreting
the temporal behavior of ambient sulfate by this type of
approach. For example, Henry and Hidy (1979, 1982)
analyzed large data sets from New York City, St. Louis,
Los Angeles, and Salt Lake City using regression on
principal components. In New York, St. Louis, and Los
Angeles, the regression of sulfate on SO_2 at the same
measurement location was not significant. Instead, the
explanatory variables, ozone, temperature, and absolute
humidity, accounted for the variability in sulfate data.
In Salt Lake City, sulfate was found to be related to
SO_2 concentrations and to conditions promoting
atmospheric mixing. Evidently, variations in ambient
SO_2 levels in themselves have only a weak influence on
variations in airborne sulfate concentrations in these
cities, compared with the influences of variations in
atmospheric processes or meteorological conditions.

Rural data from the northeastern United States have
also been examined using these techniques (Henry et al.
1980, Mueller and Hidy 1983). Regression on principal
components of rural data indicated very complicated
relationships between aerometric variables. However,
significant evidence for a direct relationship between
sulfate levels and ambient SO_2 concentrations of 3-hour
and 24-hour averaged SURE data emerged from the analysis.
With regression on principal components, a linear

proportionality between ambient SO_2 and ambient sulfate
variability at the same site can be derived objectively.
If regional SO_2 concentrations in air are proportional
to SO_2 emissions and if dry deposition is proportional
to ambient sulfur oxide concentrations at ground level,
then one can infer that reduction in SO_2 emissions
should logically result in reductions in dry deposition,
even with the dominance of meteorological factors.

Empirical Orthogonal-Function Analysis

The influence of the spatial distributions of SO_2
emissions on ambient sulfate distributions also can be
obtained using empirical orthogonal-function analysis to
analyze sulfate measurements at a number of stations over
a period of time (e.g., Henry et al., 1980, Mueller and
Hidy 1983, Peterson 1970). This technique derives eigenvectors from the estimated covariances of sulfate concentrations between pairs of stations. Use of covariances
rather than correlations gives greater weight to sites
with larger sulfate variability. Spatial patterns are
then derived for each empirical orthogonal function (or
eigenvector) separately by plotting the value of the
eigenvector for each station, interpolating the values
onto a grid, and then drawing contours from this grid.
Examples of the first two empirical orthogonal
functions derived from the SURE data for July 1978 are
shown in Figure 4.12. The patterns shown are contours of
constant values of the empirical orthogonal functions,
where stations with the same sign covary together (when
one is high, the other is high) and stations with
opposite signs vary in opposition. The shading indicates
areas of high SO_2 emission density.
The empirical orthogonal function that accounts for
the largest amount of sulfate variability in the data,
identified as the "first," is shown in Figure 4.12(a).
The function accounting for the second largest amount of
the variability, identified as the "second" function,
accounts for the next largest fraction of sulfate
variability [Figure 4.12(b)]. For the July 1978 data,
the first two empirical orthogonal functions account for
74 percent of the deviation of the sulfate concentration
from its mean value.
The spatial patterns of August 1977 were similar to
the July 1978 patterns shown in Figure 4.12; however, the
patterns obtained from the SURE data differ from season

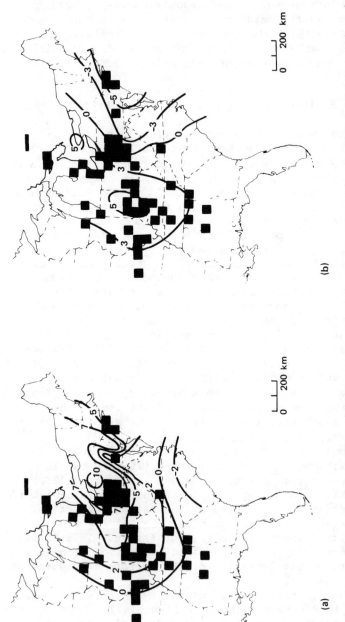

(b)

(a)

FIGURE 4.12 Empirical orthogonal functions (EOF) in the SURE data for July 1978. Shown are the lines of constant values of the EOFs in micrograms of sulfate per cubic meter. Shaded areas denote SO$_2$ emissions greater than 500 Mt/day. Of the total sulfate variance between locations where measurements were taken, the first EOF explains 54 percent (a), while the second EOF explains 20 percent (b). SOURCE: Henry et al. (1980).

to season (e.g., summer to fall or winter). In all of
these cases, the patterns can be rationalized to be
consistent with the dominant weather observed during the
periods examined. In summer, for example, the pattern of
the first empirical orthogonal function was identified
with large-scale meteorological conditions leading to
persistent mass transport of air around zones of high
barometric pressure northeastward across the Ohio River
Valley toward New England. These conditions were
identified as being most likely for long-range pollution
transport (Mueller and Hidy 1983). The second important
empirical orthogonal function in summer was identified
with large-scale air stagnation under conditions of poor
ventilation on the west side of summer anticyclones
(e.g., Vukovich et al. 1977). (It should be noted that
there were 10 days with conditions of air stagnation
during July 1978, whereas the 40-year average for July is
2 days.)

The association between SO_2 emission distributions
and the important empirical orthogonal functions found in
the SURE data is circumstantial evidence that changes in
the source patterns should induce changes in regionwide
sulfate variability over the eastern United States.
However, empirical orthogonal-function analysis and other
statistical techniques have not progressed to the stage
at which they can be used to predict consequences of
selective changes in emissions.

Elemental-Tracer Analysis

Sources often emit particulate matter that may serve as
chemical or elemental tracers of the emissions. Sampling
and analysis for the tracer substances at receptor
locations can provide evidence of the influence of
various types of sources on those locations. Because
monitoring stations are in the receiving field, these
statistical approaches to analyzing empirical data are
often called receptor models.

The technique, developed originally for studies of
urban air quality, has recently been proposed for
application to long-range transport of acidic substances. The method relies on the fact that particles
associated with different sources have different
elemental compositions (Gordon 1980, National Research
Council 1980). For example, sodium in particles
collected in aerosol monitors is usually associated with

a marine source, such as salt from windblown sea spray. Other elements in particles that are strong indications of particular sources are calcium (limestone, hence construction and demolition activities), lead (motor vehicles), vanadium (fuel oil), and zinc (municipal refuse). Particles from coal combustion are more difficult to distinguish because their elemental composition is much like that of soil. However, coal combustion produces fine particles that are significantly enriched in selenium and arsenic with respect to soil, so these elements potentially can be used to distinguish the contributions of coal combustion from those of windblown dust. The tracer method has been tested and improved for application to urban air sheds in several U.S. cities, including Washington, D.C. (Kowalczyk et al. 1982), Los Angeles (Friedlander 1973, Gartrell and Friedlander 1975), Portland, Oregon (Core et al. 1981), and St. Louis (Alpert and Hopke 1981, Dzubay 1980). The method has also been applied on the scale of global circulation to determine the sources of pollution particles collected at sites remote from human activities, such as in the Arctic (Cunningham and Zoller 1981; Heidam 1981, 1982; Rahn 1981a,b).

It has been suggested (for example by K. Rahn, University of Rhode Island, in an unpublished paper, 1982) that the elemental composition of particulate matter in aerosols might also be indicative of the origins of the aerosols after regional-scale transport and might help to resolve the question of the relative contributions of distant and local sources to acid deposition in eastern North America. Among major stationary sources that contribute substantially to the total burden of emission of SO_2, those in the eastern United States are fueled by coal and oil, whereas those in the Midwest burn mainly coal. Thus an aerosol collected in the East that is enriched in vanadium would be presumed to be of relatively local origin, whereas an aerosol characteristic of coal combustion would be taken to be of midwestern origin.

As yet there has been no systematic evaluation of the method, nor has it been applied to a number of available sets of data. Care should therefore be exercised in interpreting the results of preliminary analyses using the method, which may be more difficult to apply on a regional scale than on either an urban or a global scale.

Urban air quality is usually determined by local emissions, so the elemental composition of particulate

matter collected at urban monitoring sites can be expected
to reflect the types of sources in the urban air shed,
with the possible exception of aerosol particles, such as
sulfates, formed in the atmosphere from precursor gases
and transported long distances. Polluted air reaching
remote sites, by contrast, has usually undergone so much
mixing that its elemental composition can be considered
to be a composite representative of a large source area.
When polluted air travels from the source region to a
remote site mostly over the ocean, however, there are few
additional sources contributing particles to the air mass
during transit, and the method can give unambiguous
results. At rural continental sites, the dominant local
source of airborne particles is usually soil. Most other
types of particles come from distant sources such as
large power plants or major metropolitan areas.

There appear to be two critical problems in the
application of elemental tracer analysis to regional
transport that must be solved before the method can be
useful. The first problem arises because of the potential that the elements of interest as tracers may not be
transported at the same rates and over the same distances
as sulfates and nitrates, which are of primary concern in
acid deposition.

For example, manganese and vanadium have been proposed
as characteristic tracers of emissions from midwestern
and eastern source regions, respectively. Manganese,
most of which arises from entrained soil, is typically
found on relatively large particles that settle to the
ground close to the source; only about one third of airborne manganese is found in fine particles that would be
expected to undergo long-range transport. Elemental
tracer analysis attempts to correct for the presence of
crustal material. Because most of the crustal manganese
occurs on large particles, the technique focuses on
manganese in fine particles. The source of airborne
fine-particulate manganese is uncertain; coal is
deficient in manganese relative to soil, for example.
Almost all of the vanadium in particulate matter in the
East is associated with fine particles. Whether deposition processes for sulfates and nitrates are similar to
those of manganese and vanadium is unknown. The elements
used as tracers are normally emitted in the form of solid
particles, whereas the precursor gases SO_2 and NO_x
are transformed into particles by chemical processes
after emission. To the extent that formation of sulfate
and nitrate is dominated by in-cloud processes, it seems

likely that different processes will govern transport and deposition of the acids and the tracers. Coal combustion is predicted to be a major source of airborne selenium, appreciable fractions of which leave stacks in the vapor phase and condense on particles as the exhaust gases cool in the atmosphere. The behavior of selenium may therefore be more like that of sulfur in the atmosphere than is that of manganese.

The second, and perhaps more difficult, problem is that of differentiating contributions of midwestern sources from those of eastern cities over which an air mass originating in the Midwest may pass before it reaches sensitive receptor areas. While experience with the technique indicates that it is practical to differentiate the contributions to urban aerosols of types of sources in a local area, it may not be possible to assess the contributions to regional aerosol burdens of sources in different urban areas. It may indeed be possible to distinguish eastern air masses from midwestern air masses that are uncontaminated by emissions from eastern cities but not possible to identify a midwestern air mass that has passed over and received inputs from eastern sites. It seems likely that arrays of elements, rather than a single one, will have to be used as tracers of pollution of midwestern origin.

ANALYSIS OF HISTORICAL TRENDS

In this and the next section we analyze the relationships between emissions and wet deposition using data obtained from monitoring networks. The central question is whether the data can be used to judge whether deposition is linearly or nonlinearly related to emissions. (See Chapters 2 and 3 for discussions of the processes that might lead to a nonlinear relationship.) If the relationship is linear, a change in emissions would be reflected in a proportionate change in deposition; if it is nonlinear, the change in deposition, if any, would be disproportionate.

Data on precipitation for analysis are available from both Europe and North America. Considerably more data are available from Europe, where monitoring programs have been in existence for many years, than in North America, where systematic monitoring of precipitation chemistry is a comparatively recent development. In this section we use the body of historical data as direct evidence of the relationships, and then, owing to the paucity of these data, we assess the indirect evidence for linearity or nonlinearity in North America in the next section. More precisely, we consider whether currently available observations are consistent with conditions that should theoretically prevail if the relationship between emissions of SO_2 and NO_x and wet deposition of sulfate and nitrate in North America were strongly nonlinear.

The possibility that changes in deposition rates are not linearly related to changes in source strength initially arose in the examination of historical trends in emissions and deposition in Europe. The body of European data provides the only direct evidence for this phenomenon. For example, from about 1960 to 1975, most of the areas in the European network for precipitation sampling experienced either constant or declining deposition of sulfate (Figure 4.13) despite significant increases in SO_2 emissions during that period. This apparent nonlinearity of response has been attributed both to chemical factors and to changing climatic conditions. The data analyzed by Granat (1978) were collected using the bulk sampling technique, by which collection containers continuously exposed to the atmosphere collect combined wet and dry deposition. The observed trends are made somewhat uncertain by changes in analytical techniques, sampling techniques, and analytical laboratories throughout the sampling period. Furthermore, bulk deposition samples are subject to contamination from windblown dust, leaves, and insects, for example, and consequently are considered to be of poorer quality than current methods, which collect only wet deposition.

Extensive direct evidence, comparable with that for Europe (Kallend et al. 1983), that can demonstrate a disproportionate relationship between emissions and deposition does not exist in North America.

Efforts to reconstruct historical trends from the sporadic and disparate data that have been collected in the United States since the late 1950s (Likens and Butler 1981) are beset by large uncertainties (Hansen and Hidy 1982, Stensland and Semonin 1982). The monitoring of precipitation chemistry at the Hubbard Brook Experimental Forest appears to provide the longest continuous record of deposition data at a receptor site in the northeastern United States. These data, as in the European network, were obtained from samples of bulk deposition. The Hubbard Brook samples were collected on a weekly basis,

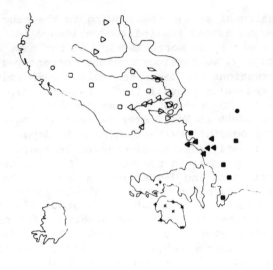

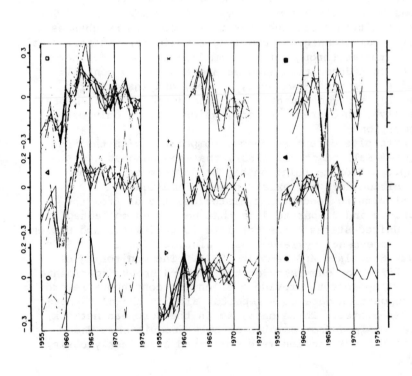

FIGURE 4.13 Long-term variation in deposition of excess sulfate for groups of stations in Europe. Symbols identify stations in each group. Scale is logarithmic. SOURCE: Granat (1978).

usually with three samples collected simultaneously. We
regard the Hubbard Brook bulk-deposition data as
reasonably reliable because samples were collected
frequently and the collection of several samples at one
site permits the detection and elimination of
contaminated data.

The Hubbard Brook data reveal several trends (Likens
et al. 1980), which are borne out at least qualitatively
in bulk deposition monitoring with relatively unreliable
quality control from nine stations in the New York State
area (Miles and Yost 1982, Peters et al. 1982):   (1)
There has been a decrease in sulfate concentration since
1964 but an increase in nitrate concentration over the
same time (Figure 4.14 and Table 4.4).   (2) The annual pH
of precipitation showed no long-term significant change
from 1964 to 1977, though several short-term changes did
occur.   (3) A linear regression equation of data points
from 1964 to 1977 indicates no statistically significant
trends in H^+ deposition from 1964 to 1977.   (4) Recent
changes in H^+ deposition correspond more with changes
in nitrate deposition than with sulfate deposition, even
though sulfuric acid is the dominant acid at Hubbard
Brook.   The contribution of NO_3^- to total acidity has
been increasing, whereas that of $SO_4^=$ has been
decreasing (Galloway and Likens 1981).   Year-to-year
changes superimposed on the long-term trend may be
related to climatological influences.

A linear-regression analysis of the Hubbard Brook data
against time indicates a decline in sulfate concentrations between 1965-1966 and 1979-1980 of about 33 ± 18
percent (95 percent confidence limits).   The regression
of sulfate concentration against time gives a slope of
-0.074 (standard error 0.016) and an intercept of 3.186
(standard error 0.138).   Using the t test, the slope is
significantly different from zero at the 0.0003 level.
The assumptions of the regression—that errors on the fit
are independent, have zero mean and a constant variance,
and follow a normal distribution—were tested and do not
appear to have been violated.

The Hubbard Brook record, unlike the European data,
appears in toto to demonstrate a reduction in sulfate
concentration similar to the general reduction in SO_2
emissions (Figure 4.15 and Table 4.5).   Between 1965 and
1978 SO_2 emissions in EPA Region I (comprising
Connecticut, Maine, Massachusetts, New Hampshire, Rhode
Island, and Vermont) declined by about 38 percent
(excluding data for Vermont, which are not reported in

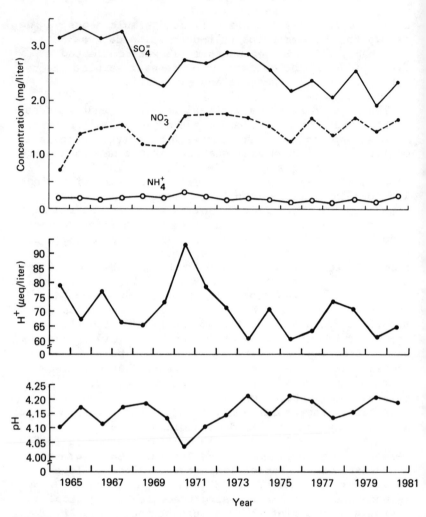

FIGURE 4.14 Annual weighted concentrations of sulfate, nitrate, ammonium, and hydrogen ions and weighted pH of precipitation at Hubbard Brook Experimental Forest from 1964 to 1977. SOURCE: Likens et al. (1980) and G.E. Likens, Cornell University, personal communication (1983).

the reference cited for Table 4.5). In Region II (New Jersey and New York) the reduction was about 40 percent, while that in Region III (Delaware, the District of Columbia, Maryland, Pennsylvania, Virginia, and West Virginia) was about 11 percent during the same period. Aggregate sulfur emissions in Region IV (Alabama,

TABLE 4.4 Annual Average Concentrations
of Sulfate and Nitrate from Weekly Bulk
Samples at Hubbard Brook Weighted by the
Annual Amount of Precipitation (mg/liter)[a]

Year	$SO_4^=$	NO_3^-
1964/65	3.16	0.70
1965/66	3.33	1.39
1966/67	3.13	1.49
1967/68	3.27	1.56
1968/69	2.42	1.18
1969/70	2.24	1.14
1970/71	2.75	1.71
1971/72	2.67	1.74
1972/73	2.87	1.74
1973/74	2.84	1.67
1974/75	2.54	1.52
1975/76	2.14	1.22
1976/77	2.20	1.66
1977/78	2.04	1.32
1978/79	2.55	1.69
1979/80	1.91	1.44
1980/81	2.36	1.66

SOURCE: G.E. Likens, Cornell University, personal
communication, January 9, 1983.
[a]Data are recorded in the water year, from June 1 through
May 31.

Florida, Georgia, Mississippi, Kentucky, North Carolina,
South Carolina, and Tennessee) increased by about 33
percent between 1965 and 1978. Region V emissions (from
Illinois, Indiana, Michigan, Minnesota, Ohio, and
Wisconsin) decreased by approximately 18 percent over the
period. Total SO_2 emissions from EPA Regions I through
V, which comprise all states east of the Mississippi
River plus Minnesota, decreased by about 8 percent
between 1965 and 1978.

The nitrate data at Hubbard Brook suggest an erratic
trend toward a maximum around 1970, followed by a
leveling off or a slight decrease. The emissions of
NO_x in the Northeast increased 26 percent between 1960
and 1970, and then decreased 4 percent by 1978. The
nitrate deposition data also appear to reflect emission
trends in the Northeast.

The data of Peters et al. (1982) were obtained from a
network operated by the U.S. Geological Survey (USGS).
The USGS network employed bulk samplers; single samples
were collected on a monthly basis.

The scatter in the USGS data is quite large. According
to Peters et al. (1982), the data do not suggest statis-

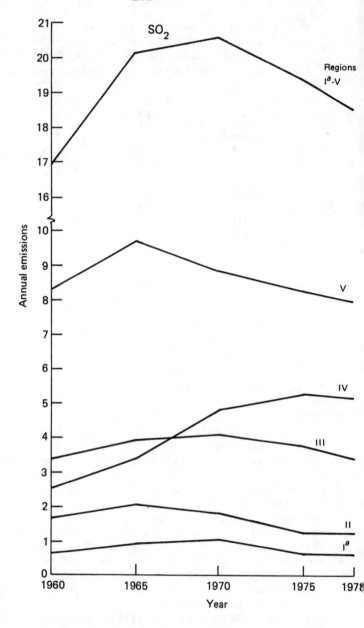

^{a}Excludes emissions from Vermont, which in 1980 amounted to about 7,500 tonnes of SO_2 and 12,000 tonnes of NO_x.

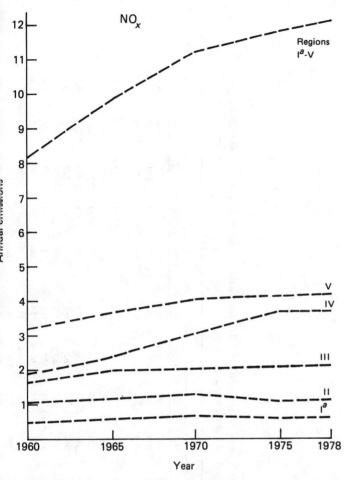

FIGURE 4.15 Annual emission of SO_2 and NO_x in the eastern United States by EPA Region, 1960-1978 (million metric tonnes). SOURCE: U.S./Canada Work Group #3B (1982).

tically significant long-term trends except for sulfate at one site (Hinckley). Although the direction of change is consistent, these results suggest weaker trends since the mid-1960s than the Hubbard Brook results. Uncertain quality control, especially during the early years of operation obviates the usefulness of these data however (Miles and Yost 1982).

Another source of data is the New York State Department of Environmental Conservation (1976, 1978,

TABLE 4.5 Annual Emissions of SO_2 and NO_x for Selected States Adjacent to or East of the Mississippi River, 1960-1978 (thousands of metric tonnes)[a]

State	SO_2					NO_x				
	1960	1965	1970	1975	1978	1960	1965	1970	1975	1978
Alabama	552	803	881	888	686	278	403	374	523	426
Connecticut	217	412	286	172	101	137	152	182	164	165
Delaware	177	196	201	174	169	46	55	65	59	64
District of Columbia	35	43	70	24	16	32	34	52[b]	33	30
Florida	307	451	776	745	617	289	379	497	660	700
Georgia	178	273	369	514	636	204	267	358	469	494
Illinois	2,208	2,512	2,256	1,756	1,573	806	957	1,008	1,016	1,017
Indiana	1,657	1,962	1,747	1,782	1,663	526	500	519	569	541
Iowa	328	397	333	283	347	195	223	279	278	289
Kentucky	568	823	1,153	1,320	1,197	251	340	447	511	507
Maine	63	87	74	61	59	44	54	68	65	69
Maryland	466	529	421	290	322	201	263	269	265	283
Massachusetts	337	399	526	326	362	229	273	324	306	328

Minnesota	353	378	406	344	341	216	246	298	353	360
Mississippi	37	40	72	174	238	136	177	274	219	246
Missouri	524	607	997	1,057	1,177	265	305	382	534	507
New Hampshire	26	37	86	68	61	28	36	57	61	60
New Jersey	434	561	531	307	291	326	395	484	416	445
New York	1,285	1,481	1,310	971	937	690	827	900	782	818
North Carolina	209	265	480	451	506	261	339	492	511	532
Ohio	2,640	2,863	2,813	2,944	2,804	864	979	1,049	1,099	1,149
Pennsylvania	2,126	2,292	2,021	1,918	1,710	918	1,029	980	984	1,009
Rhode Island	79	37	55	22	18	41	33	50	40	38
South Carolina	104	110	167	182	260	135	160	214	228	270
Tennessee	658	694	889	1,028	1,047	302	342	420	554	534
Virginia	154	169	428	343	324	234	326	390	379	392
West Virginia	477	699	882	1,098	945	203	290	312	424	416
Wisconsin	544	633	290	150[b]	597	267	331	410	410	426

SOURCE: U.S./Canada Work Group #3B (1982).

[a]Data on emissions in Vermont for this period are not contained in the referenced source and are believed to be small.

[b]Questionable datum.

TABLE 4.6 Three-Year Running Average Values for
Sulfate Concentrations in New York (μg/m^3)

Year	New York Statewide Sulfate Levels	Rochester Station 2701-01
1964	10.78	
1965	11.01	
1966	10.68	9.6
1967	9.88	9.37
1968	9.46	9.40
1969	9.04	8.83
1970	8.89	8.23
1971	8.84	8.13
1972	9.01	8.70
1973	9.00	8.77
1974	9.15	8.57
1975	9.32	8.53
1976	9.45	9.40
1977	9.02	9.8
1978	8.72	10.0
1979	8.3	9.53
1980	8.42	9.5
1981	8.42	9.65

SOURCE: Department of Environmental Conservation (1976, 1978, 1981).

1981). According to these data, statewide concentrations
of sulfate particles in New York have also decreased
about 22 percent between 1964 and 1978 (Table 4.6). This
decrease corresponds fairly closely to the decrease in
Northeast SO$_2$ emissions (Figure 4.15). However, several
stations, e.g., at Rochester, reveal more complex trends
throughout the period (Table 4.6). These data are difficult to interpret. Many of the stations are located in
urban areas so that both the statewide average and values
from specific stations should be strongly influenced by
local urban sources. In addition, controls for both
SO$_2$ and primary sulfate emissions were implemented
during this period.

Both the USGS and New York State data are therefore of
limited usefulness for assessing relationships between
emissions and deposition influenced by long-range transport to rural areas. The most reliable data available in
a continuous record are those from Hubbard Brook. These
data comprise the direct historical evidence for an
emissions-deposition relationship and show no indication
of a significant nonlinearity.

ANALYSIS OF RELATIVE BEHAVIOR OF SULFUR
AND NITROGEN EMISSIONS

Because of the small amount of direct evidence available, we resort to examining less direct empirical evidence that may reveal for eastern North America a nonlinear relationship between emissions and deposition.

An explanation of the inconsistent emission and precipitation trends observed in Europe comes from work of Rodhe et al. (1981), who hypothesized that (1) gas-phase photochemistry is significantly responsible for producing the acid incorporated in precipitation and (2) conversion of SO_2 to H_2SO_4 is indirectly influenced by emissions of NO_x in the atmosphere (see the detailed discussion in Chapter 3 on the Rodhe et al. model). In the Rodhe hypothesis, HNO_3 production is thought to be favored near the source region, while H_2SO_4 is favored in more distant regions. Nitric acid vapor formed by the reaction of NO_x and the OH radical can be rapidly removed from boundary-layer air by either dry deposition (B.B. Hicks, National Oceanic and Atmospheric Administration, private communication, 1982; Huebert 1982) or by wet deposition (Levine and Schwartz 1982). If the mechanisms suggested by Rodhe et al. significantly influence the concentrations of $SO_4^=$ and NO_3^- in precipitation, the ratio of dissolved sulfate to nitrate is expected to increase as the SO_2 oxidation becomes more efficient with increasing distance from the source region.

The occurrence of the gradient of ratios does not, however, unambiguously demonstrate the importance of this mechanism, since other factors (such as very rapid dry deposition of nitric acid vapor in relation to that of $SO_4^=$ aerosol) might yield a similar effect if gas-phase chemistry plays a dominant role. However, the absence of the gradient of ratios would be difficult to reconcile with the proposed model. Molar ratios of sulfate to nitrate in precipitation in Europe calculated from the data of Rodhe et al. (1981) do indeed increase with increasing distance from the source region in summer (Figure 4.16 and Table 4.7), when the influence of gas-phase chemistry is expected to be greatest.

The annual average of the molar ratio of sulfate to nitrate in precipitation is relatively constant through-out the area of the northeastern United States and eastern Canada in regions in which the pH precipitation is low but increases in extreme northern Canada (Figure 4.17). High values of the ratios in northeastern Canada

128

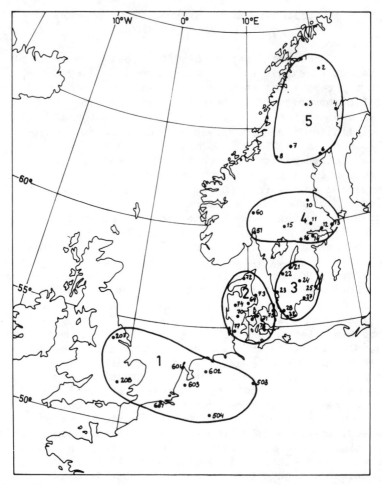

FIGURE 4.16   Locations of monitoring stations used in the data analysis of Table 4.7.
SOURCE: Rodhe et al. (1981).

and far north Europe (region 5 in Figure 4.16) may be
affected by local smelting facilities. Enhancements of
the ratio occur in areas of concentrated point-source
emissions, e.g., the Ohio and Tennessee River Valleys
(Figure 4.18) and along coastal areas, possibly because
of sea-salt sulfate. A lower ratio occurs near the Great
Lakes, a region with a relatively low ratio of SO_2 to
NO_x emissions. That the ratio is relatively constant
over a broad region appears inconsistent with the exis-

TABLE 4.7 Sulfate, Nitrate, and the Molar Ratio of Sulfate
to Nitrate in Precipitation in Europe

	Group of Stations[a]				
	1	2	3	4	5
Sulfate (μmoles/liter)					
1955–1959	62.5	50.0	28.1	25.0	15.6
1970–1974	90.6	56.3	50.0	40.6	40.6
Increase (%)	45	13	78	63	160
Nitrate (μmoles/liter)					
1955–1959	35.0	22.9	13.6	7.1	4.3
1970–1974	60.7	51.4	25.0	21.4	7.1
Increase (%)	73	125	84	200	67
Molar ratio					
1955–1959	1.8	2.2	2.1	3.5	3.6
1970–1974	1.5	1.1	2.0	1.9	5.7

SOURCE: Rodhe et al. (1981).
[a]Geographic locations of the groups of stations are indicated on Figure
4.16.

tence of a strong nonlinearity due to photochemistry and
demonstrates the apparent effectiveness of atmospheric
processes to lead to thorough mixing of pollutants over a
large spatial scale (up to 1,000 km in linear dimension).

When the ratio of sulfate to nitrate in precipitation
is considered on a monthly basis, the pattern becomes
more complex. Average ratios have been calculated for
the MAP3S stations (MAP3S/RAINE Research Community 1982)
for two seasonal periods, April to September and October
to March, for the period 1977 to 1981. The seasonal and
annual ratios are given in Table 4.8. The ratio tends to
be larger in summer than in winter, with the seasonal difference diminishing at lower latitudes (possibly because
the intensity of sunlight varies less with season) or in
maritime environments (where sea-salt sulfate may contribute to the ratio). The seasonal change may reflect a
less efficient conversion of SO_2 than of NO_x during
the cooler months than during the warmer months, when
photochemistry is more active and transport is more
vigorous. Incorporation of nitrate into precipitation is
much less sensitive to seasonal change. Emissions of
SO_2 are relatively constant throughout the year (16
percent increase in winter) (Electric Power Research
Institute 1981). Seasonal variations in emissions,
therefore, cannot account for the seasonal change in
deposition ratios. The deposition ratios, however, are
relatively invariable with respect to distance from major
source regions, particularly during the warm season (1.4

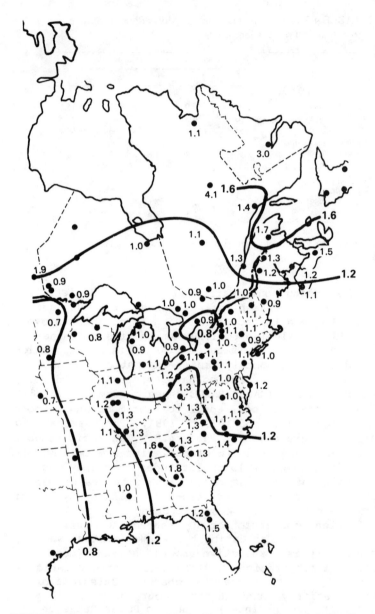

FIGURE 4.17 Average molar ratio of sulfate to nitrate in precipitation in eastern North America in 1980. SOURCE: B. Heikes, National Center for Atmospheric Research, personal communication (1983). Based on data from U.S./Canada Work Group #2 (1982).

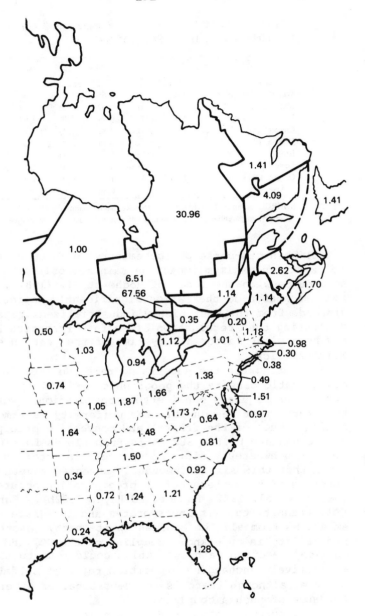

FIGURE 4.18   Average molar ratio of sulfur to nitrogen oxides in emissions in eastern North America in 1980. SOURCE: B. Heikes, National Center for Atmospheric Research, personal communication (1983). Based on data from U.S./Canada Work Group #3B (1982).

TABLE 4.8 Molar Ratios of Sulfate to Nitrate in
Precipitation in the United States, 1977–1981[a]

Station	April–September		October–March		Annual Average
Whiteface Mt., N.Y.	1.4	(29)	0.65	(26)	1.08
Ithaca, N.Y.	1.4	(25)	0.62	(21)	1.07
University Park, Pa.	1.3	(27)	0.74	(29)	1.04
Charlottesville, Va.	1.3	(27)	0.81	(21)	1.16
Urbana, Ill.	1.4	(15)	0.94	(13)	1.25
Brookhaven, N.Y.	1.12	(17)	1.0	(17)	1.10
Lewes, Del.	1.4	(18)	0.93	(18)	1.16
Oxford, Ohio	1.5	(16)	1.2	(19)	1.37
Average	1.4 ± 0.1		0.86 ± 0.19		

SOURCE: Adapted from MAP3S/RAINE Research Community (1982).
[a]Numbers in parentheses are the number of months of data in each sample.

± 0.1) when gas-phase photochemistry would be expected
to exert its maximum impact on rain composition in
accordance with the model of Rodhe et al. (1981). In
neither season do the ratios significantly increase with
distance from the most highly industrialized areas,
suggesting that retardation of sulfate formation by NO_x
may be a minor factor at any time of the year in eastern
North America.

Accounting for the surprising relative invariability
of the ratio despite the geographic variability of
conditions requires considerable speculation. Perhaps
the factor most responsible for uniformity is atmospheric
mixing. Much of the frontal and convective precipitation
in the eastern United States is from clouds forming in
moist southwesterly air. Air-parcel trajectories have
shown that this air is frequently advected several
hundreds of kilometers before precipitation occurs
(Lazrus et al. 1982, Raynor and Hayes 1982b). During
this transit, the air incorporates and integrates the
emissions from all the sources in its path. Another
possibility is an unknown coupling of the NO_x and SO_2
chemical reaction pathways, which could tend to maintain
a relatively unchanging deposition ratio of sulfate to
nitrate, although there is no theoretical or experimental
evidence of such a coupling.

It is surprising to note in addition that the annually
averaged molar ratios in precipitation at the MAP3S,
NADP, and CANSAP sites in eastern North America resemble
the molar ratios of SO_2 to NO_x emissions averaged on
an annual basis by state or province (Figure 4.18). The
emission ratios are somewhat higher than the deposition

ratios in the Ohio and Tennessee River Valleys. Similarly, the emission ratios are somewhat lower than the
low deposition ratios immediately north of Lake Ontario.
The smaller spatial variability of the deposition ratios
compared with the emission ratios near large point sources
is probably related to a rapid rate of atmospheric mixing
relative to the rates of wet-deposition processes. Alternatively, a preferential dry deposition of SO_2 relative
to that of NO_2 may occur in the near-source region, or
the differences may indeed represent a suppression of SO_2
oxidation within a short distance from the emission
sources. However, the general similarity of wet-
deposition and emission ratios on an annual average basis
suggests that the incorporation of sulfuric acid in rain
is not retarded with respect to the conversion and
incorporation of nitric acid in rain.

A small body of data suggests that the historical
trends in the molar deposition ratios of sulfate to
nitrate correspond to the trend in molar SO_2/NO_x
emission ratios. The historical trends of the molar
deposition ratio $SO_4^=/NO_3^-$ at Hubbard Brook and
the molar ratios of SO_2 to NO_x in emissions in the
eastern United States by EPA Region are plotted in Figure
4.19. Also shown in the figure is the result of a linear-
regression analysis of the Hubbard Brook deposition ratio
against time. Between 1965-1966 and 1979-1980, the linear
regression gives a slope of -0.039 (standard error 0.006)
and an intercept of 1.444 (standard error 0.054). The
significance level of the slope is 0.0001.

Another perspective is provided by comparing the
annually averaged regional wet-deposition rates with
regional emissions. Integrating wet deposition over the
northeast quadrant of the United States [using the
isopleths of Pack (1980) based on MAP3S and SURE data],
Golomb (1983) found sulfate deposition to be 4.3×10^{10}
(±20 percent) moles/year and total nitrate deposition
to be 3.9×10^{10} (±25 percent) moles/year. Golomb
calculated regional emissions from U.S./Canada (1981) to
be 2.2×10^{11} moles/year of SO_2 and 1.8×10^{11}
moles/year of NO_x. The emissions were similarly
distributed geographically. Golomb also indicates that
19 ± 6 percent of the sulfur and 20 ± 7 percent of
the nitrogen emitted in this region are wet deposited
there. It is interesting to note that Golomb found the
molar ratio of sulfate to nitrate wet deposited annually
in this region to be 1.1, compared with an annually
averaged molar ratio of emissions for SO_2 and NO_x of

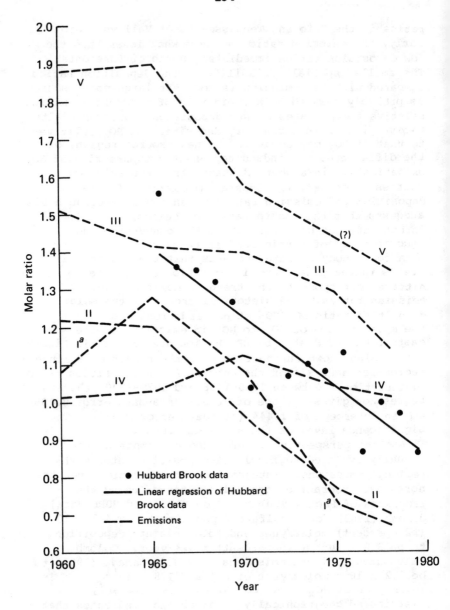

^a Excludes Vermont.

FIGURE 4.19 Molar ratio of SO_2 to NO_x in emissions in the eastern United States by EPA Region and molar ratio of sulfate to nitrate in deposition at Hubbard Brook, New Hampshire, 1960-1978. SOURCE: Adapted from Tables 4.4 and 4.5.

1.2.   The similarity of the ratios suggests that, whatever
specific processes are taking place, there is a tendency
on an annual basis for sulfate and nitrate to be deposited
in the northeast quadrant of the United States with no
significant loss of one component relative to the other.
Two inferences result from this observation. On an
annual basis there is no indication that within the
region the oxidation of SO_2 is retarded relative to the
oxidation of NO_x. Laboratory evidence, field observations, and photochemical theory predict a relatively
rapid gas-phase oxidation of NO_x to HNO_3, especially
in summer (see Appendix A). If the rate of SO_2
oxidation (as a result of both gas-phase and aqueous
reactions) is comparable with that of NO_x, then no
significant retardation of sulfate formation is likely to
occur within the eastern region of high acidity as a
result of limited availability of gaseous or aqueous
oxidants.

On the basis of Golomb's calculation and our own
analysis, we conclude, therefore, that (1) on an annual
average basis there is no evidence that the hypothesis of
Rodhe et al. (1981) applies in the northeastern United
States and further that (2) the net oxidation of SO_2
leading to sulfate in precipitation is as efficient as
the net oxidation of NO_x leading to nitrate in
precipitation in this region.

The first conclusion relates to a specific mechanism;
the second to the core of the critical issue of nonlinearity. If SO_2 oxidation were limited by the
availability of oxidant, it would be possible for SO_2
emissions to overwhelm, or saturate, the capacity of the
atmosphere to oxidize all or most of the SO_2 within the
region of eastern North America. Oxidation over the
ocean and in remote land regions would presumably convert
the remaining SO_2. Since a large fraction of emitted
SO_2 could conceivably be in excess of the oxidant
available in North America, a significant reduction in
SO_2 might not yield a significant reduction in wet
deposition of sulfate in North America, including, of
course, the regions deemed sensitive to acid deposition.
The second conclusion, however, suggests that the
oxidative capacity is not saturated in the northeast.
Furthermore, the similarity of deposition ratios in
eastern Canada both to those in the eastern United States
and to emission ratios upwind also tend to support this
implication. The indirect evidence agrees with the small
amount of reliable North American historical data, which

suggest that there is no evidence of a strong nonlinearity in the relationship between $SO_4^=$ deposition
and SO_2 emissions.

The inference to be drawn from these data is that if
the average ratio of SO_2 to NO_x in emissions in the
region of high acidity covered by the data (consisting of
the northeast quadrant of the United States, bounded by
the southwest corner of Tennessee, the northwest corner
of Illinois, New Hampshire, and the southeastern corner
of North Carolina) were changed by changing SO_2
emissions while keeping NO_x emissions constant, that
change should be reflected in a similar change in the
ratio of sulfate to nitrate in wet deposition, all other
emissions and conditions remaining unchanged. If dry
deposition is linearly proportional to emissions, as
suggested in the section on principal-component analysis,
then the annual average ratio in bulk deposition in the
area should also respond to changes in the emission
ratio. The analysis is limited in its applicability to
circumstances in which the spatial distribution of
emissions is unchanged. Without greater confidence in
the results of deterministic models, we cannot judge the
consequences of emission reductions in a smaller region,
such as the Midwest, for deposition in another region,
such as the Adirondacks or southern Ontario (Chapter 3).

It is now necessary to consider the uncertainties
inherent in the application of these observations to the
problem of nonlinearity.

Our present inability to model the processes linking
air composition with the composition of precipitation is
revealed by a comparison of wet-deposition ratios with
the molar ratios of SO_x to NO_x observed in air by the
SURE network (Table 4.9). Precipitation scavenges
airborne material both in and below clouds at some time
after the injection of emissions into the atmosphere.
Molar ratios in precipitation are expected therefore to
reflect concentrations in ambient air. The ratios
observed in ambient air at ground level in the SURE
network, however, are not similar to those either in
precipitation or in emissions. In ambient air at ground
level, SO_x is depleted with respect to NO_x in the sum
of gas and particle phases but is enriched in particles
alone. The depletion of SO_x in ambient air relative to
NO_x may be related to inherent nonlinearities in the
processes involved, to differences in dry-deposition
rates, to differences in the composition of air at ground
and cloud levels, or to the comparison of ambient

137

TABLE 4.9 Molar Ratios of SO_x to NO_x in the Region of the SURE Experiment in 1977–1978[a]

Emissions	Ambient Air[b]		Precipitation
1.2 (all sources)	Gas and particles	0.66	0.73 to 1.3
2.2 (utilities)[c]	Gas only	0.42	
	Particles only	24.0	

SOURCE: Adapted from Mueller and Hidy (1983).
[a]Ratio taken as SO_2/NO_2.
[b]Sampled at ground level.
[c]Weighted heavily toward emissions injected at altitudes at or above 300 m.

concentration data that are averaged over all meteorological conditions with precipitation data that reflect only those conditions specifically favoring precipitation.

As indicated earlier, the ratio of sulfate to nitrate in precipitation varies regularly throughout the year, being high in summer and low in winter (see Table 4.8). It is unclear how the specific processes controlling these variations average out annually to be so uniform spatially and so similar to the emission ratios over much of the area when ambient atmospheric conditions apparently are so variable.

Conceivably the uniformity of the annual $SO_4^=/NO_3^-$ ratio in precipitation in the northeastern United States could reflect some as yet unidentified stoichiometric linkage between the atmospheric chemistry of SO_2 and NO_2. For example, the reaction of $HOSO_2$ radicals formed in reactions of HO with SO_2 could lead to $HOSO_2O_2$ and $HOSO_2O$ radicals, which could ultimately react preferentially with NO_2 to form a one-to-one, S to N compound such as $HOSO_2ONO_2$ (see Chapter 3 and Appendix A). This compound would in principle hydrolyze in cloud water to form equal amounts of H_2SO_4 and HNO_3 acids. However, this reaction mechanism would have to be the dominant source of $SO_4^=$ and NO_3^- to maintain the ratio of moles of $SO_4^=$ to moles of NO_3^- near unity, as observed in precipitation in the eastern United States. There is no laboratory or field evidence of the significant participation of nitryl sulfuric acid ($HOSO_2ONO_2$) or other similar compounds in acid development; in fact, the existing limited evidence appears to discount the possible importance of such a product as an intermediate in the $HO-SO_2-NO_x$ reaction system (see Appendix A).

From theoretical considerations (e.g., Chapter 2 and Appendix A), we expect that ambient concentrations of oxidants (such as H_2O_2, O_3, and HO) will have maximum values in summer and minimum values in winter. It is expected, therefore, that in winter in areas at higher latitudes, aqueous oxidation is slow because of low concentrations of oxidants in cloud water. Hence, if a limitation of oxidant were to lead to nonlinearity in the relationship between emissions and deposition, that effect would be most likely to be observed in winter. The evidence that the relationship between average annual emissions and deposition is not strongly nonlinear may result from the fact that most of the total annual deposition of acid occurs during the warm months.

Comparison of emission ratios and deposition ratios constitutes indirect evidence for the absence of a serious nonlinearity even though we lack an adequate understanding of the intermediate chemical and meteorological processes to predict, either qualitatively or quantitatively, the deposition ratios from the emission ratios. It is our opinion that the necessary data and theoretical understanding to model the effects of emissions from distant sources on the composition of precipitation reliably and accurately will not be available in the near future. The evidence based on the empirical observations appears incompatible with a seriously nonlinear system in terms of our current knowledge. Unquestionably, however, it would be preferable to reinforce such conclusions with a more complete understanding of atmospheric chemistry and meteorology under conditions prevalent in eastern North America.

Additional uncertainties are introduced by the inherent variabilities of natural processes (e.g., annual rates of precipitation), imprecision of measurements, and errors in determining emission rates. For example, the ratios of emissions and wet depositions in the Northeast are uncertain by a factor of about 30 percent. The concentrations of sulfate and nitrate in the MAP3S study are estimated to be uncertain by a factor of 15 to 20 percent.

A significant reduction in the uncertainties related to natural variability, measurement imprecision, or establishment of emission rates is unlikely in the near future.

FINDINGS AND CONCLUSIONS

Although data from which to assess the relationships between emissions and deposition are relatively sparse in

North America, analysis of the available empirical data
provides insight into the nature of the atmospheric
proceeses involved in acid deposition.

Nonlinearity

Data indicate that the variability in ambient concentrations of SO_2 vapor and $SO_4^=$ particles do not
necessarily correlate with the variability in SO_2
emissions but are predominantly controlled by meteorological conditions. Variations in sulfate particle
concentrations tend to correlate with variations in SO_2
vapor concentrations in rural areas of the Northeast.
Regression on principal components and empirical
orthogonal-function analysis suggest that if other
factors were constant, ambient concentrations of SO_2
and $SO_4^=$ would be determined largely by patterns of
SO_2 emissions.

Direct evidence of a strongly nonlinear relationship
between wet deposition of sulfate and SO_2 emissions is
limited to extensive historical data in Europe. The
continuous historical record of reasonably reliable data
in North America, at Hubbard Brook, indicates no evidence
for a strongly nonlinear relationship between annual
depositions and annual emissions in the Northeast.
Indirect evidence based on patterns of the ratio of
$SO_4^=$ to NO_3^- in annual deposition do not support the
hypothesis of a strongly nonlinear relationship between
SO_2 emissions and sulfate deposition in eastern North
America. Differences in the relationship between emissions and deposition in Europe and eastern North America
may be the consequence of differences in meteorology,
latitude, or other factors, such as the spatial distribution of sources, between the two regions.

On the basis of currently available empirical data and
within the limits of uncertainty associated with the data
and with estimating emissions, we therefore conclude that
there is no evidence for a strong nonlinearity in the
relationships between long-term average emissions and
depositions in eastern North America.

The conclusion that nonlinearity is probably not significant for annual average deposition in eastern North
America is clouded by three types of uncertainties.
First, direct evidence based on long-term time-series
data is severely limited in North America to only ten
stations, all of which have collected bulk deposition

data. The data from only one site, Hubbard Brook, are considered to be reasonably reliable. Therefore we have relied on the historical record at only one station combined with the indirect evidence provided by data on sulfate and nitrate deposition compared with SO_2 and NO_x emissions. Second, there remain uncertainties in our detailed understanding of the meteorological, physical, and chemical processes that relate emissions to deposition. Third, the unknown influences natural variability in the composition and occurrence of precipitation, imprecision in sampling and analysis, and uncertainties in estimation of emissions further limit confidence in this conclusion.

Influence of Local and Distant Sources

Both observational and theoretical evidence exists for the long-range transport of pollutants leading to acid deposition. It is apparent that any receptor site will be influenced to one degree or another by both local and distant sources. The issue of concern is the extent of this zone of influence for sensitive ecological areas, including the relative contributions to deposition of nearby and distant sources.

In the case of the Hubbard Brook data, the trends in concentrations of sulfate and nitrate appear to reflect general trends in emissions. Analyses of the trajectories of precipitating systems delivering acidic precipitation to Whiteface Mountain and Ithaca in New York and south central Ontario in Canada indicate that most of the acidity in precipitation at these sites--as well as most of the precipitation--is associated with air masses that have passed over source regions to the south and southwest.

The spatial distribution of the annual average molar ratios of pollutants in emissions and deposition suggest that atmospheric processes in eastern North America lead to a thorough mixing of pollutants over a wide geographic area, making it difficult to distinguish between the effects of distant and local sources.

On the basis of currently available empirical data, we cannot in general determine the relative importance for the net deposition of acids in specific locations of long-range transport from distant sources or more direct influences of local sources. We regard the problem of relating emissions from a given region to depositions in a given receptor region to be of primary importance and recommend that high priority be given to research relevant to its solution.

The SURE data have indicated a relatively small zone of influence (of the order of 300 to 600 km) on ambient sulfate concentrations in general, with occasional long-range influence during ducting situations involving southwesterly flows of air. Similarly, warm frontal precipitation may involve cloud formation in air parcels advected for hundreds of kilometers from a southwesterly direction. The relative contributions of such long-range effects and of more local regional effects are currently unknown and cannot be reliably estimated using currently available models.

REFERENCES

Allard, D.W., I.H. Tombach, H. Mayrsohn, and C.V. Mathai. 1982. Aerosol measurements: western regional air quality studies. Presented at the 75th Annual Meeting of the Air Pollution Control Association, New Orleans.

Alpert, D.J., and P.K. Hopke. 1981. A determination of the sources of airborne particles collected during the regional air pollution study. Atmos. Environ. 15:675-688.

Altshuller, A.P. 1980. Seasonal and episodic trends in sulfate concentrations (1963-1978) in the eastern United States. Environ. Sci. Technol. 14:1337-1348.

Altshuller, A.P., and R.A. Linthurst, ed. 1982. Critical Assessment Document: The Acidic Deposition Phenomenon and Its Effects. Draft. Prepared for the U.S. Environmental Protection Agency. Raleigh, N.C.: The North Carolina State University Acid Precipitation Program.

Atkinson, R., A.C. Lloyd, and L. Winges. 1982. An updated chemical mechanism for hydrocarbon/NO_x/SO_2 photooxidations suitable for inclusion in atmospheric simulation models. Atmos. Environ. 16:1341-1356.

Belsley, D.A., E. Kuh, and R.E. Welsch. 1980. Regression Diagnostics: Identifying Influential Data and Sources of Collinearity. Pp. 85-191. New York: John Wiley and Sons.

Chamberlain, J., H. Foley, D. Hammer, G. MacDonald, D. Rothaus, and M. Ruderman. 1981. The Physics and Chemistry of Acid Precipitation. Technical Report JSR-81-25. Arlington, Va.: SRI International.

Core, J.E., P.L. Hanrahan, and J.A. Cooper. 1981. Air particulate control strategy development: a new approach using chemical mass balance methods. In Atmospheric Aerosol: Source/Air Quality Relationships, E.S. Macias and P.K. Hopke, eds. ACS Symposium Series No. 167. Washington, D.C.: American Chemical Society.

Cunningham, W.C., and W.H. Zoller. 1981. The chemical composition of remote area aerosols. J. Aerosol Sci. 12:367-384.

Department of Environmental Conservation. 1976. New York State Air Quality Report. DAR-77-1. Albany, N.Y.: New York State Department of Environmental Conservation.

Department of Environmental Conservation. 1978. New York State Air Quality Report. DAR-79-1. Albany, N.Y.: New York State Department of Environmental Conservation.

Department of Environmental Conservation. 1981. New York State Air Quality Report. DAR-82-1. Albany, N.Y.: New York State Department of Environmental Conservation.

Dzubay, T.G. 1980. Chemical element balance method applied to dichotomous sampler data. In Aerosols: Anthropogenic and Natural, Sources and Transport, T.J. Kneip and P.J. Lioy, eds. Annal 338. New York: New York Academy of Sciences.

Eichenlaub, V.L. 1979. Weather and Climate in the Great Lakes Region. South Bend, Ind.: The University of Notre Dame Press.

Electric Power Research Institute. 1981. EPRI Sulfate Regional Experiment: Results and Implications. EPRI EA-2165-SY-LD. Palo Alto, Calif.

Fisher, B.E.A. 1982. The transport and removal of sulphur dioxide in a rain system. Atmos. Environ. 16:775-784.

Friedlander, S.K. 1973. Chemical element balances and identification of air pollution sources. Environ. Sci. Technol. 7:235-240.

Galloway, J.N., and G.E. Likens. 1981. Acid precipitation: the importance of nitric acid. Atmos. Environ. 15:1081-1086.

Gartrell, G. Jr., and S.K. Friedlander. 1975. Relating particulate pollution to sources: the 1972 California aerosol characterization study. Atmos. Environ. 9:279-300.

Golomb, D. 1983. Acid deposition-precursor emission relationship in the northeastern U.S.A.: the effectiveness of regional emission reduction. Atmos. Environ. In press.

Gordon, G.E. 1980. Receptor models. Environ. Sci. Technol. 14:792-800.

Granat, L. 1978. Sulfate in precipitation as observed by the European atmospheric chemistry network. Atmos. Environ. 12:413-424.

Hansen, D.A., and G.M. Hidy. 1982. Review of questions regarding rain acidity data. Atmos. Environ. 16:2107-2126.

Hegg, D.A., and P.V. Hobbs. 1981. Cloud water chemistry and the production of sulfates in clouds. Atmos. Environ. 15:1597-1604.

Hegg, D.A., and P.V. Hobbs. 1982. Measurements of sulfate production in natural clouds. Atmos. Environ. 16:2663-2668.

Heidam, N.Z. 1981. On the origin of the Arctic aerosol: a statistical approach. Atmos. Environ. 15:1421-1428.

Heidam, N.Z. 1982. Atmospheric aerosol factor models, mass and missing data. Atmos. Environ. 16:1923-1932.

Henderson, R.G., and K. Weingartner. 1982. Trajectory analysis of MAP3S precipitation chemistry data at Ithaca, New York. Atmos. Environ. 16:1657-1655.

Henry, R.C., and G.M. Hidy. 1979. Multivariate analysis of sulfate and other air quality variables by principal components. Part I. Annual data from Los Angeles and New York. Atmos. Environ. 13:1581-1596.

Henry, R.C., and G.M. Hidy. 1982. Multivariate analysis of particulate sulfate and other air quality variables by principal components. Part II. Salt Lake City and St. Louis. Atmos. Environ. 16:929-943.

Henry, R.C., G.M. Hidy, P.K. Mueller, and K.K. Warren. 1980. Assessing sources, dispersion, chemical transformation and deposition of atmospheric sulfur over eastern North America by advanced multivariate data analysis. Paper presented at the 5th International Clean Air Congress, Buenos Aires. International Union of Air Pollution Prevention Associations.

Hidy, G.M., P.K. Mueller, and E.Y. Tong. 1978. Spatial and temporal distributions of airborne sulfate in parts of the United States. Atmos. Environ. 12:735-752.

Huebert, B.J. 1982. Measurements of the dry-deposition flux of nitric acid vapor to grasslands and forest. Paper presented at the 4th International Conference on Precipitation Scavenging Dry Depositions and Resuspension, November 29 to December 3, 1982. Santa Monica, Calif.

Junge, C.E. 1958. The distribution of ammonia and nitrate in rain water over the United States. Trans. Am. Geophys. Union 39:241-248.

Junge, C.E., and R.T. Werby. 1958. The concentrations of sulfur dioxide, sodium, potassium, calcium and sulphate in rainwater over the United States. J. Meteorol. 15:417-425.

Kallend, A.S., A.R. Marsh, J.H. Pickles, and M.V. Proctor. 1983. Acidity of rain in Europe. Atmos. Environ. 17:127-137.

King, W.J., and F.M. Vukovich. 1982. Some dynamic aspects of extended pollution episodes. Atmos. Environ. 16:1171-1182.

Kowalczyk, G.S., G.E. Gordon, and S.W. Rheingrover. 1982. Identification of atmospheric particulate sources in Washington, D.C., using chemical element balances. Environ. Sci. Technol. 16:79-90.

Kurtz, J., and W.A. Schneider. 1981. An analysis of acidic precipitation in south-central Ontario using air parcel trajectories. Atmos. Environ. 15:1111-1116.

Lazrus, A.L., P.L. Haagenson, G.L. Kok, B.J. Huebert, C.W. Kreitzberg, G.E. Likens, V.A. Mohnen, W.E. Wilson, and J.W. Winchester. 1983. Acidity in air and water in a case of warm frontal precipitation. Atmos. Environ. 17:581-591.

Levine, S.Z., and S.E. Schwartz. 1982. In-cloud and below-cloud scavenging of nitric acid vapor. Atmos. Environ. 16:1725-1734.

Likens, G.E., and T.J. Butler. 1981. Recent acidification of precipitation in North America. Atmos. Environ. 15:1103-1110.

Likens, G.E., F. H. Bormann, and J.S. Eaton. 1980. Variations in precipitation and streamwater chemistry at the Hubbard Brook Experimental Forest during 1964-1977. Pp. 443-464 in Effects of Acid Precipitation on Terrestrial Ecosystems. T.C. Hutchinson and M. Havas, eds. New York: Plenum Press.

Lioy, P.J., and M.T. Morandi. 1982. Source-related winter and summer variations in SO_2, $SO_4^=$ and vanadium in New York City for 1972-1974. Atmos. Environ. 16:1543-1550.

MAP3S/RAINE Research Community. 1982. The MAP3S/RAINE Precipitation Chemistry Network: statistical overview for the period 1976-1980. Atmos. Environ. 16:1603-1631.

Mardia, K.V., J.T. Kent, and J.M. Bibby. 1979. Multivariate Analysis. Pp. 213-246. London: Academic Press.

Miles, L.J., and K.J. Yost. 1982. Quality analysis of USGS precipitation chemistry data for New York. Atmos. Environ. 16:2889-2898.

Mueller, P.K., and G.M. Hidy (principal investigators). 1983. The Sulfate Regional Experiment: Report of Findings. Report EA-1901(3). Palo Alto, Calif.: Electric Power Research Institute.

Mueller, P.K., G.M. Hidy, K. Warren, T.F. Lavary, and R.L. Baskett. 1980. The occurrence of atmospheric aerosols in the northeastern United States. In Aerosols: Anthropogenic and Natural, Sources and Transport. T.J. Kneip and P. J. Lioy, eds. Annal 338, pp. 463-482. New York: New York Academy of Sciences.

National Research Council. 1980. Controlling Airborne Particles. Washington, D.C.: National Academy Press.

Niemann, B.L. 1982a. Proceedings of the American Chemical Society, Division of Environmental Chemistry, Symposium on Acid Precipitation, Las Vegas, Nevada, 29 March 1982. In press.

Niemann, B.L. 1982b. Regional Relationships between H^+, $SO_4^=$, NO_3^-, NH_4^+, Wet Depositions. Washington, D.C.: Office of Technology Assessment.

Niemann, B.L. 1982c. The 1980 Data Set for Further Evaluation of Regional Air Quality/Acid Deposition Simulation Models. Washington, D.C.: U.S. Environmental Protection Agency.

Pack, D.H. 1980. Precipitation chemistry patterns: a two network data set. Science 208. 1143-1145.

Peters, N.E., R. Schroeder, and D. Troutman. 1982. Temporal trends in the acidity of precipitation and surface waters of New York. U.S. Geol. Surv. Water Supply Paper 2188.

Peterson, J.T. 1970. Distribution of sulfur dioxide over metropolitan St. Louis, as described by empirical eigenvectors and its relation to meteorological parameters. Atmos. Environ. 4:501-518.

Pitchford, A.M., W. Malm, R. Flocchini, T. Cahill, and E. Walther 1981. Regional analysis of factors affecting visual air quality. Atmos. Environ. 15:2043-2054.

Rahn, K.A. 1981a. Relative importance of North American and Eurasia as sources of Arctic aerosol. Atmos. Environ. 15:1447-1456.

Rahn, K.A. 1981b. The Mn/V ratio as a tracer of large-scale sources of pollution aerosol for the Arctic. Atmos. Environ. 15:1457-1464.

Raynor, G.S., and J.V. Hayes. 1982a. Variation in chemical wet deposition with meteorological conditions. Atmos. Environ. 16:1647-1656.

Raynor, G.S., and J.V. Hayes. 1982b. Effects of varying air trajectories on spatial and temporal precipitation chemistry patterns. Water, Air, Soil Pollut. 18:173-190.

Rodhe, H., P. Crutzen, and A. Vanderpol. 1981. Formation of sulfuric and nitric acid in the atmosphere during long-range transport. Tellus 33:132-141.

Semonin, R.G. 1981. State Water Survey Division, Atmospheric Chemistry Section at the University of Illinois, SWS Contract Report, Vol. 252, p. 56.

Shaw, R.W., and H. Rodhe. 1983. Nonphotochemical oxidation of SO_2 in regionally polluted air during winter. Atmos. Environ. In press.

Snedecor, G.W., and W.G. Cochran. 1967. Statistical Methods. 6th ed. Ames, Iowa: Iowa State University Press.

Stensland, G.J., and R.G. Semonin. 1982. Another interpretation of the pH trend in the United States. Bull. Am. Meteorol. Soc. 63:1277-1284.

Tanner, R.L., and B.P. Leaderer. 1982. Seasonal variations in the composition of ambient sulfur-containing aerosols in the New York area. Atmos. Environ. 16:569-580.

U.S./Canada. 1981. Memorandum of Intent on Transboundary Air Pollution, Interim Working Paper, 2-14. Washington, D.C.: U.S. Department of State.

U.S./Canada Work Group #2. 1982. Atmospheric Sciences and Analysis. Final Report. H.L. Ferguson and L. Machta, cochairmen. Washington, D.C.: U.S. Environmental Protection Agency.

U.S/Canada Work Group #3B. 1982. Emissions, Costs and Engineering Assessment. Final Report. M.E. Rivers and K.W. Riegel, Cochairmen. Washington, D.C.: U.S. Environmental Protection Agency.

Vukovich, F.M., W. Bach, B. Crissman, and W. King. 1977. On the relationship between high ozone in the rural surface layer and high pressure systems. Atmos. Environ. 11:967-984.

White, W., S. Heisler, R. Henry, and G.M. Hidy. 1978. The same-day impact of power plant emissions on sulfate levels in the Los Angeles air basin. Atmos. Environ. 12:779-784.

Wilson, J.W., V.A. Mohnen, and J.A. Kadlecek. 1982. Wet deposition variability as observed by MAP3S. Atmos. Environ. 16:1667-1676.

Wisniewski, J., and J. Kinsman. 1982. An overview of acid rain monitoring activities in North America. Bull. Am. Meteorol. Soc. 63:598-618.

Wolff, G.T., N.A. Kelly, and M.A. Ferman. 1981. On the sources of summertime haze in the eastern United States. Science 211:703-704.

Wolff, G.T., N.A. Kelly, and M.A. Ferman. 1982. Source regions of summertime ozone and haze episodes in the eastern U.S.A. Water, Air, Soil Pollut. 18:65-82.

5 Research Needs

In light of existing uncertainties in knowledge about the relationships between emissions and the deposition of acid-forming materials, it is appropriate to consider the research that may help to clarify understanding. In this chapter we describe research that we believe is needed to answer the most central questions in the shortest amount of time.

A considerable amount of research is currently being performed with funding from both governmental and private sources in the United States and Canada as well as in Europe. We have not attempted to review these research programs. As a consequence, some of the research activities that we recommend may be--indeed we know they are--included in current programs. Others, to the best of our knowledge, are not. On the basis of our review of current knowledge as described in this report, we are convinced that the research that we recommend forms essential elements in an integrated effort to improve understanding of the phenomenon of acid deposition.

We do not know whether incorporation into a model of all the chemical and physical processes currently thought to be important would account for the concentrations of sulfate observed in precipitation during both winter and summer. No quantitative determination of the relative contributions to the production of sulfate and nitrate by gas- and liquid-phase processes has been attempted to date; such attempts are probably several years in the future.

The details of dispersion of pollutants from sources, their chemical transformation and transport over long distances, and their contribution to acids in precipitation systems are sufficiently unknown and complex that a great deal of research will probably be required before they can be described with precision by models. Furthermore, many years of precipitation measurements will be needed to establish a reliable data set from which tests of models can be made.

We believe that extensive laboratory, field, and modeling studies should be continued, to establish the physical and chemical mechanisms governing acid deposition. However, it appears to us that useful information about the delivery of acids to rural areas by transport and transformation processes can be determined fairly quickly by direct empirical observation in the field. Although the results of such field studies may not yield complete detailed descriptions of the interactions of all the processes involved, they are likely to provide basic phenomenological evidence with sufficient reliability to form the basis for improving the near-term strategy for dealing with the problem of acid deposition. The data are essential for enhancing theoretical understanding and developing improved deposition models. In the long term, the ultimate strategy for dealing with acid deposition will depend on the application of realistic, validated models.

FIELD STUDIES

Field research capable of testing the possibility of a nonlinear relationship between emissions and wet deposition in North America is needed. A limitation of oxidant in winter is believed to be a principal cause of such nonlinearity. Because the nature and availability of oxidants is sensitively dependent on the general composition of polluted air, field studies should be conducted in regions particularly involved in the problem of acid precipitation. Of primary importance are those experiments directed at discovering the rates of production of $SO_4^=$ and NO_3^- in clouds and the detection of possible limitations of those rates. Supplemental information, such as the availability of oxidants under varying conditions, is also important.

We describe below some of the more important issues currently amenable to study in the field.

Cloud Processes

As described in Chapter 2 and Appendix A, the conversion of SO_2 to H_2SO_4 in certain cloud systems may be relatively complete and rapid, but we lack conclusive information on either oxidation rate constants or the completeness of conversion in acidified clouds typical of the eastern United States and Canada. Most of the mass of sulfuric acid already existing in aerosols in dry air is efficiently incorporated into cloud droplets as the air is cooled and the cloud forms. Models indicate that nitric acid vapor is also incorporated with high efficiency into cloud droplets. If, during the warm months, when the rate of wet deposition of acid is highest, complete in-cloud conversion of SO_2 and NO_2 occurs, and if the rainout of HNO_3 vapor and sulfate aerosol is highly efficient, then the assumption that the wet deposition of sulfuric and nitric acid is linearly related to ambient concentrations of SO_2 and NO_x would be justified. This could be readily checked experimentally with available technology in appropriate storm systems, such as warm fronts and summertime convective storms.

Studies of Chemical Mechanisms

A number of key problems relating to the linearity of the dependence of acid production rates on precursor concentrations provide the focus for studies of chemical mechanisms. (1) The efficiency and seasonal dependence of transformation processes are no doubt different for gas-phase and aqueous-phase processes. What are the relative contributions of these two pathways? (2) What are the lifetimes of NO_2 and SO_2 in clouds? Do rates of acid production differ markedly in various cloud types? Are the production rates inhibited by increasing acidity or limited by availability of oxidants? (3) What are the dominant reaction paths for SO_2 oxidation in clouds? (4) Do competing reactions with other atmospheric constituents (e.g., formaldehyde), especially in polluted air, seriously inhibit or lower the effectiveness of H_2O_2 reactions with SO_2 in clouds? (5) What processes govern the apparently significant rate of production of HNO_3 in clouds? (6) What are the relative rates of production of HNO_3 and H_2SO_4 in clouds? (7) What is the role of ultraviolet light in the production of free radicals and H_2O_2 in the gas phase in clouds?

Dry Deposition

We noted in Chapter 3 that dry deposition of SO_2 and NO_2 may account for about one half of the total deposition of sulfur and nitrogen oxides and acids in eastern North America. Accurate evaluations of the extent of dry deposition of these pollutants is required to obtain a quantitative measure of their fates in the environment. The eddy-correlation technique, in which simultaneous measurements of the vertical wind speeds and specific pollutant concentrations are measured with rapid-response instrumentation, should prove to be a valuable tool in the study of the deposition of SO_2, NO_2, HNO_3, and other pollutant gases. The development, refinement, and application of this method and possibly other new and accurate methods of measuring dry deposition, which are suitable for monitoring applications, are important research objectives. The challenge of developing accurate methods is great, adapting as they must to the great variety of surfaces that cover the Earth and to the effects of humidity, temperature, sunlight intensity, and other factors.

Tracers

To develop the most cost-effective strategy for ameliorating the problem of acid deposition, it is necessary to know the relative impacts of specific source regions on specific sensitive receptor regions. In view of the uncertainties inherent in the calculation of trajectories, especially during storm conditions, it seems important to develop tracer techniques that can yield experimental tracking of air parcels. Such techniques could be applied especially during the meteorological conditions that are currently believed to provide the greatest opportunity for long-range transport of acidic materials. The use of insoluble and chemically inert gaseous tracers, as is currently planned, provides a promising approach. Admixtures of insoluble and unique materials that are subject to reaction with oxidants such as the HO radical would in addition provide information on possible limitations of the oxidation of SO_2 and NO_x.

In light of the potential advantages of the method of elemental tracers for understanding the relationships between source and receptor regions, a considerable amount of research and development to establish and test the method on a regional scale is warranted. The first priority in this work should be to measure the detailed patterns of elemental composition of particulate matter collected at various rural sites in both source and receptor regions in eastern North America. The particles should be segregated by size, and a large number of elements and species should be measured for each size range. Other data should also be collected concurrently, including ambient concentrations of pollutant gases, wind speed and direction, relative humidity, precipitation, and optical qualities of the atmosphere. High-quality back trajectories should be calculated for the air masses sampled for each sampling period to aid in interpretation of the results. Few of these data now exist.

Research using elemental tracers would attempt to determine if particles coming from major source regions have distinctive chemical signatures that can be used to identify the origins of polluted air masses at long distances from source regions. For example, are there elemental tracers that can be used as clear indicators of coal combustion, in the same way as vanadium apparently can be used to indicate oil combustion? Can observed patterns of elemental composition in rural areas be resolved into linear combinations of known composition released by certain types of sources? Does the behavior of fine particles, particularly those bearing soluble species, provide insight into the behavior of sulfate and nitrate species that are of primary concern in acid deposition?

Building on the first stage of research, it may also be necessary to conduct studies of the elemental composition of particles in clear air and clouds as well as on the ground during specific episodes, such as the ducting of a polluted air mass from the midwestern United States toward the northeast along a southward-moving warm front. Changes in elemental composition as a specific parcel of air moves on a regional scale should help to identify the types of changes in the characteristics of the air during transport and as it passes over additional sources. Much of this work could be conducted in conjunction with the field studies described in the previous section.

Meteorological Studies

The climatology of storm movements in North America is well developed. The development of a quantitative relationship between storm type and acid deposition over eastern North America would help to estimate the long-range transport of pollutants. Despite the associated uncertainties, statistical studies of air parcel trajectories associated with the types of storm systems responsible for depositing acid precipitation would also be extremely valuable.

An important requirement in the experimental determination of rates of SO_2 and NO_x oxidation in clouds is a realistic evaluation of the airflow in the vicinity of clouds to establish the quantity of materials processed by and the time resident in the cloud.

All of these field studies bear directly on the critical question of the zone of influence of sources on receptor regions.

Our suggestion that these field measurements be a first priority is not meant to imply that pertinent laboratory and modeling studies should not also proceed. However, in view of the complexities in these systems, we believe that well-conceived field studies may answer many of the outstanding questions in a shorter time than that required for a complete molecular and dynamic description of the phenomenon.

LABORATORY STUDIES

The direct measurement of elementary rate constants for the many apparently important reactions related to the chemistry of acid deposition has been restricted largely to conditions that are not typical of the lower troposphere. The direct determination of the rate constants for reactions of the HO radical with SO_2 and NO_2 should be made at pressures (near 1 atm of air) and temperatures characteristic of the troposphere, as should other measurements.

Many aspects of cloud chemistry can and should be examined quantitatively under controlled laboratory conditions. For example, the mechanisms of the development of H_2O_2 from O_3 and other unidentified reactants should be established. Other important parameters amenable to laboratory measurement that are required for the quantitative description of cloud processes are the

sticking coefficients of the gas-phase reactants such as
HO, HO_2, and NO_3 on collision with cloud droplets.

In general, laboratory studies are needed to establish
the mechanistic detail that is required in the development of the chemically and physically sound models of
acid precipitation.

DEVELOPMENT OF THEORETICAL MODELS

Models of the development, transport, and deposition of
acids and acid-forming materials based on physical and
chemical principles are under development. These models
will serve as a useful framework with which the latest
data from field and laboratory studies can be combined to
provide a suitable test of theory and improved planning
for further field studies. State-of-the-art theoretical
models that treat quantitatively the complicated gas-
phase chemistry, cloud processes, transport, and
deposition will require many years for development. The
ultimate test of our understanding of the chemistry and
physics of the processes of acid rain is the successful
development and use of such models.

Appendix A The Chemistry of Acid Formation

It is well established that the oxides of sulfur--sulfur dioxide (SO_2) and sulfur trioxide (SO_3)--and nitrogen-- nitric oxide (NO) and nitrogen dioxide (NO_2)--are converted (oxidized) in the troposphere to sulfuric acid (H_2SO_4) and nitric acid (HNO_3), respectively. However, the most prevalent end products of these reactions--H_2SO_4, HNO_3, ammonium bisulfate (NH_4HSO_4), ammonium nitrate (NH_4NO_3), etc.--give few clues about which of several oxidizing pathways are important. Yet for the development of scientifically sound, predictive models of acid deposition that define theoretical source-receptor relationships, knowledge of the elementary chemical steps that are involved (among many other quantities such as deposition rates and transport rates) is required. The rates of these reactions follow well-defined mathematical laws (rate expressions) that relate reaction rates to concentrations of the reactants, temperature, pressure, and other variables. Considerable progress has been made in developing an understanding of the nature of this chemistry, but a large number of uncertainties still remain in key areas.

As our knowledge of the atmospheric chemistry of SO_2, NO, and NO_2 continues to grow, it has become increasingly clear that many different pathways exist for generating H_2SO_4 and HNO_3 in the troposphere. Reactions can occur in the gas phase; in the solution phase in cloud water and rainwater, for example; and in reactions on surfaces of solid particles in the atmosphere. Thus we are concerned with the rates of exchange of gaseous reactants and their reaction products between liquids and the surfaces of solids as well as the rates of interactions among gaseous molecules, aqueous phase molecules and ions, and species adsorbed on solid

155

surfaces. For current purposes, it is not necessary to review and evaluate every detail of these chemical processes, but we should note the nature of the many chemical processes, the variety of interactions among the many species involved, and the current state of knowledge related to the chemistry of acid deposition. We consider first the homogeneous gas-phase chemistry that results in oxidation of SO_2, NO, and NO_2 in the troposphere (Calvert and Stockwell 1983, Calvert et al. 1978).

GAS-PHASE REACTIONS LEADING TO GENERATION OF ACID IN THE TROPOSPHERE

Oxidation by Stable Atmospheric Molecules

The thermodynamic properties of the oxides of sulfur indicate that sulfur dioxide has a strong tendency to react with oxygen in the air under normal tropospheric conditions:

$$2SO_2 + O_2 \rightarrow 2SO_3. \tag{1}$$

Thus the ratio of $[SO_3]/[SO_2]$ is about 8×10^{11} at equilibrium in air at 1 atm and 25°C. (Square brackets indicate the concentration of the species inside the brackets.) Thermodynamic arguments also tell us that at humidities normally encountered in the lower troposphere, the SO_3 produced by reaction (1) will be converted efficiently to sulfuric acid, $H_2SO_4(aq)$, according to

$$SO_3 + H_2O \rightarrow H_2SO_4(aq). \tag{2}$$

The reaction of SO_3 with H_2O is so fast that any process in which SO_3 is formed in the moist troposphere can be considered equivalent to the formation of H_2SO_4. Certain metal ions (Mn^{2+}, Fe^{3+}, etc.) in aqueous solutions of SO_2 (HSO_3^-) can catalyze the overall sequence of reactions (1) and (2). However, thermodynamics tells us nothing about the rates of chemical reactions, and the rate of reaction (1) is so slow under catalyst-free conditions in the gas phase that it can be neglected as a source of sulfuric acid in the atmosphere.

The thermal oxidation of NO and NO_2 in the gas phase is also slow. Pathways that are thermodynamically favored are

$$2NO + O_2 \rightarrow 2NO_2, \qquad\qquad (3)$$

$$2NO_2 + H_2O \rightarrow HNO_3 + HONO. \qquad\qquad (4)$$

Reaction (3) requires literally days for the conversion of a significant fraction of the nitric oxide present [at concentrations of the order of parts per billion (ppb)] in the troposphere, and the homogeneous gas-phase reaction (4) is immeasurably slow (Schwartz and White 1982). Obviously, other, seemingly less direct reactions must be invoked to account for the observed rates of SO_2, NO, and NO_2 oxidation, which are between 1 and 100 percent/h.
A number of the more complex pathways involve photochemistry. In one, sulfur dioxide absorbs light in the ultraviolet region of the solar radiation incident in the troposphere, and, in principle, excited states of SO_2 generated in this fashion could lead to SO_2 oxidation in the troposphere. Figure A.1 indicates that there is a significant overlap between the actinic flux incident in the lower troposphere and two distinct absorption regions of SO_2. Excitation in the "forbidden," long-wavelength band forms the excited $SO_2(^3B_1)$ species, while excitation at the wavelengths below 330 nm generates higher excited states, presumably the $^1A_2$ and $^1B_1$ species.

$$SO_2(\tilde{X}^1A_1) + h\nu(340 < \lambda < 400 \text{ nm}) \rightarrow SO_2(^3B_1), \qquad (5)$$

$$SO_2(\tilde{X}^1A_1) + h\nu(240 < \lambda < 330 \text{ nm}) \rightarrow SO_2(^1A_2, {}^1B_1). \qquad (6)$$

These excited states of SO_2 are nondissociative; only quanta of light at wavelengths below 218 nm (which do not penetrate to the troposphere) provide sufficient energy to allow photodissociation:

$$SO_2(\tilde{X}^1A_1) + h\nu(\lambda < 218 \text{ nm})$$

$$\rightarrow SO_2(^1B_2) \rightarrow O(^3P) + SO(^3\Sigma^-). \qquad (7)$$

The lower excited singlet states of $SO_2(^1A_1, {}^1B_1)$ appear to be very short lived in air at 1 atm, and they are rapidly converted by collisional perturbations to $SO_2(^3B_1)$ molecules, possibly $SO_2(^3A_2)$ and $SO_2(^3B_2)$ molecules, and ground-state $SO_2(\tilde{X}^1A_1)$ molecules. The rate of excitation of SO_2 through absorption of sunlight can be very significant. If this excitation were the rate-determining step in the photooxidation of SO_2, that is, if every molecule of SO_2 that is photoexcited were oxidized

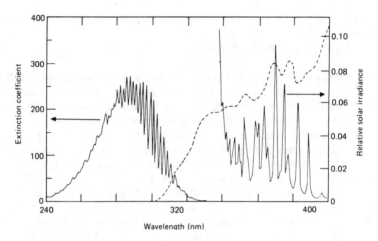

FIGURE A.1 Comparison of the extinction coefficients (liters mole^{-1} cm^{-1}, base 10) of SO_2 within the first allowed band (left), the "forbidden" band (right), and a typical distribution of the flux of solar quanta (relative) at ground level (dashed curve). SOURCE: Calvert et al. (1978).

through subsequent reaction with O_2 or other reactants, then the lifetime of SO_2 in the lower troposphere with overhead sun should be as low as 52 minutes (Sidebottom et al. 1972). Of course, this is not the case. The SO_2 ($^3B_1$) species appears to be one of the most favored states, which is ultimately populated through absorption of sunlight and collisional processes in the lower atmosphere. The reactions of this species with various atmospheric gases and many atmospheric impurities have been studied extensively (Calvert et al. 1978). Quenching of SO_2 ($^3B_1$) by atmospheric gases is expected to be the dominant process. In air at 1 atm, 25°C, and 50 percent relative humidity, quenching by N_2, O_2, H_2O, and Ar will occur 45.7, 41.7, 12.2, and 0.3 percent of the time, respectively. Quenching by impurity gases is highly improbable. Even when impurities are present at concentrations of the order of ppm, the rates of SO_2 conversion by such species are very slow (Calvert et al. 1978). All available evidence suggests that the only significant chemical result of the major quenching reactions of SO_2 ($^3B_1$) occurs with O_2, and this does not lead efficiently to any overall chemical change in the SO_2, but low-lying, excited electronic states of molecular oxygen, $O_2(^1\Delta_g)$ and $O_2(^1\Sigma_g^+)$, are formed by energy transfer.

$$SO_2(^3B_1) + O_2(^3\Sigma_g^-) \rightarrow SO_2(\tilde{X}^1A_1) + O_2(^1\Sigma_g^+) \qquad (8)$$

$$\rightarrow SO_2(\tilde{X}^1A_1) + O_2(^1\Delta_g). \qquad (9)$$

These reactions are not unique sources of excited oxygen, species that are also formed in other atmospheric reactions at much higher rates (Calvert et al. 1978). Thus we conclude that photooxidation of SO_2 is also not an important source of acids in the atmosphere.

Similar conclusions are reached from a study of the photochemistry of NO and NO_2. NO does not absorb solar radiation in the wavelength range available near the Earth's surface, so its photochemistry is not important in the troposphere. NO_2 does absorb radiation over a wide range, and absorption of the wavelengths below 430 nm leads to dissociation ($NO_2 + h\nu \rightarrow O + NO$). However, formation of acid as a direct result of the photochemistry of NO and NO_2 is unimportant.

Reactive Transient Species in the Troposphere

Most of the gas-phase tropospheric chemistry of SO_2, NO, NO_2, and other impurity molecules involves reactions with a variety of reactive excited molecules, atoms, and free radicals (neutral fragments of stable molecules) formed by absorption of sunlight by trace gases in the atmosphere. As a background to discussions to come we review briefly here some of this important chemistry since it enters directly or indirectly into many of the reaction pathways that lead to the formation of acids in the troposphere.

In the polluted troposphere, NO_2 is dissociated by sunlight absorption ($\lambda < 430$ nm) to form reactive, ground state oxygen atoms, $O(^3P)$, and NO, while the oxygen atom reacts rapidly to form ozone (O_3):

$$NO_2 + h\nu\,(\lambda < 430 \text{ nm}) \rightarrow O(^3P) + NO, \qquad (10)$$

$$O(^3P) + O_2(+M) \rightarrow O_3(+M). \qquad (11)$$

Ozone can reoxidize NO to NO_2 in (12) or react with alkenes to give highly reactive ozonides in (13) and Criegee intermediates in (14):

$$O_3 + NO \rightarrow O_2 + NO_2 \tag{12}$$

$$O_3 + RHC=CHR \rightarrow \overset{\displaystyle O-O-O}{\underset{\displaystyle}{RHC}\!\!-\!\!-\!\!CHR} \tag{13}$$

$$\overset{\displaystyle O-O-O}{RHC\!\!-\!\!-\!\!CHR} \rightarrow RCHO_2 + RCHO \tag{14}$$

Ozone can also oxidize NO_2 to the reactive transient NO_3 in (15), and this can lead to N_2O_5 in (16):

$$O_3 + NO_2 \rightarrow O_2 + NO_3, \tag{15}$$

$$NO_3 + NO_2 (+M) \rightleftharpoons N_2O_5 (+M). \tag{16}$$

The photodecomposition of ozone may generate electronically excited oxygen atoms, $O(^1D)$, and excited molecular oxygen with absorption in the short-wavelength region of the spectrum:

$$O_3 + h\nu(290-306 \text{ nm}) \rightarrow O(^1D) + O_2(^1\Delta_g), \tag{17}$$

$$O_3 + h\nu(290-350 \text{ nm})$$
$$\rightarrow O(^1D) + O_2 \text{ or } O + O_2(^1\Delta_g, {}^3\Sigma_g^-), \tag{18}$$

$$O_3 + h\nu(450-700 \text{ nm}) \rightarrow O + O_2. \tag{19}$$

The $O(^1D)$ species formed in (17) is much more reactive than the ground-state oxygen atoms $[O(^3P)$, often simply symbolized by O]. $O(^1D)$ reacts efficiently when it collides with a water molecule to form a highly important transient in atmospheric chemistry, the hydroxy radical, HO:

$$O(^1D) + H_2O \rightarrow 2HO. \tag{20}$$

This radical, unlike many molecular fragments formed from carbon-containing molecules, is unreactive toward oxygen, and it survives to react with most atmospheric impurities such as the hydrocarbons, aldehydes, NO, NO_2, SO_2, and CO. Its reactions with carbon monoxide and the hydrocarbons (RH) lead to another important class of reactive transients, the peroxy radicals:

$$HO + CO \rightarrow H + CO_2, \tag{21}$$

$$H + O_2(+M) \rightarrow HO_2(+M),\tag{22}$$

$$HO + RH \rightarrow H_2O + R,\tag{23}$$

$$R + O_2(+M) \rightarrow RO_2(+M).\tag{24}$$

Here R represents the alkyl groups such as methyl (CH_3), ethyl (C_2H_5), or another larger group derived from the parent hydrocarbons, methane (CH_4), ethane (C_2H_6), or larger hydrocarbon (RH), respectively. The reaction of the HO radical with aldehydes (RCHO) forms the acyl (RCO) and acylperoxy ($RCOO_2$) radicals in similar reactions:

$$RCHO + HO \rightarrow RCO + H_2O,\tag{25}$$

$$RCO + O_2(+M) \rightarrow RCOO_2(+M).\tag{26}$$

The peroxy radicals react rapidly with NO to form NO_2 and other classes of reactive species. In the case of the HO_2-NO reaction, HO is regenerated, while with the RO_2 and $RCOO_2$ radicals, alkoxy (RO) and acyloxy (RCO_2) radicals, respectively, are formed:

$$HO_2 + NO \rightarrow HO + NO_2,\tag{27}$$

$$RO_2 + NO \rightarrow RO + NO_2,\tag{28}$$

$$RCOO_2 + NO \rightarrow RCO_2 + NO_2.\tag{29}$$

The most common fate of the smaller alkoxy radicals in the lower atmosphere is reaction with oxygen, leading to HO_2 radicals and a carbonyl compound. For example, with the simplest alkoxy radical, methoxy (CH_3O), the following reaction occurs:

$$CH_3O + O_2 \rightarrow HO_2 + CH_2O.\tag{30}$$

The RCO_2 radicals are of short lifetime, decomposing to form an alkyl radical (R) and CO_2, with the subsequent generation of another peroxyalkyl radical:

$$RCO_2 \rightarrow R + CO_2,\tag{31}$$

$$R + O_2(+M) \rightarrow RO_2(+M).\tag{32}$$

Reactions (14)-(26) combined with reactions (27)-(32) form a chain reaction. That is, a single initial HO radical

oxidize CO, hydrocarbon, or aldehyde, and additional
radicals will be regenerated as NO is oxidized to NO_2;
tne subsequent steps occur again and again in a repeating
cycle of events. The peroxy radicals and ozone are the
principal oxidizing agents for NO in the lower atmosphere
[reactions (15) and (27)-(29)] (Demerjian et al. 1974).

When peroxy radicals react with NO_2, an additional
class of highly reactive compounds is generated--the
peroxynitrates:

$$HO_2 + NO_2 \rightleftharpoons HO_2NO_2, \tag{33}$$

$$RO_2 + NO_2 \rightleftharpoons RO_2NO_2, \tag{34}$$

$$RCOO_2 + NO_2 \rightleftharpoons RCOO_2NO_2. \tag{35}$$

Peroxynitric acid formed in (33), and the alkyperoxy-
nitrates formed in (34) are relatively unstable in the
lower troposphere at temperatures common in summer
months; they dissociate readily to reform peroxy radicals
and NO_2. However, during the cold winter months or in
the stratosphere they can act as temporary sinks for
HO_2 and RO_2 radicals and NO_2. Peroxyacylnitrates
($RCOO_2NO_2$), of which peroxyacetylnitrate ($CH_3COO_2NO_2$) is
the most common, have longer lifetimes and can be the
source of radical generation even during the nighttime
hours.

The excited singlet delta molecular oxygen, $O_2(^1\Delta_g)$, a
product of reactions (17) and (18), and excited singlet
sigma oxygen, $O_2(^1\Sigma_g^+)$, the product of reaction (8), can
also be created by direct absorption of sunlight by
atmospheric O_2 and by energy transfer reactions from
other photoexcited species such as $NO_2(\lambda > 430$ nm) and
excited triplet aromatic hydrocarbons.

For current purposes the complex array of interactions
that occur among the reactive species outlined here and
with the various atmospheric impurities need not be
considered. It suffices to say that many aspects of
tropospheric chemistry, including photochemical "smog"
and ozone generation, depend on these happenings
(Demerjian et al. 1974). It is important, however, to
evaluate the potential significance of the many highly
reactive transient species of the atmosphere for reactions
that oxidize SO_2 and NO_2 to acids.

Atmospheric Oxidation of SO_2
by Reactive Transient Species

There are a large number of potentially significant
gas-phase reactions of the reactive transients leading to
oxidation of SO_2 in the troposphere. The potential
candidate reactions, summarized in Table A.1, have the
thermodynamic potential to occur as measured qualitatively
by the sign of the change in enthalpy ($\Delta H°$) for the
overall reaction. Rate constants for most of these
elementary reactions have been determined. These data,
coupled with estimates of the concentrations of the
transients in the atmosphere, allow us to evaluate the
significance of each reactant in oxidizing SO_2 in the
atmosphere. Such evaluations have shown that reactions
(42), (50), (56), and (59) or (61) are potentially significant sources of SO_2 oxidation; some of these
reactions are only important for certain peculiar
atmospheric conditions. By far the most important of the
gas-phase reactions is the reaction of HO radicals with
SO_2:

$$HO + SO_2(+M) \rightarrow HOSO_2(+M). \tag{56}$$

The rate-constant data for reaction (56) have been
reviewed recently (Calvert and Stockwell 1983) and are
summarized in Figure A.2. The theoretical effect on this
rate constant of the altered temperature and pressure of
the atmosphere at various altitudes is shown in Figure
A.3.

The "best" values of $k_{II(56)}$ suggested from the
analysis given by Calvert and Stockwell (1983) are
somewhat larger than those chosen by Moortgat and Junge
(1977), Zellner (1978), and the values recommended for
use by the CODATA (1980) group and the NASA (1979) panel
and are more consistent with currently available information on the $HO-SO_2$ reaction. One must, however,
retain considerable pessimism about the accuracy of all
these data; realistic confidence limits should include
±50 percent of the suggested value.

Evidence is good that $HOSO_2$ formed in reaction (56)
ultimately leads to the generation of sulfuric acid
aerosol. However, $HOSO_2$ is not a stable molecule; it
is a free radical that is probably highly reactive with
several atmospheric compounds. It is not now clear what
elementary reaction pathways are important in its
conversion to H_2SO_4. Although there has been a great

TABLE A.1 Enthalpy Changes and Recommended Rate Constants for Potentially Important Reactions of Ground State SO_2 and SO_3 Molecules in the Lower Atmosphere

Reaction	$-\Delta H^{ou}$ (kcal/mole) (25°C)	k^b (cm³/molec-sec)
(36) $O_2\,(^1\Delta_g) + SO_2 \rightarrow SO_4$ (biradical; cyclic)	~25, ~28	
(37) $O_2\,(^1\Delta_g) + SO_2 \rightarrow SO_3 \rightarrow O\,(^3P)$	−13.5	3.9×10^{-20}
(38) $O_2\,(^1\Delta_g) + SO_2 \rightarrow O_2\,(^3\Sigma_g) + SO_2$	22.5	
(39) $O_2\,(^1\Sigma_g^+) + SO_2 \rightarrow SO_4$ (biradical; cyclic)	~40, ~43	
(40) $O_2\,(^1\Sigma_g^+) + SO_2 \rightarrow SO_3 + O\,(^3P)$	1.5	6.6×10^{-16}
(41) $O_2\,(^1\Sigma_g^+) + SO_2 \rightarrow SO_2 + O_2\,(^1\Delta_g)$	15.0	
(42) $O\,(^3P) + SO_2\,(+M) \rightarrow SO_3\,(+M)$	83.0	5.7×10^{-14}
(43) $O_3 + SO_2 \rightarrow O_2 + SO_3$	57.6	$<8 \times 10^{-24}$
(44) $NO_2 + SO_2 \rightarrow NO + SO_3$	9.9	8.8×10^{-30}
(45) $NO_3 + SO_2 \rightarrow NO_2 + SO_3$	32.6	$<7 \times 10^{-21}$
(46) $ONOO + SO_2 \rightarrow NO_2 + SO_3$	~30	$<7 \times 10^{-21}$
(47) $N_2O_5 + SO_2 \rightarrow N_2O_4 + SO_3$	24.0	$<4 \times 10^{-23}$
(48) $HO_2 + SO_2 \rightarrow HO + SO_3$	16.7	$<1 \times 10^{-18}$
(49) $HO_2 + SO_2\,(+M) \rightarrow HO_2SO_2\,(+M)$	~7	
(50) $CH_3O_2 + SO_2 \rightarrow CH_3O + SO_3$	~27	$<1 \times 10^{-18}$
(51) $CH_3O_2 + SO_2\,(+M) \rightarrow CH_3O_2SO_2\,(+M)$	~31	$\sim1.4 \times 10^{-14c}$
(52) $(CH_3)_3CO_2 + SO_2 \rightarrow (CH_3)_3CO + SO_3$	~26	$<7.3 \times 10^{-19}$
(53) $(CH_3)_3CO_2 + SO_2 \rightarrow (CH_3)_3CO_2SO_2$	~30	
(54) $CH_3COO_2 + SO_2 \rightarrow CH_3CO_2 + SO_3$	~33	$<7 \times 10^{-19}$
(55) $CH_3COO_2 + SO_2 \rightarrow CH_3COO_2SO_2$	~37	
(56) $HO + SO_2\,(+M) \rightarrow HOSO_2\,(+M)$	~37	1.1×10^{-12}
(57) $CH_3O + SO_2\,(+M) \rightarrow CH_3OSO_2\,(+M)$	~24	$\sim5.5 \times 10^{-13}$
(58) $\overset{O-O-O}{\underset{}{RCH-CHR}} + SO_2 \rightarrow 2RCHO + SO_3$	~69	See text
$\overset{O\cdot\quad O-O\cdot}{\underset{}{RCH-CHR}} + SO_2 \rightarrow 2RCHO + SO_3$	~89	See text
(59) $RCHOO\cdot + SO_2 \rightarrow RCHO + SO_3$	~79	$k_{59}/k_{60} \sim 6 \times 10^{-5}$
(60) $RCHOO\cdot + H_2O \rightarrow RCOOH + H_2O$	~121	(R = CH_3)
(61) $\overset{O\cdot}{\underset{}{RCHO\cdot}} + SO_2 \rightarrow RCHO + SO_3$	~58	$k_{61}/k_{62} \cong 4$
(62) $\overset{O\cdot}{\underset{}{RCHO\cdot}} + CH_2O \rightarrow \overset{O\cdot}{\underset{}{RCHOCH_2O\cdot}}$	~12	(R = H)
(63) $SO_3 + H_2O \rightarrow H_2SO_4$	24.8	9.1×10^{-13}

[a]Enthalpy change estimates were derived from the data of Benson (1978), Harding and Goddard (1978), and Domalski (1971).
[b]Rate constants are expressed as second-order reactions for 1 atm of air at 25°C; see Calvert and Stockwell (1983) for the references to the original literature.
[c]The reverse reaction is so fast that the rate of oxidation of SO_2 via (51) is very dependent on alternate fates of the $CH_3O_2SO_2$ species.

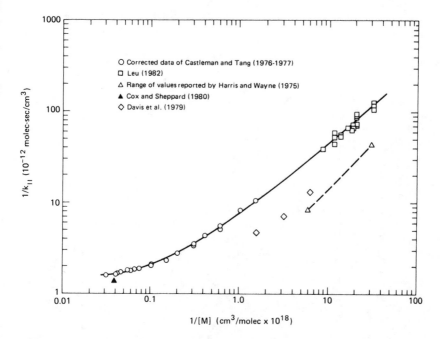

FIGURE A.2 Comparison of the experimental data for the effective second-order rate constants for the reaction (56) with $M = N_2$. SOURCE: Calvert and Stockwell (1983).

amount of speculation in this regard, there is little experimental evidence to help determine the relative importance of suggested alternative routes. Thus Cox (1974-1975, 1975), Calvert and McQuigg (1975), Calvert et al. (1978), Davis et al. (1979), Friend et al. (1980), Leu (1982), Benson (1978), and others have suggested that the $HOSO_2$ radical may participate in a variety of radical-radical and radical-molecule reactions. These reactions are summarized in Figure A.4. Although the mechanism of H_2SO_4 generation following (56) is unclear today, it is probable that reaction (56) is the rate-determining step in the sequence. Recent evidence suggests that the concentration of HO in photooxidizing mixtures of HONO, NO, NO_2, and CO is insensitive to even large additions of SO_2 (Stockwell and Calvert 1983). Thus the following sequence of reactions seems favored:

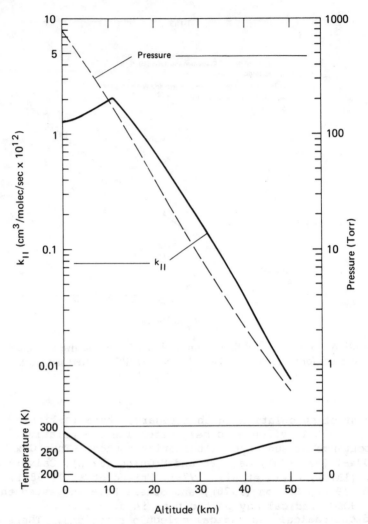

FIGURE A.3 Pressure, temperature, and the apparent second-order rate constant for the reaction (56) as a function of altitude. SOURCE: Adapted from Calvert and Stockwell (1983) for the conditions of pressure and temperature defined for the standard atmosphere (Valley 1965).

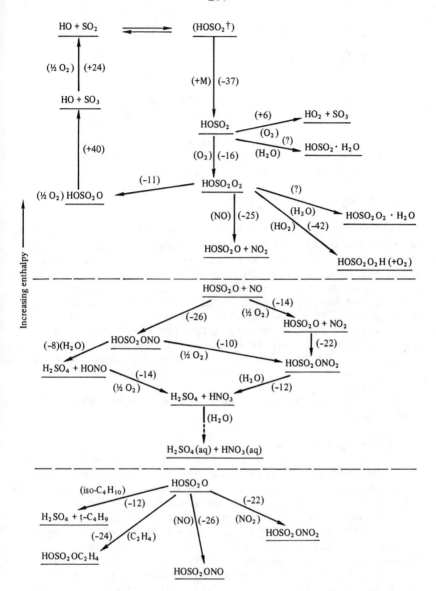

FIGURE A.4 Enthalpy relationships between various possible products of the HO-radical reaction with SO_2. SOURCE: Calvert and Stockwell (1983).

$$HOSO_2 + O_2 \rightarrow HO_2 + SO_3,$$
$$HO_2 + NO \rightarrow HO + NO_2,$$
$$SO_3 + H_2O \rightarrow H_2SO_4.$$

Atmospheric Oxidation of NO_2
by Reactive Transient Species

A great variety of experimental evidence supports the view that a very important pathway for the tropospheric oxidation of NO_2 is similar to reaction (56) for SO_2:

$$HO + NO_2 (+M) \rightarrow HONO_2 (+M). \tag{64}$$

Estimates of $k_{II(64)}$ are well characterized for conditions applicable to the troposphere ($M = N_2$). Using Anderson's (1980) recommendations for this rate constant, we have made the plot shown in Figure A.5 of the effective second-order rate constant for various altitudes and conditions of [M] and temperature characteristics of the standard atmosphere. The same general trends in $k_{II(64)}$ with altitude are seen as observed for the $HO-SO_2$ reaction rate constant $k_{II(56)}$ in Figure A.3, but the values of $k_{II(56)}$ are roughly a factor of 10 larger than those for $k_{II(64)}$ for a given temperature and pressure in the troposphere.

A second homogeneous mode of potential formation of $HONO_2$ involves N_2O_5:

$$H_2O + N_2O_5 \rightarrow 2HONO_2. \tag{65}$$

It is difficult to distinguish the homogeneous component of this reaction in laboratory experiments, since reaction at a moist cell wall can be more important than the homogeneous reaction for most experimental systems. Morris and Niki (1973) reported an upper limit for this constant of 1.3×10^{-20} cm^3 $molec^{-1}$ sec^{-1}. For theoretically expected concentrations of N_2O_5 in the troposphere ($\leq 2.5 \times 10^9$ molec cm^{-3}) during the daylight hours (Demerjian et al. 1974), the rate of (65) is insignificant compared with that of (64). In theory the reaction (65) can be important at night when NO_3 and N_2O_5 concentrations may rise.

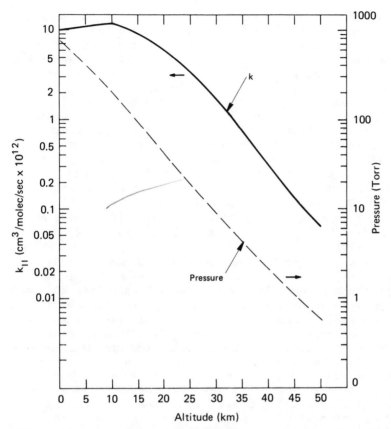

FIGURE A.5 Pressure and the apparent second-order rate constant for reaction (64) as a function of altitude. SOURCE: Calculated from the equations of Anderson (1980) for the conditions of pressure and temperature defined for the standard atmosphere (Valley 1965) (from Calvert and Stockwell 1983).

Theoretical Rates of SO_2 and NO_2 Conversion to H_2SO_4 and $HONO_2$ Through Gas-Phase Reactions in the Troposphere

The major homogeneous processes for SO_2 and NO_2 conversion to H_2SO_4 and $HONO_2$, respectively, are governed by the rates of reactions (56) and (64), which depend on the concentrations of the reactants, NO_2, SO_2, and HO. Knowledge of [HO] in the troposphere is needed to make theoretical estimates of the minimum rates of the gas-phase generation of the H_2SO_4 and $HONO_2$.

Although the direct measurement of [HO] in the troposphere is very difficult, there are both theoretical and experimental estimates of this quantity that are of some value for our considerations here (Table A.2). We have excluded early estimates based on laser-induced fluorescence of the HO radical because the generation of HO by the probing laser beam itself was then an unrecognized problem. Data based on the ^{14}CO-chemical method of Campbell et al. (1979), the laser-induced fluorescence data of Wang et al. (1981), and the HO absorption data of Perner et al. (1976) are all reasonably consistent both among themselves and with theoretical estimates based on computer models of the complex atmospheric chemistry (Calvert and McQuigg 1975, Chang et al. 1977, Crutzen and Fishman 1977, Demerjian et al. 1974, Graedal et al. 1976, Hecht and Seinfeld 1972, Höv and Isaksen 1979, Levy 1974, Niki et al. 1972).

We may use these data to make reasonable estimates of the gas-phase oxidation rates for SO_2 and NO_2 in order to judge the potential of reactions (56) and (64) to develop the ingredients for "acid rain." At high concentrations of HO radical characteristic of midday sunny summer skies in a polluted atmosphere (about 9×10^6 molec cm^{-3}), we anticipate from our selected rate constant data that SO_2 will oxidize at a rate of $\approx 3.7 \pm 1.9$ percent/h through HO-radical attack in reaction (56). Rates of $HONO_2$ generation in (64) are expected to be about 34 ± 17 percent/h for these conditions. Somewhat lower midday solar intensities typical of the winter months in a polluted atmosphere for which the maximum noontime value for [HO] $\approx 2.4 \times 10^6$ molec cm^{-3} lead to rates of about 1 ± 0.5 percent/h for SO_2 and about 18 ± 9 percent/h for NO_2. Of course, these rates are expected to track [HO] during the day and thus to drop rapidly to near zero at night.

Computer simulations (Stockwell and Calvert 1983) show that for conditions that provide a maximum summertime value for [HO] of about 9.1×10^6 molec cm^{-3}, the 24-h average [HO] is about 1.7×10^6 molec cm^{-3}, leading to a daily average rate of SO_2 oxidation through reaction (56) of 0.7 percent/h or 16.4 percent/24-h period. The equivalent NO_2 oxidation rate in (64) is 6.2 percent/h or 150 percent/24-h period. The average wintertime rate of oxidation is much lower: for SO_2 it is about 0.12 percent/h or 3 percent/24-h period and for NO_2 about 1.1 percent/h or 25 percent/24-h period. In theory therefore we expect the homogeneous gas-phase conversion of SO_2

TABLE A.2 Measured and Theoretical Estimates of the [HO] in the Troposphere

	[HO], molec/cm^3
A. Experimental Methods	
1. Chemical Method: $^{14}CO + HO \rightarrow \ ^{14}CO_2 + H$ (Campbell et al. 1979)	
Pullman, Wash. (46.7° N, 770 m)	$(5.8 \pm 2.4) \times 10^6$
New Zealand (44° S, 1030 m)	$(1.1 \pm 0.4) \times 10^6$
Tennessee (rural, 36° N, 270 m)	$(1.9 \pm 0.7) \times 10^6$
Los Angeles (34° N, 270 m)	$(5.0 \pm 1.2) \times 10^6$
Arizona (desert, 37° N, 2300 m)	$(57 \pm 23) \times 10^6$
2. Laser-Induced Fluorescence, 282.07-nm excitation, 309-nm emission, 302-nm N_2 Raman (Wang et al. 1981)	
Niwot Ridge, Colo. (3048 m)	$(40 \text{ to } 6) \times 10^6$
Los Angeles, Calif. (11,886 m)	$(2.5 \pm 2) \times 10^6$
San Bernadino, Calif. (10,057 m)	$(20 \pm 6) \times 10^6$
San Diego, South, Calif. (10,667 m)	$(10 \pm 4) \times 10^6$
Denver, Colo. (10,057 m)	$(1 \pm 3) \times 10^6$
3. Absorption Spectroscopy, 307.995 nm (Perner et al. 1976)	
Jülich, Germany (51° N)	$(11 \pm 6) \times 10^6$ usually $(6 \pm 3) \times 10^6$
B. Theoretical Estimates: Typical recent estimates found in simulations of Calvert, Demerjian, Seinfeld, Niki, Graedel, Isaksen, etc., and their co-workers (values are sensitive to ultraviolet solar irradiance at point of interest, levels of impurities, NO, NO_2, RH's, RCHO's, etc.):	
Daytime maximum (summer, 40° N)	$\sim (9 \pm 4) \times 10^6$
Daytime maximum (winter, 40° N)	$\sim (2 \pm 1) \times 10^6$
Nighttime	$\lesssim 2 \times 10^5$

SOURCE: Calvert and Stockwell (1983).

and NO_2 to provide a significant quantity of sulfuric and nitric acids in the troposphere. When making such estimates for the gas-phase SO_2 oxidation rates, we must remember that other as yet unevaluated contributions from the alkene-O_3 products, the CH_3O_2-radical, and probably other unidentified reactants may contribute to the total SO_2 oxidation rate as well. For example, the reaction (42) of $O(^3P)$ with SO_2 can be significant in a highly NO_2-polluted atmosphere such as that present in the early stages of dilution of a stack gas plume. In this case, reaction (42) can theoretically account for an initial burst of SO_3 (H_2SO_4) formation at a maximum rate of about 1.4 percent/h (Calvert et al. 1978). In addition, the methylperoxy (CH_3O_2) and possibly other primary alkylperoxy

radicals may lead to SO_2 oxidation in the troposphere
under special circumstances of high $NO-NO_2$ pollution,
although the evidence for the significance of reaction
(50) is not unambiguous (Kan et al. 1981). Also the
products of the ozone-alkene reactions oxidize SO_2
readily, presumably through the Criegee intermediates in
reactions (59) or (61). The aldehydes, water vapor,
carbon monoxide, NO, and possibly other atmospheric
impurities compete with SO_2 for these reactive intermediates, and the effectiveness of the competition of
SO_2 with NO remains unclear today (Calvert and Stockwell
1983). If the rate constants for these $RCHO_2$ species
with NO and SO_2 are similar and/or the ratio of $[NO]/[SO_2]$
is low, then SO_2 oxidation by Criegee intermediates can
be significant in alkene-ozone-containing atmospheres (a
few tenths of a percent per hour).

Although the contribution of SO_2 oxidation from
reactions other than (56) may be important, the theoretical rates given here for the HO reactions alone appear
to match reasonably well those observed in relatively dry
SO_2 urban plumes (Eatough et al. 1981, Forrest et al.
1981, Garber et al. 1981, Gillani and Kohli 1981, Hegg
and Hobbs 1980, McMurry and Rader 1981, Meagher et al.
1981, Newman 1981, Williams et al. 1981, Wilson and
McMurry 1981, Zak 1981). Certainly a large part of the
observed oxidation in the cloud-free ambient troposphere
during the daylight hours arises from the $HO-SO_2$ reaction.
The conversion rates for NO_2 observed by recent workers
are also consistent with the estimates presented here
(Spicer 1982).

THE SOLUTION-PHASE OXIDATION OF SO_2 IN THE TROPOSPHERE

In recent years the role of the aqueous-phase reactions in
the development of acid precipitation has received increase
attention. The results of field and laboratory studies
suggest that the relative importance of the gas-phase and
solution-phase pathways may vary depending on a variety of
meteorological conditions such as the extent of cloud cover
relative humidity, the amount of precipitation, the intensi
of the solar radiation, and the presence and concentration
of various pollutants. Rates of oxidation of SO_2 through
gas-phase reactions are relatively slow (a few percent per
hour during daylight), whereas in theory those for the
solution-phase pathways may be as high as 100 percent/h for
seemingly realistic concentrations of the reactants in clou

water. Despite the considerable difference in rates of oxidation pathways in the aqueous and gaseous phases, both must be regarded as participating in the development of acid deposition in the eastern United States because air masses in this region are more likely to be free of clouds and precipitation a large fraction of the time (Niemann 1982). There appears to be little question that both gas-phase and liquid-phase processes can contribute significantly to the formation of H_2SO_4 in the atmosphere. Clear evidence of the relative importance of the two processes for various conditions is not now available.

Field data taken in dilute stack plumes in relatively cloud-free atmospheres show that formation of sulfuric acid at night is very slow and that during daylight hours the rate correlates with solar intensity (Hegg and Hobbs 1980, Wilson and McMurry 1981). Measurements are not frequently made at night, however, and there is no clear test, of which we are aware, of cloud chemistry for conditions that minimize the possible gas-phase processes. Wilson and McMurry (1981) have observed the evolution of aerosol size distributions as a result of gas-to-particle conversion. From these data they suggest that droplet-phase conversion of SO_2 to sulfate may be important at high humidities (>50 percent), where up to 20 percent of the growth of the aerosol was attributed to this reaction. Aerosol from gas-phase processes dominated the plume chemistry for low humidities and in the absence of clouds.

The clearest evidence of the major input from precipitating clouds was derived in the Acid Precipitation Experiment (APEX) (Lazrus et al. 1983). The investigators found significant production of both sulfuric and nitric acids in clouds. The acidity of dry air south of a warm front was measured before it ascended and produced a large area of warm frontal precipitation. Comparison of the chemical composition of the dry air and precipitation at the base of the cloud showed that rapid production of acid had occurred in the cloud. Rates of HNO_3 and H_2SO_4 production in the cloud were estimated to be of the order of 0.5 ppb/h and 1.2 ppb/h, respectively. Although analyses of H_2O_2 in the cloud suggested the involvement of this oxidizing agent, unidentified interferences in the luminol method for H_2O_2 detection discovered later leave the nature of the oxidizing agent unresolved.

Recently, transport and transformation chemistry of SO_2 were studied in an air mass tagged with SF_6 along the coast of southern California (Cass and Shair 1983). These studies appear to provide the first unambiguous evidence

174

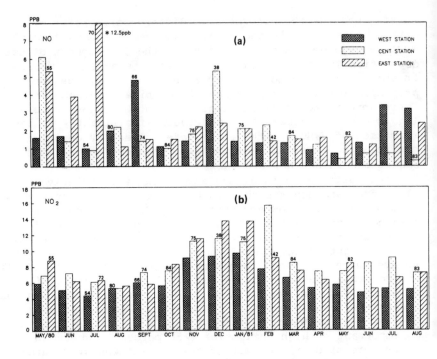

that SO_2 conversion to sulfate can occur at a measurable
rate at night (about 10 percent/h) in low stratus clouds
over the coastal waters.

Strong seasonal variations in the sulfate and SO_2 in
precipitation (measured as H^+ and HSO_3^-) have been
seen in the MAP3S precipitation chemistry data (Henderson
and Wingartner 1980). A more direct and quantitative
measure of seasonal variations in the atmospheric chemistry involved in acid deposition has been presented by
Shaw and Paur (1983). They measured gaseous NO, NO_2,
SO_2, and airborne sulfate aerosol at three sites in the
Ohio River Valley during the period May 1980 to August
1981 (Figure A.6). Two sites (labeled west and central
in the figure) were located in rural farming communities
(Union County, Kentucky, and Franklin County, Indiana),
and the third (east) was in a forest clearing (Ashland
County, Ohio). Figure A.6 indicates that while monthly
average concentrations of NO remained relatively constant
throughout the period of the experiment, NO_2 and SO_2
(gas-phase sulfur) increased in the winter months.
Airborne particulate sulfate shows a minimum in winter,
as does the percentage of total sulfur existing as

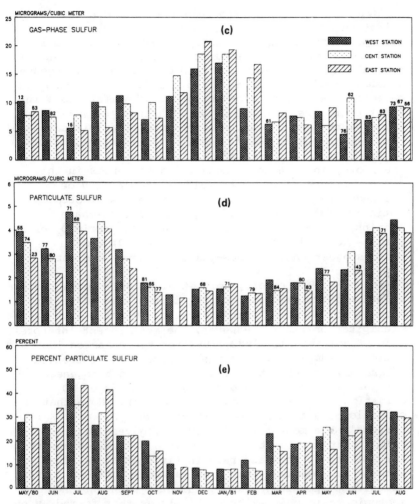

FIGURE A.6 Average monthly concentrations of airborne pollutants at three sites in the Ohio River Valley in 1980-1981. (a) NO, (b) NO_2, (c) SO_2, (d) sulfate aerosol, (e) percentage of total airborne sulfur existing as sulfate aerosol. SOURCE: Shaw and Paur (1983).

sulfate aerosol. The results are consistent with the general expectations of pathways for oxidation in both the gaseous and the aqueous phases since higher concentrations of the oxidizing agents (HO, H_2O_2, and O_3) are anticipated in the summer months when both solar intensities and the length of daylight are higher.

Evidence concerning the relative contributions to SO_2 oxidation by gaseous- and liquid-phase processes has also been presented using the isotopic abundance of ^{18}O in sulfate formed in various laboratory experiments and in natural samples of snow and rain collected near Argonne, Illinois (Holt et al. 1981, 1982, 1983). The method is based on the fact that oxygen atoms in sulfate ions in water do not reach equilibrium with those of water for temperatures and times characteristic of acidification of droplets in acid precipitation. Differences also exist in the ratios of ^{18}O to ^{16}O in O_2 and H_2O. Therefore, in principle, measured differences in the ^{18}O content of cloud-water sulfate should provide evidence of the origins of the sulfate. An isotopic ratio typical of H_2O would suggest solution-phase origins, whereas a ratio typical of atmospheric O_2 or species derived from the O_2 (O_3, H_2O_2, HO, etc.) would suggest that gaseous-phase chemistry was more important.

In their 1981 report, Holt et al. (1981) suggested that solution-phase oxidation of SO_2 is the more effective pathway for observed isotopic ratios in sulfate in natural precipitation water. However, the most recent interpretation of Holt et al. (1982, 1983) of the measured isotopic ratios in samples of atmospheric water and controlled laboratory experiments using a variety of homogeneous gas- and liquid-phase and catalytic methods of SO_2 oxidation is more ambiguous (Figure A.7). Note that the oxidation reaction of SO_2 with aqueous H_2O_2, currently the most favored reaction, gives isotopic ratios that are least in accord with the values observed for sulfate in snow and rain in field studies. The authors now conclude that some currently unknown source of sulfate (possibly primary sulfate) with enriched isotopic ratio must exist that dilutes that formed by one or more of the currently known reactions leading to sulfate formation.

An abundance of data based largely on laboratory measurements points to the probable involvement of the oxidizing agents hydrogen peroxide and/or ozone in the formation of acids in cloud water. Also suggested in theory as having a possible influence in these transformations are free radicals such as HO and HO_2 either formed in the gas phase and transported to the liquid phase or formed in the liquid phase. Soot (carbonaceous

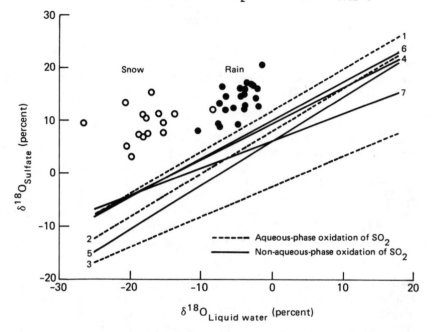

1. Aqueous air oxidation with Fe^{3+} catalyst
2. Aqueous air oxidation with charcoal catalyst
3. Aqueous H_2O_2 oxidation
4. Electric spark in humidified air
5. NO_2 in humidified air
6. Gamma irradiation in humidified air
7. Water vapor, air, and SO_2 absorbed on charcoal at 22° C

$\delta^{18}O = \left[(f/f_{smow}) -1\right] \times 10^3$, where $f = {}^{18}O/{}^{16}O$ and $f_{smow} = 2.005 \times 10^{-3}$, the isotopic ratio of standard mean oceanic water.

FIGURE A.7 Comparison of isotopic ratios in sulfates formed by seven laboratory methods and in precipitation water. SOURCE: Holt et al. (1982, 1983).

material) and various metal ions (copper, iron, manganese, and vanadium, for example) can act as catalysts for solution-phase oxidation of SO_2. Ammonia and other basic materials (such as metal carbonates) can influence the rates of solution-phase oxidation of SO_2 through alteration of the pH of the solution. We will review briefly here the kinetic results that define the reaction rates for the solution-phase oxidation of SO_2. Many of the rate data presented here are from the review of sulfite oxidation by Martin (1983).

Aqueous-Phase Oxidation of SO_2

Sulfur dioxide dissolves in water to form the hydrate $SO_2 \cdot H_2O$ and the ions HSO_3^-, H^+, and $SO_3^=$; these species are related by the following equilibria, which are established rapidly:

$$SO_2(g) + H_2O(liq) \rightleftharpoons SO_2 \cdot H_2O(aq),$$
$$K_H = 1.23 \text{ M/atm } (25°C),$$

$$SO_2 \cdot H_2O(aq) \rightleftharpoons H^+(aq) + HSO_3^-(aq),$$
$$K_1 = 1.7 \times 10^{-2} \text{ M},$$

$$HSO_3^-(aq) \rightleftharpoons H^+(aq) + SO_3^=(aq),$$
$$K_2 = 6.0 \times 10^{-8} \text{ M}.$$

The concentration of undissociated H_2SO_3 is insignificant. The solubility of gaseous SO_2 or the concentration of total dissolved $S(IV)$ equilibrated in water at some specified pH with gaseous SO_2 at a pressure P_{SO_2} is given by the relation:

$$[S(IV)] = P_{SO_2} K_H (1 + K_1/[H^+] + K_1 K_2/[H^+]^2),$$

where $[S(IV)] = [SO_2 \cdot H_2O(aq)] + [HSO_3^-(aq)] + [SO_3^=(aq)]$. It is apparent that SO_2 becomes less soluble as the acidity of a solution increases. Freiberg and Schwartz (1981) and Schwartz and Freiberg (1981) have shown that the times required to establish gas-liquid equilibrium between SO_2 and aqueous aerosol droplets are usually much shorter than the chemical reaction times for reactant concentrations typical of the ambient atmosphere. The expected $[S(IV)]$ in aqueous aerosols of pH 3.0 in equilibrium with SO_2 in the atmosphere ranges from about

10^{-9} M for 0.2 ppb SO_2 to about 10^{-6} M for 200 ppb SO_2.
At pH 4.0, [S(IV)] varies from about 10^{-8} M for 0.2 ppb
to about 10^{-5} M for 200 ppb SO_2. At pH 5.0, [S(IV)] is
about 10^{-7} M for 0.2 ppb SO_2 and 10^{-4} M for 200 ppb SO_2.

Of the various oxidizing agents that can oxidize S(IV)
in solution, two species, H_2O_2 and O_3, appear to be
especially significant for ambient levels of impurities.
The possibility that H_2O_2 and O_3 could be important
oxidants was first suggested by Penkett et al. (1979).
The overall reactions are represented by

$$H_2O_2(aq) + HSO_3^-(aq) \rightarrow SO_4^=(aq) + H^+(aq) + H_2O(aq), \quad (66)$$

$$O_3(aq) + HSO_3^-(aq) \rightarrow SO_4^=(aq) + H^+(aq) + O_2(aq). \quad (67)$$

Martin and Damschen (1981) have summarized all the
pertinent rate data for reaction (66) (Figure A.8) and
derived the following rate equation for the reaction:

$$\frac{d[S(IV)]}{dt} = \frac{8 \times 10^4 [H_2O_2][SO_2 \cdot H_2O(aq)]}{0.1 + [H^+(aq)]} \quad (68)$$

$$\text{mole liter}^{-1} \text{ sec}^{-1}.$$

When $[H^+] \ll 0.1$ M, the expression (68) shows the
oxidation rate to be independent of pH. This system is
unique in this regard among the potential oxidizers of
SO_2 in solution, and hence the H_2O_2 oxidation
mechanism is favored at low values of pH common in
atmospheric aerosols and precipitation water. The
apparent activation energy for the reaction is about 8
kcal/mole in the pH range 4-6 and 6.7 kcal/mole at pH 2
(no buffer).

Hoffman and Edwards (1975) proposed a mechanism
consistent with rate law (68):

$$HSO_3^-(aq) + H_2O_2(aq) \rightarrow HOSOO_2^-(aq) + H_2O(aq),$$

$$HOSOO_2^-(aq) + HA(aq) \rightarrow 2H^+(aq) + SO_4^=(aq) + A^-(aq),$$

in which HA is H_3O^+ or a suitable weak acid in the
solution. According to this mechanism, the reaction
occurs via a nucleophilic displacement by H_2O_2 on $HOSO_2^-$
to form a peroxymonosulfurous acid intermediate,
$HOOSO_2^-(aq)$, which then undergoes a rate-determining
rearrangement assisted by H_3O^+ or HA.

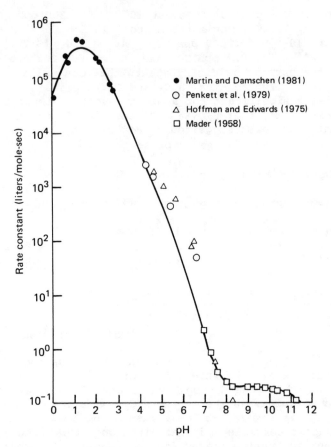

FIGURE A.8 Rate constant (k) for the bisulfite H_2O_2 reaction as a function of pH from the expression $-d[S(IV)]/dt = k[H_2O_2][S(IV)]$ with the effect of buffer removed and all results converted to $25°C$. SOURCE: Martin (1983).

The oxidation of SO_2 by O_3 in solution has been studied by several groups. Available data are summarized in Figure A.9 (Martin 1983). For aqueous solutions with pH in the range 1-3, the rate of oxidation via ozone follows

$$-\frac{d[O_3(aq)]}{dt} = 1.9 \times 10^4[H^+(aq)]^{-1/2}[O_3(aq)][S(IV)]$$

$$\text{mole liter}^{-1}\text{ sec}^{-1}. \quad (69)$$

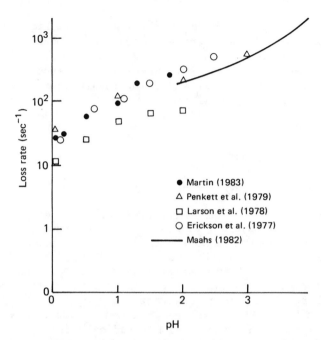

FIGURE A.9 Pseudo-first-order rate constant for ozone loss as a function of pH in the reaction of ozone with bisulfite with initial $[O_3] = 1 \times 10^{-4}$ and $[S(IV)] = 5 \times 10^{-4}$ M. SOURCE: Martin (1983).

For pH between 3 and 6.5, the rate law has this form:

$$-\frac{d[O_3\,(aq)]}{dt} = 4.19 \times 10^5 [O_3\,(aq)]\,[S(IV)]$$

$$+ 100\,[H^+(aq)]^{-1}\,[S(IV)]\,[O_3(aq)]$$

$$\text{mole liter}^{-1}\,\text{sec}^{-1}. \quad (70)$$

The inverse dependence of the rate of the O_3-S(IV) reaction on $[H^+(aq)]$ decreases the potential importance of the O_3 reaction with acidic solutions characteristic of the troposphere. The apparent activation energy of the reaction is 8.2 kcal/mole (Maahs 1982).

There is a strong link between the gas-phase and liquid-phase atmospheric chemistry related to acid precipitation. The major contributors to solution-phase oxidation of S(IV) in the troposphere, presumed to be H_2O_2 and O_3, are both products of the homogeneous gas-phase reactions. Hence the gas-phase and liquid-phase pathways of acid generation in the troposphere are not independent.

The major sources of H_2O_2 in the gaseous troposphere are reactions (71) and (73), involving the HO_2 and the hydrated HO_2 radicals:

$$2HO_2 \rightarrow H_2O_2 + O_2, \tag{71}$$

$$HO_2 + H_2O \rightarrow H_2O \cdot HO_2, \tag{72}$$

$$H_2O \cdot HO_2 + HO_2 \rightarrow H_2O_2 + H_2O + O_2. \tag{73}$$

The complex $H_2O \cdot HO_2$ represents only a few percent of the total HO_2 concentration in the atmosphere; however, the inequality of the rate constants, $k_{73} > k_{71}$, ensures that its role in H_2O_2 formation is not insignificant. Of course, if water is available to gaseous H_2O_2, this compound will end up largely in the aqueous phase because of the large Henry's law constant for H_2O_2 in water (7.1 x 10^4 M atm^{-1} at 25°C). In the polluted atmosphere, there is strong competition for the HO_2 radicals through reactions (27), (74), (75), and possibly others.

$$HO_2 + NO \rightarrow HO + NO_2, \tag{27}$$

$$HO_2 + CH_2O \rightleftharpoons HO_2CH_2O \rightarrow O_2CH_2OH, \tag{74}$$

$$HO_2 + O_2CH_2OH \rightarrow HO_2CH_2OH + O_2. \tag{75}$$

The efficiency with which H_2O_2 is generated in the polluted troposphere through (71) and (73) is a complex function of [NO] and [CH$_2$O] as well as the concentrations of CO, hydrocarbons (RH), and aldehydes (RCHO) from which the HO_2 radical is derived. [See reactions (21), (22), (30), and others.] Simulations of the complex chemistry of the polluted troposphere show that with a highly polluted air mass ([NO$_x$] = 100 ppb, [RH] = 500 ppb, [RCHO] = 30 ppb), H_2O_2 development is unimportant until the [NO] has been depleted late in the day (Calvert and Stockwell 1983). However, for conditions more typical of a nonurban air mass, where it is not uncommon to have very low NO$_x$ levels (≈ 1 ppb) and somewhat higher RH and RCHO concentrations in this circumstance, generation of H_2O_2 can begin early in the day. When [NO$_x$] is very high (100 ppb) and [RH] and [RCHO] are low initially ([RH] = 50 ppb, [RCHO] = 3 ppb), little H_2O_2 forms. However, with small levels of all the pollutants, H_2O_2 formation in (71) and (73) can be significant. In theory, the amount of H_2O_2 formed in these gas-phase reactions can often be sufficient to oxidize a large fraction of the bisulfite usually present in cloud water and precipitation.

Production of O_3 in the troposphere follows a somewhat different pattern. Its development is most favored in the highly polluted atmosphere, although significant concentrations (0.08 ppm) may be generated in relatively clean urban and rural atmospheres. The Henry's law constant for O_3 in water (0.01 M atm^{-1} at 25°C) and normal concentrations of O_3 found in the relatively clear troposphere (30-60 ppb) ensure that the aqueous droplets in the atmosphere will contain ozone concentrations of about 3-6 x 10^{-10} M.

These limited observations suggest that conditions of low NO_x and high hydrocarbon and aldehyde impurity levels favor H_2O_2 formation, whereas those of relatively high NO_x, RH, and RCHO favor high O_3 generation rates. If homogeneous air masses containing preformed H_2O_2 and O_3 encounter cloud water, rainwater, high aqueous acid aerosol levels, etc., then solution phase pathways for H_2SO_4 formation are favored as well. Because the optimum impurity levels for generation of O_3 can be very different from those for H_2O_2 formation, the components of the H_2O_2 and O_3 oxidation of $SO_2(HSO_3^-)$ in the solution phase will not proceed exactly in phase with fixed fractions occurring by each pathway. The rates of SO_2 oxidation in the homogeneous gas-phase oxidation and those of the solution-phase reactions will depend on two quite different, complex functions of pollutant concentrations. However, it is clear that the two processes are not entirely independent.

Other interesting possibilities for the oxidation of S(IV) to sulfuric acid in cloud water have been suggested but remain untested. Chameides and Davis (1982) studied theoretically the contribution to aqueous-phase chemistry of HO and HO_2 gas-phase radicals scavenged by and incorporated into cloud droplets during daylight. A seemingly logical series of chemical reactions involving these species and their products was suggested, and calculations indicated that for small water droplets (<20 μm) this scavenging process may be a very important source of oxidant (H_2O_2, HO_2, HO) for acid development provided that the sticking coefficient for these species impinging upon the water droplets is greater than about 10^{-3}.

Both NO_2 and HONO can oxidize sulfite in dilute
solutions. Martin et al. (1981) report that $HONO_2$ does
not react significantly for any of the pH conditions
employed. Lee and Schwartz (1981) and Schwartz and White
(1982) estimated the oxidation rate of NO_2(aq) with
S(IV) to be 2.4 x 10^{-6} M/h at pH 5 and P_{SO_2}= P_{NO_2}=
10 ppb in the gas phase. Their preliminary data suggest
that for relatively high NO_2 and SO_2 pollution levels the
acid generation from the NO_2-SO_2-H_2O(liq) system could be
significant.

In laboratory experiments, nitrous acid can oxidize
sulfite at a reasonably fast rate. Martin et al. (1981)
report that the rate law for this system is given by

$$\frac{d[S(IV)]}{dt} = 142[H^+(aq)]^{1/2}[N(III)][S(IV)]$$

$$\text{mole liter}^{-1} \text{ sec}^{-1}. \quad (76)$$

The rate constant data are summarized in Figure A.10.
The nitrogen-containing product of the reaction up to pH
3.5 was N_2O. Above this pH, hydroxylamine disulfonate
$(HON(SO_3)_2^=)$ forms. Further reactions of this species
occur to give hydroxylamine and nitrous acid. At the
very low levels of gaseous HONO in the atmosphere both
anticipated theoretically and observed experimentally, it
is not likely that this reactant will contribute significantly to the oxidation of S(IV) in the troposphere.

The truly uncatalyzed oxidation of sulfite solutions
by oxygen is very slow, and there is some question
whether small impurities of metal ions such as iron may
not account entirely for observed "uncatalyzed" rates
(Martin 1983).

$$O_2(aq) + 2HSO_3^-(aq) \rightarrow 2H^+(aq) + 2SO_4^=(aq).$$

Both iron (Fe^{3+}) and manganese (Mn^{2+}) ions catalyze the
reaction effectively. The rate law derived by Martin
(1983) for the iron-containing system is given by

$$\frac{d[SO_4^=(aq)]}{dt} = 0.82[H^+(aq)]^{-1}[Fe^{3+}][S(IV)]$$

$$\text{mole liter}^{-1} \text{ sec}^{-1}. \quad (77)$$

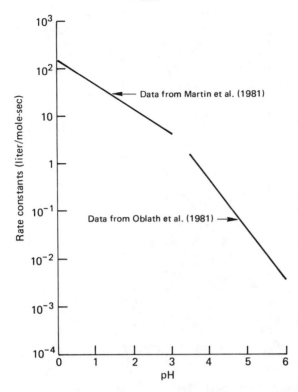

FIGURE A.10   Rate constant (k) for the reaction of HONO with bisulfite as a function of pH in the expression $d[S(IV)]/dt = k\,[N(III)]\,[S(IV)]$. SOURCE: Martin (1983).

These data and those of previous workers are summarized in Figure A.11. An increase in $[H^+(aq)]$ typical of a reacting mixture in cloud water or rainwater tends to lower the rate of this reaction pathway. In the high pH range (pH > 4), a condition uncommon for most natural atmospheric situations, ferric ion concentrations fall below 10^{-8} M, and the inconsistency of the rate laws observed in the laboratory for this system in this pH range may reflect the unrecognized problem of iron ion removal as a precipitate.

Catalysis of sulfite oxidation by manganese ions is governed by different mechanisms and rate laws depending on the concentrations of reactants. Martin (1983) reported the following expression for the [S(IV)] regime from 10^{-3} to 10^{-4} M:

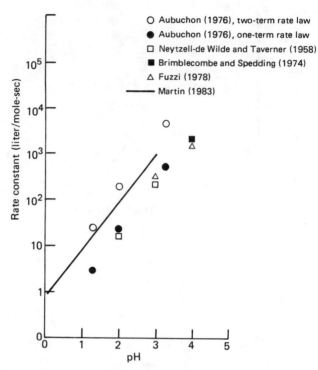

FIGURE A.11 Rate constant (k) for the iron-catalyzed oxidation of bisulfite as a function of pH in the expression $d[S(IV)]/dt = k [Fe^{3+}] [S(IV)]$ with data corrected to 25°C. SOURCE: Martin (1983).

$$\frac{d[SO_4^=(aq)]}{dt} = 4.7[H^+(aq)]^{-1}[Mn^{2+}(aq)]^2$$

mole liter^{-1} sec^{-1}.　(78)

The rate is independent of [S(IV)] for these conditions. This and other estimates of the rate constants for this reaction are shown as a function of pH in Figure A.12. The activation energy for the reaction is 27.3 kcal/mole (Hoather and Goodeve 1934a,b). At low [S(IV)] (less than 10^{-6} M), Martin (1983) finds the rate law to be

$$\frac{d[S(IV)]}{dt} = 25[H^+(aq)]^{-1}[Mn^{2+}] [S(IV)]$$

mole liter^{-1} sec^{-1}.　(79)

For the entire range of concentration the rate is inversely proportional to $[H^+(aq)]$. For the conditions of relatively high concentrations of Fe^{3+} or Mn^{2+} in atmospheric precipitation or urban fogs (Hoffman and Jacob 1983), the rates of S(IV) oxidation through these catalyzed reactions can be very significant. These conditions usually do not prevail in the free troposphere in nonurban areas, however.

Barrie and Georgii (1976) found that when both Fe^{3+} and Mn^{2+} ions are present in sulfite solutions, the rate of sulfite oxidation is enhanced over that expected from the sum of the Mn^{2+}-S(IV) and Fe^{3+}-S(IV) rates alone. Martin (1983) studied these systems in detail and confirmed this finding. For the combined Fe-Mn system the rate of sulfite oxidation was typically 3 to 10 times faster than that anticipated from the sum of the independent rates in the individual systems.

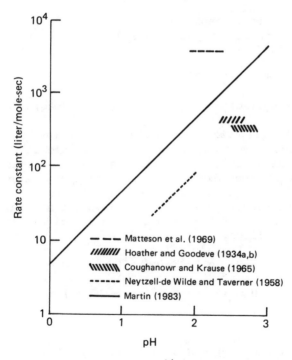

FIGURE A.12 Rate constant (k) for the Mn^{2+} ion catalyzed oxidation of bisulfite as a function of pH in the expression $d[S(VI)]/dt = k[Mn^{2+}]^2$. SOURCE: Martin (1983).

Ions of other metals such as Cu and Co are much less effective catalysts for sulfite oxidation than Mn and Fe for the low values of pH encountered in the environment (Martin 1983).

The mechanism by which the catalytic reactions occur has been studied by Hoffman et al. (1982) and others. A basic hypothesis being tested by Hoffman et al. is that the presence of certain trace metals in atmospheric aerosols and the occurrence of H_2O_2, HO, and HO_2 may be interrelated phenomena. Hoffman et al. illustrate this notion by the following mechanistic sequence:

$$Cu^+(aq) + O_2(aq) \rightleftharpoons CuO_2^+(aq),$$

$$CuO_2^+(aq) + H^+(aq) \rightarrow Cu^{2+}(aq) + HO_2(aq),$$

$$Cu^+(aq) + HO_2(aq) \rightarrow Cu^{2+}(aq) + HO_2^-(aq),$$

$$H^+(aq) + HO_2^-(aq) \rightleftharpoons H_2O_2(aq).$$

For example, in atmospheric systems, Cu^+, Co^+, and Fe^{2+} could be generated by photoinduced reduction of Cu^{2+}, Co^{2+}, and Fe^{3+}.

$$Cu^{2+}(aq) + h\nu \rightarrow Cu^{2+*}(aq),$$

$$Cu^{2+*} + H_2O(aq) \rightarrow Cu^+(aq) + HO(aq) + H^+(aq).$$

Subsequent catalytic decomposition of the intermediate hydrogen peroxide may produce HO and additional HO_2 free radicals. This hypothetical sequence of reactions could proceed according to the mechanism suggested for the classical "Fenton's reagent" reaction for the decomposition of H_2O_2. This mechanism involves a catalytic couple for Fe^{2+} and Fe^{3+} and a chain reaction that may proceed as follows:

$$Fe^{2+}(aq) + H_2O_2(aq) \rightarrow Fe(OH)^{2+}(aq) + HO(aq),$$

$$Fe^{3+}(aq) + H_2O_2(aq) \rightarrow Fe^{2+}(aq) + H^+(aq) + HO_2(aq),$$

$$Fe^{3+}(aq) + HO_2(aq) \rightarrow Fe^{2+}(aq) + H^+(aq) + O_2(aq),$$

$$HO(aq) + H_2O_2(aq) \rightarrow H_2O(aq) + HO_2(aq),$$

$$HO_2(aq) + H_2O_2(aq) \rightarrow O_2(aq) + H_2O(aq) + HO(aq).$$

The reactive intermediates, H_2O_2, HO, and HO_2, in the overall sequence described in the equations may oxidize S(IV) to H_2SO_4.

Hoffman et al. (1982) postulate another mechanism by which transition metals or transition metal complexes may catalyze the autooxidation of trace-level hydrocarbons present in cloud droplets. For a generalized hydrocarbon (RH) a possible sequence is as follows:

$$M^{2+}(aq) + O_2(aq) \rightarrow M^{3+}-O_2^-(aq),$$

$$M^{3+}-O_2^-(aq) + RH(aq) \rightarrow M^{2+}(aq) + R(aq) + HO_2(aq),$$

$$M^{3+}-O_2^-(aq) + RH(aq) \rightarrow M^{3+}-O_2H + R^-(aq),$$

$$M^{3+}-O_2^-(aq) + RH(aq) \rightarrow M^{3+}(aq) + R^-(aq) + HO_2(aq)$$
$$\rightarrow M^{3+}(aq) + R(aq) + HO_2^-(aq),$$

$$HO_2^-(aq) + H^+(aq) \rightarrow H_2O_2(aq).$$

There is some field evidence for such reactions in the atmosphere; rainwater analyses from Norway show significant numbers of alkanes, polycyclic aromatic hydrocarbons, fatty acids and esters, benzoic acids, etc.

The interesting facet of this reaction scheme is the production of peroxide in the cloud droplet by gas-phase scavenged hydrocarbon products. Hoffman et al. suggest that complexes of $Fe^{2+,3+}$, $Mn^{2+,3+}$, $Co^{2+,3+}$, and $V^{3+,4+}$ should be the most effective catalysts for the autooxidation of sulfite. They also showed that certain metal-catalyzed reactions of sulfite may be enhanced by a photoassisted pathway. The coupling of photolytic and metal-catalyzed processes is also consistent with the observed difference between daytime and nighttime SO_2 conversion rates. Dissolved organic molecules can act as competitive complexing agents for metals. The presence of complexing agents of this type will accelerate the dissolution of Fe_2O_3 and MnO_2, which are the likely sources of soluble iron and manganese in cloud droplets.

Another area of potential importance is that involving solid catalysts. Chang et al. (1978, 1981), and Brodzinski et al. (1980) showed that carbon can be an important catalyst for oxidation of sulfite in urban smog. They found from experiments at 20°C that the following rate expression decribed their results:

$$\frac{d[S(IV)]}{dt} = 1.69 \times 10^{-5}[C_x] [O_2(aq)]^{0.69}$$

$$\cdot \left\{ \frac{1.5 \times 10^{12}[S(IV)]^2}{1 + 3.06 \times 10^6 [S(IV)] + 1.5 \times 10^{12}[S(IV)]^2} \right\} \cdot (80)$$

The rates are in moles per liter per second, concentrations of soluble species are in moles per liter, and $[C_x]$ is the suspended carbon content of the suspension in grams per liter. The rate is independent of the pH (1.45-7.5), but it is dependent on the type of carbon (graphite, soot, etc.) employed and the surface area of the carbon particles.

Comparison of Reaction Pathways for Solution-Phase Oxidation of SO_2

The potential contributions of the various reactions we have considered here to solution-phase chemistry are compared for several realistic atmospheric conditions in Figure A.13. To obtain Figure A.13, Martin (1983) assumed that there are no limitations due to mass transport rates. The rate constants presented here have been applied to a hypothetical cloud containing 1 ml of liquid water per cubic meter of air at 25°C. The concentrations of reactants assumed in the calculation are listed on the figure. Rates for other impurity concentrations may be derived easily by recognizing that most of the processes are first order in each species so that the conversion rates are independent of SO_2 and linearly proportional to the concentration of the other species. The following are exceptions: (1) The mechanism for manganese-catalyzed oxidation changes at pH 4, because the equilibrium concentration of S(IV) goes above 10^{-6} M at higher values of pH; the shape of the curve is sensitive to the assumed concentrations of sulfur and manganese. (2) Oxidation by carbon catalysis is nonlinear in sulfur, and the curve of Figure A.13 will change in shape if the sulfur concentration is changed. (3) The upper portion of the curve for iron catalysis is nonlinear in sulfur.

For iron-catalyzed oxidation at high pH, Martin used the result of Brimblecombe and Spedding (1974) converted to a third-order rate. The dotted part of the curve between pH 3 and 5 is very uncertain, and there may be local maxima in this region. No synergism has been included in deriving the data of Figure A.13.

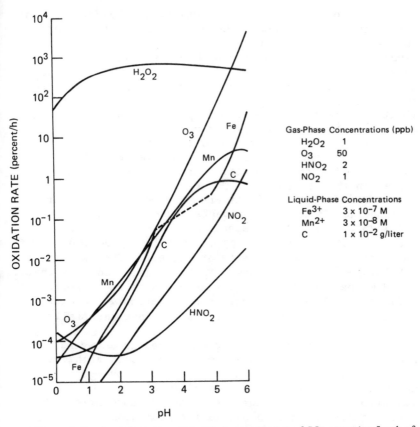

FIGURE A.13 Theoretical rates of liquid-phase oxidation of SO_2 assuming 5 ppb of SO_2, 1 ml/m^3 of water in air, and concentrations of impurities as shown. SOURCE: Martin (1983).

For the hydrated NO_2 oxidation, Martin used the Henry's law constant and rate constants of Schwartz and White (1982) and Lee and Schwartz (1981). The concentrations of NO_2, HONO, and NO are related thermodynamically, and so they may not be regarded as completely independent.

The trend of rising conversion rates with increased pH results from either the rising equilibrium concentration of sulfur (IV) or the sensitivity of the rate constants to pH, or both. The relative conversion rate increases with rising concentrations of sulfur (IV) in this range of pH values and for this amount of liquid. H_2O_2, which undergoes an acid-catalyzed reaction, is the only oxidant for which the rate dependence on $[H^+]$ compensates for the decreased solubility of SO_2 with increased $[H^+]$; hence the rate of oxidation by H_2O_2 is nearly independent of pH.

Martin emphasized that the relative positions of the curves in Figure A.13 differ for differing assumptions about the composition of the droplets. For example, an aerosol with less liquid water will, as a rule, have higher concentrations of nonvolatile species such as carbon, iron, and manganese but a smaller reaction volume, so the relative positions of the curves may differ in this situation. The relative positions will also be different at night, when the concentration of photochemically derived oxidants drops. At temperatures below 25°C, the rate constants are lower in accordance with the activation energies of the given reactions, and the Henry's law coefficients are higher. The two effects act in opposite directions. In some cases, such as that for H_2O_2 oxidation, the net rate rises as the temperature falls (at constant gas-phase concentration). In other cases, such as for iron catalysis, the net rate falls with temperature.

The concentrations of reactants used in deriving Figure A.13 are representative of those that might be anticipated in the atmosphere. Oxidation by H_2O_2 dominates all reactions for conditions of low pH. Oxidation rates can be greater than 100 percent/h. The rate for oxidation by ozone varies from about 10 percent/h at pH 4.5 to about 1 percent/h at pH 4.0. The contributions from the Fe^{3+}, Mn^{2+}, and carbon-catalyzed reactions (for the conditions specified) are below 1 percent/h for solutions of pH < 4.5. Oxidation by NO_2 and HONO is even less significant under these conditions.

An additional influence on the rates and kinetics of sulfite oxidation in clouds and rainwater can arise from the complexation of HSO_3^-(aq) by aldehydes scavenged by the droplets. The common gaseous products of atmospheric oxidation of the impurity hydrocarbons, CH_2O, CH_3CHO, CH_3COCH_3, and possibly other carbonyl compounds, can in principle complex with HSO_3^-. The result of such interactions could increase the solubility of SO_2 in the droplet and conceivably retard sulfite oxidation by the oxidants.

Although there is significant theoretical evidence that H_2O_2 may be the most important oxidizing agent for acid generation in cloud water and rain, unambiguous experimental measurements of H_2O_2 levels in air and in cloud water and rain have not been possible to date. Peroxide development in the sampling train and lack of selectivity of the luminol technique employed in previous work prevents a firm conclusion about the H_2O_2 levels in the atmosphere and in precipitation (Heikes et al. 1982, Zika and Saltzman 1982). Theoretical simulations of the atmospheric chemistry of a mixture of reactants in a highly diluted urban atmosphere show that H_2O_2 generation through reactions (71) and (73) can provide substantial levels of H_2O_2, CH_3O_2H, and other hydroperoxides in the gas phase. If these species are absorbed into aqueous aerosols, cloud water, or rainwater, for example, as is probable for H_2O_2 in view of its very high Henry's law constant, then oxidation by H_2O_2, and possibly other peroxides, can be most significant. The possible roles for peroxyacetylnitrate, peroxynitric acid, CH_3O_2H, and other peroxides in solution-phase sulfite oxidation remain to be evaluated.

Solution-Phase Generation of Nitric and Nitrous Acids in the Troposphere

There is some evidence of the formation of $HONO_2$ in clouds and rainwater. Recently, both theory and experiment suggest that $HONO_2$ may be formed rapidly from a combined gas-phase/liquid-phase process. Through simulation, Heikes and Thompson (1981) have suggested that N_2O_5 generated by O_3-NO_2 reactions (15) and (16) may be scavenged effectively by H_2O droplets to form $HONO_2$ in clouds or rainwater.

$$NO_2 + O_3 \rightarrow O_2 + NO_3, \qquad (15)$$

$$NO_3 + NO_2 (+M) \rightleftharpoons N_2O_5 (+M), \qquad (16)$$

$$N_2O_5 + H_2O(liq) \rightarrow 2H^+(aq) + 2NO_3^-(aq) \qquad (81)$$

A preliminary report of the experimental observation of this process has been made recently by Gertler et al. (1982). An evaluation of the contribution of this mechanism to the total [$HONO_2$] found in the atmosphere is not now possible.

Lee and Schwartz (1981) studied the rates of reactions (82) and (83):

$$2NO_2(g) + H_2O(liq) \rightarrow 2H^+(aq) + NO_3^-(aq) + NO_2^-(aq), \quad (82)$$

$$NO(g) + NO_2(g) + H_2O(liq) \rightarrow 2H^+(aq) + 2NO_2^-(aq). \quad (83)$$

They found that at the lower partial pressures characteristic of moderately polluted atmospheres the rates are
slow (10^{-9} to 10^{-8} M/h). Unless high partial pressures
(e.g., 10^{-7} atm) of these gases are maintained in contact
with liquid water for substantial periods of time (tens of
hours), reactions (82) and (83) cannot represent a substantial source of atmospheric acidity. Lee and Schwartz found
no evidence of Fe^{2+} ion catalysis of these reactions, but
they suggested that possibly other metal ions could enhance
these rates. There is no evidence related to this possibility now available.

Although significant uncertainty remains concerning the
source of HNO_3 in clouds and rainwater, the limited
evidence currently available favors the probable importance
of the formation of N_2O_5 in (15) and (16) followed by its
reaction in water droplets to form HNO_3.

SUMMARY

In summary, a wide variety of interrelated homogeneous
gas-phase, solution-phase, and heterogeneous chemistry may
result ultimately in oxidation of SO_2 to sulfuric acid and
NO_x to nitric acid. The homogeneous gas-phase oxidation of
SO_2 by the HO radical and the solution-phase oxidation of
S(IV) through H_2O_2, O_3, and possibly other species appear to
be the major sources of H_2SO_4. In the cloud-free,
ambient, sunlight-irradiated troposphere, nitric acid is
probably generated largely by the reaction of HO radicals
with NO_2. Both $HONO_2$ and H_2SO_4 produced in the gas phase
can be scavenged effectively by cloud water and precipitation. NO_2 may be oxidized to $HONO_2$ if sufficient O_3 and
NO_2 are present. Following its gas phase generation, N_2O_5
may be scavenged effectively by water droplets to form
$HONO_2$.

All the various pathways that lead to the oxidation of
SO_2 and NO_x are coupled by common products and reactants
that can directly and indirectly influence the rates of
reaction by the other pathways. In general the homogeneous gas-phase reactions can lead to maximum daylight
rates of acid formation of a few percent per hour for
SO_2 and 20-30 percent/h for NO_2. Solution-phase
reactions involving H_2O_2 and O_3 can in principle

convert SO_2 to H_2SO_4 in cloud water and precipitation at much higher rates (as high as 100 percent/h) for concentrations of H_2O_2 and O_3 in the troposphere that reasonably could result from the normal homogeneous reactions characteristic of atmosphere chemistry during daylight.

REFERENCES

Anderson, L.G. 1980. Absolute rate constants for the reaction of OH with NO_2 in N_2 and He from 225 to 389K. J. Phys. Chem. 84:2152-2155.

Aubuchon, C. 1976. The rate of iron catalyzed oxidation of sulfur dioxide by oxygen in water. Ph.D. thesis. The Johns Hopkins University, Baltimore, Md.

Barrie, L.A., and H.W. Georgii. 1976. An experimental investigation of the absorption of sulfur dioxide by water drops containing heavy metal ions. Atmos. Environ. 10:743-749.

Benson, S.W. 1978. Thermochemistry and kinetics of sulfur-containing molecules and radicals. Chem. Rev. 78:23-35.

Brimblecombe, P., and D.J. Spedding. 1974. The reaction order of the metal ion catalyzed oxidation of sulfur dioxide in aqueous solution. Chemosphere 1:29-32.

Brodzinski, R., S.G. Chang, S.S. Markowitz, and T. Novakov. 1980. Kinetics and mechanism for the catalytic oxidation of sulfur dioxide on carbon in aqueous suspension. J. Phys. Chem. 84:3354-3358.

Calvert, J.G., and R.D. McQuigg. 1975. The computer simulation of rates and mechanisms of photochemical smog formation. Int. J. Chem. Kinet. Symp. 1:113-154.

Calvert, J.G., and W.R. Stockwell. 1983. The mechanism and rates of the gas phase oxidations of sulfur dioxide and nitrogen oxides in the atmosphere. In Acid Precipitation: SO_2, NO, and NO_2 Oxidation Mechanisms: Atmospheric Considerations. Ann Arbor, Mich.: Ann Arbor Scientific Publications. In press.

Calvert, J.G., F. Su, J.W. Bottenheim, and O.P. Strausz. 1978. Mechanism of the homogeneous oxidation of sulfur dioxide in the troposphere. Atmos. Environ. 12:197-226.

Campbell, M.J., J.C. Sheppard, and B.F. Au. 1979. Measurements of hydroxyl concentration in the boundary layer air by monitoring CO oxidation. Geophys. Res. Lett. 6:175-178.

Cass, G.R., and F.H. Shair. 1983. Sulfate accumulation in a sea breeze/land breeze circulation system. J. Geophys. Res. In press.

Castleman, A.W., Jr., and I.N. Tang. 1976-1977. Kinetics of the association reaction of SO_2 with hydroxyl radical. J. Photochem. 6:349-354.

Chameides, W.L., and D.D. Davis. 1982. The free radical chemistry of cloud droplets and its impact upon the composition of rain. J. Geophys. Res. 87:4863-4877.

Chang, J.S., P.J. Wuebbles, and D.D. Davis. 1977. A theoretical model of global tropospheric OH distributions. UCRL-78392. Livermore, Calif.: Lawrence Livermore Laboratory. 22 pp.

Chang, S.G., R. Brodzinsky, R. Toossi, S.S. Markowitz, and T. Novakov. 1978. Catalytic oxidation of SO_2 on carbon in aqueous suspensions. Pp. 122-130 in Conference on Carbonaceous Particles in the Atmosphere. Berkeley, Calif.: Lawrence Berkeley Laboratory.

Chang, S.G., R. Roosi, and T. Novakov. 1981. The importance of soot particles and nitrous acid in oxidizing SO_2 in atmospheric aqueous droplets. Atmos. Environ. 15:1287-1292.

CODATA. 1980. Evaluated kinetic and photochemical data in atmospheric chemistry. J. Phys. Chem. Ref. Data 9:295-471.

Coughanowr, D.R., and R.E. Krause. 1965. The reaction of SO_2 and O_2 in aqueous solutions of $MnSO_4$. Ind. Eng. Chem. Fundam. 4:61-66.

Cox, R.A. 1974-1975. The photolysis of nitrous acid in the presence of carbon monoxide and sulfur dioxide. J. Photochem. 3:291-304.

Cox, R.A. 1975. The photolysis of gaseous nitrous acid--a technique for obtaining kinetic data on atmospheric photooxidation reactions. Int. J. Chem. Kinet. Symp. 1:379-398.

Cox, R.A., and D. Sheppard. 1980. Reaction of OH radicals with gaseous sulfur compounds. Nature 284:330-331.

Crutzen, P.J., and J. Fishman. 1977. Average concentrations of OH in the troposphere and the budgets of methane, carbon monoxide, molecular hydrogen and 1,1,1-trichloromethane. Geophys. Res. Lett. 4:321-324.

Davis, D. D., A. R. Ravishankara, and S. Fischer. 1979. SO_2 oxidation via the hydroxyl radical: atmospheric fate of HSO_x radicals. Geophys. Res. Lett. 6:113-116.

Demerjian, K.L., J.A. Kerr, and J.G. Calvert. 1974. The mechanism of photochemical smog formation. Adv. Environ. Sci. Technol. 4:1-262.

Domalski, E.S. 1971. Thermochemical properties of peroxyacetyl (PAN) and peroxybenzoyl nitrate (PBN). Environ. Sci. Technol. 5:443-444.

Eatough, D.J., B.E. Richter, N.L. Eatough, and L.D. Hansen. 1981. Sulfur chemistry on smelter and power plant plumes in the western U.S. Atmos. Environ. 15:2241-2253.

Erickson, R.E., L.M. Yates, R.L. Clark, and D. McEwen. 1977. The reaction of sulfur dioxide with ozone in water and its possible atmospheric significance. Atmos. Environ. 11:813-817.

Forrest, J., R.W. Garber, and L. Newman. 1981. Conversion rates in power plant plumes based on filter pack data: the coal-fired Cumberland plume. Atmos. Environ. 15:2273-2282.

Freiberg, J.E., and S.E. Schwartz. 1981. Oxidation of SO_2 in aqueous droplets: mass-transport limitation in laboratory studies and the ambient atmosphere. Atmos. Environ. 15:1145-1154.

Friend, J.P., R.A. Barnes, and R.M. Vasta. 1980. Nucleation by free radicals from the photooxidation of sulfur dioxide in air. J. Phys. Chem. 84:2423-2436.

Fuzzi, S. 1978. Study of iron (III) catalyzed sulfur dioxide oxidation in aqueous solution over a wide range of pH. Atmos. Environ. 12:1439-1442.

Garber, R.W., J. Forrest, and L. Newman. 1981. Conversion rates in power plant plumes based on filter pack data: the oil fired Northport plume. Atmos. Environ. 15:2283-2292.

Gertler, A.W., D.F. Miller, D. Lamb, and U. Katz. 1982. SO_2 and NO_2 reactions in cloud droplets. Pp. 112-115 in Abstracts, American Chemical Society, 185th National Meeting, Las Vegas, Nev.

Gillani, N.V., and S. Kohli. 1981. Gas-to-particle conversion of sulfur in power plant plumes. I. Parameterization of the conversion rate for dry, moderately polluted ambient conditions. Atmos. Environ. 15:2293-2313.

Graedel, T.E., L.A. Farrow, and T.A. Wicker. 1976. Kinetic studies of the photochemistry of the urban troposphere. Atmos. Environ. 10:1095-1117.

Harding, L.B., and W.A. Goddard III. 1978. Mechanism of gas phase and liquid phase ozonolysis. J. Am. Chem. Soc. 100:7180-7188.

Harris, G.W., and R.P. Wayne. 1975. Reaction of hydroxyl radicals with NO, NO_2, and SO_2. J. Chem. Soc., Faraday Trans. 1. 71:610-617.

Hecht, T.A., and J.H. Seinfeld. 1972. Development and validation of generalized mechanism for photochemical smog. Environ. Sci. Technol. 6:47-57.

Hegg, D.A., and P.V. Hobbs. 1980. Measurements of gas-to-particle conversion in the plumes from five coal-fired electric power plants. Atmos. Environ. 14:99-116.

Heikes, B.G., and A.M. Thompson. 1981. Nitric acid formation in-cloud gas phase, photochemical and/or heterogeneous paths. Eos 62:884.

Heikes, B.G., A.L. Lazrus, G.L. Kok, S.M. Kunen, B.W. Gandrud, S.N. Gitlin, and P.D. Sperry. 1982. Evidence for aqueous phase hydrogen peroxide synthesis in the troposphere. J. Geophys. Res. 87:3045-3051.

Henderson, R., and K. Wingartner. 1980. Analysis of MAP3S Precipitation Chemistry Data. MTR-80W00265. McLean, Va.: MITRE Corporation.

Hoather, R.C., and C.F. Goodeve. 1934a. The oxidation of sulphurous acid. I. The dilatometric technique. Trans. Faraday Soc. 30:626-629.

Hoather, R.C., and C.F. Goodeve. 1934b. The oxidation of sulphurous acid. III. Catalysis by manganese sulfate. Trans. Faraday Soc. 30:1149-1156.

Hoffman, M.R., and J.O. Edwards. 1975. Kinetics of the oxidation of sulfite by hydrogen peroxide in acidic solution. J. Phys. Chem. 79:2096-2098.

Hoffman, M.R., and D.J. Jacob. 1983. Kinetics and mechanisms of catalytic oxidation of dissolved sulfur dioxide in aqueous solution: an application to nighttime fog water chemistry. In Acid Precipitation: SO_2, NO, NO_2 Oxidation Mechanisms: Atmospheric Considerations. Ann Arbor, Mich.: Ann Arbor Science Publishers, Inc. In press.

Hoffman, M.R., S.D. Boyce, A. Hong, and L. Moberly. 1982. Catalysis of the autooxidation of aquated sulphur dioxide by homogeneous and heterogeneous transition metal complex. In Heterogeneous Catalysis: Its Importance in Atmospheric Chemistry, D. Schryer, ed. AGU Monograph. Washington, D.C.: American Geophysical Union.

Holt, B.D., P.T. Cunningham, and R. Kumar. 1981. Oxygen isotopy of atmospheric sulfates. Environ. Sci. Technol. 15:804-808.

Holt, B.D., R. Kumar, and P.T. Cunningham. 1982. Primary sulfates in atmospheric sulfates: estimation of oxygen isotope ratio measurement. Science 217:51-53.

Holt, B.D., P.T. Cunningham, A.G. Engelkemeir, D.G. Graczyk, and R. Kumar. 1983. Oxygen-18 study of nonaqueous-phase oxidation of sulfur dioxide. Atmos. Environ. 17:625-632.

Höv, O., and T.S.A. Isaksen. 1979. Hydroxy and peroxy radicals in the polluted tropospheric air. Geophys. Res. Lett. 6:219-222.

Kan, C.S., J.G. Calvert, and J.H. Shaw. 1981. Oxidation of sulfur dioxide by methylperoxy radicals. J. Phys. Chem. 85:1126-1132.

Larson, T.V., N.R. Horike, and H. Harrison. 1978. Oxidation of sulfur dioxide by oxygen and ozone in aqueous solution: a kinetic study with significance to atmospheric rate processes. Atmos. Environ. 12:1597-1611.

Lazrus, A., P. L. Haagenson, G. L. Kok, C. W. Kreitzberg, G.E. Likens, V.A. Mohnen, W.E. Wilson, and J.W. Winchester. 1983. Acidity in air and water in frontal precipitation. Atmos. Environ. 17:581-591.

Lee, Y.N., and S.E. Schwartz. 1981. Reaction kinetics of nitrogen dioxide with water at low partial pressure. J. Phys. Chem. 85:840-848.

Leu, M.-T. 1982. Rate constants for the reaction of OH with SO_2 at low pressures. J. Phys. Chem. 86:4558-4562.

Levy, H. 1974. Photochemistry of the troposphere. Acta Photochem. 9:369-375.

Maahs, H.G. 1982. The importance of ozone in the oxidation of sulfur dioxide in nonurban tropospheric clouds. In Second Symposium on the Composition of the Nonurban Troposphere, Williamsburg, Va. Boston, Mass.: American Meteorological Society.

Mader, P.M. 1958. Kinetics of the hydrogen peroxide-sulfite reaction in alkaline solution. J. Am. Chem. Soc. 80:2634-2639.

Martin, L.R. 1983. Kinetic studies of sulfite oxidation in aqueous solutions. In Acid Precipitation: SO_2, NO, and NO_2 Oxidation Mechanisms: Atmospheric Considerations. Ann Arbor, Mich.: Ann Arbor Scientific Publications. In press.

Martin, L.R., and D.E. Damschen. 1981. Aqueous oxidation of sulfur dioxide by hydrogen peroxide at low pH. Atmos. Environ. 15:1615-1621.

Martin, L.R., D.E. Damschen, and H.S. Judeikes. 1981. The reactions of nitrogen oxides with SO_2 in aqueous aerosols. Atmos. Environ. 15:191-195.

Matteson, M.J., W. Stöber, and H. Luther. 1969. Kinetics of the oxidation of sulfur dioxide by aerosols of manganese sulfate. Ind. Eng. Chem. Fundam. 8:667-687.

McMurry, P.H., and D.J. Rader. 1981. Studies of aerosol formation in power plant plumes. I. Growth laws for secondary aerosols in power plant plumes: implications for chemical conversion mechanisms. Atmos. Environ. 15:2315-2327.

Meagher, J.F., L. Stockburger III, R.J. Bonanno, and M. Luria. 1981. Cross-sectional studies of plumes from a partially SO$_2$-scrubbed power plant. Atmos. Environ. 15:2263-2272.

Moortgat, G.K., and C.E. Junge. 1977. The role of SO$_2$ oxidation for the background stratospheric layer in the light of new reaction rate data. Pure Appl. Geophys. 115:759-774.

Morris, E.D., and H. Niki. 1973. Reaction of dinitrogen pentoxide with water. J. Phys. Chem. 77:1929-1932.

NASA. 1979. The Stratosphere--Present and Future. R.D. Hudson and E.I. Reed, eds. Report 1049. Greenbelt, Md.: National Aeronautics and Space Administration.

Newman, L. 1981. Atmospheric oxidation of sulfur dioxide: a review as viewed from powerplant and smelter plume studies. Atmos. Environ. 15:2231-2239.

Neytzell-de Wilde, F.G., and L. Taverner. 1958. Experiments relating to the possible production of an oxidizing acid leach liquor by autooxidation for the extraction of uranium. Pp. 303-317 in Proceedings of the Second U.N. International Conference on the Peaceful Uses of Atomic Energy, vol. 3. Geneva: United Nations.

Niemann, B.L. 1982. Analysis of wind and precipitation data for assessment of transboundary transport and acid deposition between Canada and the United States. Pp. 35-36 in Abstracts, American Chemical Society, 185th National Meeting, Las Vegas, Nev.

Niki, H., E.E. Daby, and B. Weinstock. 1972. Mechanisms of smog reactions. Chapter 2 in Photochemical Smog and Ozone Reactions, Adv. Chem. Ser. 113:16-57.

Oblath, S.B., S.S. Markowitz, T. Novakov, and S.G. Chang. 1981. Kinetics of the formation of hydroxylamine disulfonate by reaction of nitrite with sulfites. J. Phys. Chem. 85:1017-1021.

Penkett, S.A., B.M.R. Jones, K.A. Brice, and A.E.J. Eggleton. 1979. The importance of atmospheric ozone and hydrogen peroxide in oxidizing sulfur dioxide in cloud and rainwater. Atmos. Environ. 13:123-137.

Perner, D., D.H. Ehhalt, H.W. Pätz, V. Platt, E.P. Rö
and A. Volz. 1976. OH radicals in the lower
troposphere. Geophys. Res. Lett. 3:466-468.

Schwartz, S.E., and J.E. Freiberg. 1981. Mass-transport
limitation to the rate of reaction of gases in liquid
droplets: Application to oxidation of SO_2 in
aqueous solutions. Atmos. Environ. 15:1129-1144.

Schwartz, S.E., and W.H. White. 1982. Kinetics of
reactive dissolution of nitrogen oxides into aqueous
solution. Adv. Environ. Sci. Technol. 12.

Shaw, R.W., and R.J. Paur. 1983. Measurements of sulfur
in gases and particles during sixteen months in the
Ohio River Valley. Atmos. Environ. In press.

Sidebottom, H.W., C.C. Badcock, G.E. Jackson, J.G.
Calvert, G.W. Reinhardt, and E.K. Damon. 1972.
Photooxidation of sulfur dioxide. Environ. Sci.
Technol. 6:72-79.

Spicer, C.W. 1982. Nitrogen oxide reactions in the urban
plume of Boston. Science 215:1095-1097.

Stockwell, W.R., and J.G. Calvert. 1983. The mechanism of
the $HO-SO_2$ reaction. Atmos. Environ. In press.

Valley, S.L. 1965. U.S. Standard Atmosphere. In Handbook
of Geophysics and Space Environment. New York:
McGraw-Hill.

Wang, C.C., L.I. Davis, Jr., P.M. Selzer, and R. Munos.
1981. Improved airborne measurements of OH in the
atmosphere using the technique of laser induced
fluorescence. J. Geophys. Res. 86:1181-1196.

Williams, D.J., J.N. Carras, J.W. Milne, and A.C. Heggie.
1981. The oxidation and long-range transport of sulfur
dioxide in a remote region. Atmos. Environ.
15:2255-2262.

Wilson, J.C., and P.H. McMurry. 1981. Studies of aerosol
formation in power plant plumes. I. Secondary aerosol
formation in the Navajo generating station plume.
Atmos. Environ. 15:2329-2339.

Zak, B.D. 1981. Lagrangian measurements of sulfur dioxide
to sulfate conversion rates. Atmos. Environ.
15:2583-2591.

Zellner, R. 1978. Recombination reactions in atmospheric
chemistry. Ber. Bunsenges Phys. Chem. 82:1172-1179.

Zika, R.G., and E.S. Saltzman. 1982. Interaction of ozone
and hydrogen peroxide in water: implications for
analysis of H_2O_2 in air. Geophys. Res. Lett.
9:231-234.

Appendix **B** Transport and Dispersion Processes

Over the past 50 years scientists have been concerned with the transport of materials by the atmosphere. Substances of interest have included volcanic debris, Saharan dust, radioactive fallout, and industrial pollutants. The type of gas or particulate matter, their physical and chemical properties, the vigor of the atmospheric flow, and other factors help to determine how and where the material is finally deposited on the Earth's surface. This appendix treats only the physical transport of materials in the atmosphere. The effects of chemical transformation and scavenging by clouds and aerosols are discussed in detail in other appendixes.

The term transport encompasses the processes by which a substance or quantity is carried past a fixed point or across a fixed plane. In the atmosphere, the substances or quantities of interest include air parcels, gaseous impurities, suspended particles, and moisture (Huschke 1959).

CLASSIFICATION OF TRANSPORT PHENOMENA

Atmospheric motion and transport phenomena are extremely complex in both horizontal and vertical dimensions, with thermal layering, shear turbulence, convection, variation of boundary characteristics, and so on. Because of these complexities, meteorologists have devised an ordering of the various atmospheric phenomena. One way to approach the ordering is on the basis of spatial scale. After release, a given material diffuses during transport, coming under the influence of larger-scale motions as it moves farther from the source. To classify transport behavior, four scales have been defined: local, meso,

synoptic, and global. The local scale is defined as
being on the order of the vertical dimension of the
planetary boundary layer within which pollutants are
typically emitted. This dimension is the order of a
kilometer, and the time scale on which phenomena take
place on this dimension is on the order of tens of
minutes. The next largest spatial scale is the mesoscale,
which extends up to several hundred kilometers and has an
associated time scale of the order of a day (about the
time needed for a mean horizontal transport of several
hundred kilometers). Mesoscale effects include the
diurnal variability of the planetary boundary layer and,
therefore, the dynamics of plumes. In the mesoscale, an
individual plume from a power plant or urban complex of
sources loses its identity by mixing with other plumes or
by diluting into the background. The synoptic scale is
on the order of 1000 km, with transport times of about 1
to 5 days. The hemispheric or global scale reflects
intercontinental transport with times on the order of a
week. The term "long-range transport" commonly refers to
transport on the synoptic and global scales.

The prevailing winds in the lower troposphere transport
and disperse atmospheric pollutants. A combination of the
rotation of the Earth (Coriolis effect) and the existence
of synoptic-scale pressure gradients in the atmosphere
maintain the planetary or geostrophic winds. A number of
perturbances near the Earth's surface, such as surface
roughness, heat, and moisture fluxes, influence the local
winds. The perturbed layer, called the planetary boundary
layer (PBL), is of variable height ranging typically up
to 3 km. Because most atmospheric pollutants are released
in this layer, study of the PBL is vital to understanding
local or mesocale transport as opposed to synoptic or
global transport.

LOCAL AND MESOSCALE TRANSPORT

Mesoscale transport is usually confined to the planetary
boundary layer or the lowest 3 km. Embedded in this
region and closest to the ground is a highly dynamic
layer termed the mixing layer. Here the local effects of
mechanical and thermal turbulence can predominate. It is
called the mixing layer because within it atmospheric
turbulence very effectively and quickly mixes and dilutes
any concentrated release of mass, momentum, or heat. In
other parts of the atmosphere, dilution may be slow. The

mixing layer typically undergoes a diurnal cycle rising
to heights of 1 to 2 km in the day and lowering to 100 to
300 m at night. Thermal convection dominates in the day,
and small-scale mechanical turbulence at night. Because
of the efficient mixing during the day, pollutants are
quickly moved to all areas of the mixing layer including
the ground. On the other hand, elevated releases at night
may be above the shallow mixing layer and can be transported independently.

In general, mesoscale mean winds dominate plume transport, but, depending on the strength of local turbulent
eddies, the plume may also be spread horizontally and
vertically. Another factor affecting pollutant transport
is wind shear. Since thermal convection and mechanical
drag of the ground diminishes with height, the geostrophic
balance of forces varies with height and is maintained by
increasing wind speed and veering in wind direction.
Thus, vertically adjacent layers of air move at different
speeds in different directions (shear). Wind shear may
cause dispersion and dilution of atmospheric pollutants
and becomes increasingly important as the range of
transport increases.

In addition to this general picture of local and
mesoscale transport, there are significant diurnal and
seasonal variations in the boundary layer that affect
transport on these two scales. Figure B.1 shows the
different patterns for winter and summer. The major
feature to notice is that for both periods there is a
very stable nocturnal layer that extends to 300 m.
However, during the daytime, mixing heights are much
greater in summer than in winter when an elevated daytime
inversion hinders vertical mixing.

Another factor to be considered on the mesoscale is
the vertical profile of the horizontal wind speed.
Diurnal and seasonal variations in the profile are
affected by the vigor of the synoptic-scale flow. Winter
is a period of frontal passages, whereas in summer weak
anticyclonic systems tend to prevail (Figure B.2).

From the above discussion, it can be seen that the
mesoscale transport and dilution of a given pollutant
depend on whether its source is elevated or on the
ground. For example, while most of the SO_2 is emitted
from elevated point sources, NO_x emissions are more
evenly distributed between elevated and ground-level
sources. Thus on the average, elevated releases spend
more of their mesoscale transport time decoupled from the
ground, while near-ground releases maintain continuous

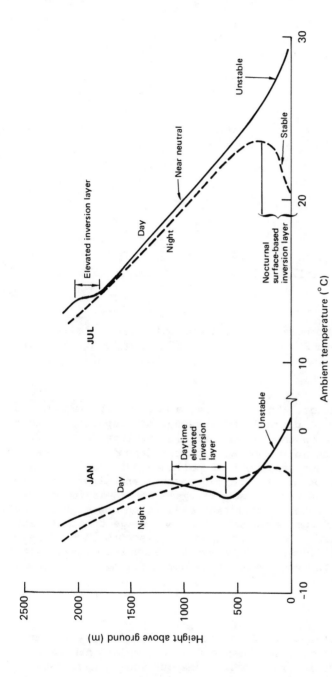

FIGURE B.1 Monthly average diurnal and seasonal variations of the vertical thermal structure of the planetary boundary layer at a rural site near St. Louis, Missouri, based on 1976 data. SOURCE: N. Gillani, Washington Unversity, St. Louis, Missouri, personal communication (1982).

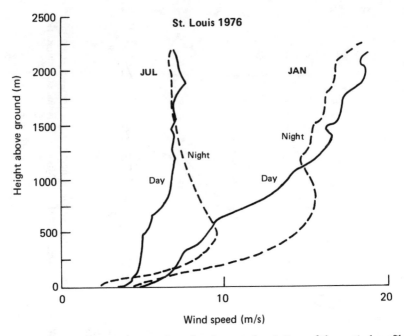

FIGURE B.2 Monthly average diurnal and seasonal variations of the vertical profiles of wind speed near St. Louis, Missouri, based on 1976 data. SOURCE: N. Gillani, Washington University, St. Louis, Missouri, personal communication (1982).

ground contact. This fact has a direct relation to the importance of diurnal and seasonal dry deposition and to some degree on the wet-deposition patterns.

While the main emphasis in acid deposition has been on the long-range transport of pollutants to remote areas, consideration of mesoscale transport and dispersion of pollutants of varying source types under varying flow conditions have an important bearing on how much of the emissions become available for long range transport and in what form. Important mesoscale factors such as release height and diurnal and seasonal variabilities must not be neglected in long-range transport modeling.

SYNOPTIC- OR CONTINENTAL-SCALE TRANSPORT

Synoptic transport of pollutants, especially acids and acid precursors, has been one of the major thrusts in acid deposition research. Numerous models have been devised to investigate the long-range transport (up to 1000 km) and are reviewed in Chapter 5 of this report. Each of the models simulates the transport, transformation, and deposition of a given substance (sulfur compounds in this case). The main transport module in the models uses the synoptic winds, which are measured in the vertical dimension every 12 h by balloon soundings. The network of balloon soundings produces, unfortunately, sparser data coverage than the precipitation chemistry measurements over eastern North America. Considering the spacing of upper-air measurements, it is optimistic to expect the knowledge of the direction of the prevailing wind at an arbitrary location in space and time to be known to better than 5° about the "actual" advecting wind. When one calculates forward or back trajectories from these winds, there is an uncertainty in the crosswind direction of 15 to 20 percent of the trajectory length for every timestop forward or backward in time. One would hope that such uncertainties and errors would cancel out when trajectories are calculated over many days and a climatology is established.

There are several key factors that determine the transmission of pollutants on the synoptic scale specifically over the North American continent. Already mentioned is the wind field. Clear patterns can be seen from a summary of the 1975-1977 data (R. Husar, Washington University, St. Louis, MO, personal communication, 1982). Conclusions are that (1) the general flow is from west to east with an important component northward from the Gulf of Mexico, (2) winter and fall have the highest speeds, (3) the southeastern United States is within a region of low mean velocity during late spring and summer, and (4) the Midwest exhibits very strong shear during summer and spring (Figure B.3).

It is important to note that winds above 1 to 2 km are not always important in the transport of surface releases, depending on the mixing depth. Also, well-mixed aged pollutants in the nocturnal stable layers may not always be re-entrained into the mixing layer the next morning. Contours of mixed depths (Figure B.4) provide some insight into the gross interaction of advecting winds and the depth of the mixing layer. However, synoptic temporal and spatial scales of interaction may be at least as important as the seasonal averages in determining the net transport of emissions. It is important to note that some of the well-mixed aged pollutants will ride over the

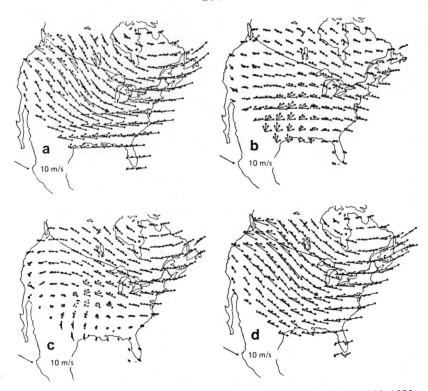

FIGURE B.3   Averages for 1975-1977 of winds in the layers 0-500, 500-1000, 1000-2000, and 2000-3000 m agl for the 000 and 1200 GMT soundings. Lower-level winds generally lie to the left and are of lower speed. a, January through March; b, April through June; c, July through September; and d, October through December.
SOURCE: R. Husar, Washington University, St. Louis, Missouri, personal communication (1982).

daytime mixed layer when moving either from south to north or from west to east owing to lowered mixed depths along the trajectory. Parameterization of the vertical structure in the models is important for simulation of continental-scale transport over several days and thousands of kilometers.

Other vertical motions must be taken into account in long-range transport, although these are difficult to simulate properly. Vertical motions are important, for example, in transmission of pollutants across major physical barriers (for example, the Rocky Mountains), along warm and cold fronts, and near simple convective cells or clusters of cells. Also the vigor of motion of

209

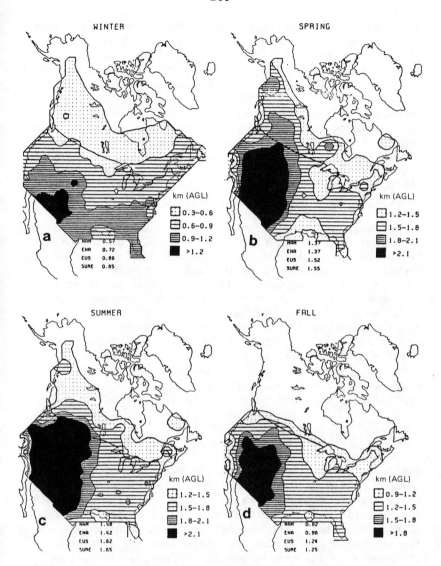

FIGURE B.4 Contour plots of maximum afternoon mixing depths by season, indicating qualitative patterns only. a, January through March; b, April through June; c, July through September; and d, October through December. SOURCE: Holzworth (1972) and Portelli (1977).

both cyclonic and anticyclonic systems can have an impact on accumulation of emissions. Korshover (1976) has pointed out that the south central United States is particularly subject to stagnating anticyclones, leading to lower ventilation of local and advected emissions.

Another factor critical to long-range transport is precipitation, which removes pollutants in a sporadic way. Trajectories from a source to a receptor will not establish the total mass transported if the air mass is likely to experience precipitation along the way. This removal depends on the type, intensity, and frequency of the precipitation. At present, the precipitation removal process is difficult to quantify over long transport paths.

Recently another important factor has been pointed out by Draxler and Taylor (1982). The authors showed that the spreading of emissions is dominated by the action of vertical wind shear acting in combination with the diurnal cycle of daytime mixing and nighttime layering of the atmosphere. Further work on the importance of this factor is being pursued.

In understanding the synoptic-scale transport, one should not lose sight of the fact that both local and mesoscale influences are important in continental transport. Thus to model the regional transport, the mesoscale must be adequately parameterized even if not explicity nested with that scale's simulation.

HEMISPHERIC OR GLOBAL TRANSPORT

Besides the fact that hemispheric transport involves greater distances and times than regional transport, there are important differences between the two scales. One fact is that the bulk of the global transport takes place over water. Because of the small changes in oceanic surface temperature, the planetary boundary layer over the oceans is relatively constant. Besides this, the oceans can be considered a homogeneous surface over large areas. Thus there are broad stretches of strong atmospheric inversions over cold water and other well-mixed regions over relatively warm water. One can expect that pollution within the boundary layer is subject to dry removal and that pollution that has been transported above the boundary layer will remain there until removed by precipitation processes or by large-scale subsidence.

The study of the movement of acidic and preacidic
material from sources in North America to other receptor
regions in the northern hemisphere has been undertaken in
several cases. However, because of the lack of chemical
and meteorological data over large stretches of the ocean,
only crude estimates of this transport can be made. For
example, the high acidity found in precipitation on the
island of Hawaii could be partially explained by long-
range transport from the west, where Japan would be the
major source (Miller and Yoshigana 1981, Dittenhoefer
1982). In this study, a single trajectory model was
useful in evaluating the transport patterns.

Another area of interest is the contribution of North
American sources to Arctic haze. This issue has been
raised more in reference to visibility or the modification
of the radiation balance, since the Canadian and U.S.
Arctic areas are deserts (100 mm per annum) with little
wet deposition. The major transport path from eastern
North America is a track around Greenland. Concentrations
of pollutant aerosols in the Arctic show a definite
wintertime peak when removal mechanisms are most inactive.
Rahn and McCaffrey (1980) indicate that residence times
of aerosol particles in the Arctic range from 2 to 3
weeks in the winter.

The transport of materials across the Atlantic has also
been a topic of interest though not firmly established.
Early estimates were made that North American contribution
to sulfate in rain in Norway could be important. More
recent studies in Bermuda indicate that trans-Atlantic
transport of acid precursors is important to the acidity
of precipitation on the island (Jickells et al. 1982).
Further studies of this transport are being continued
under a joint U.S.-Canada-Bermuda effort.

Recent studies of precipitation in remote areas of
both the northern and southern hemisphere have shown the
acidity of rain to be on the average lower than pH 5.0
(Galloway et al. 1982). The degree to which natural
sources or long-range transport of man-made pollutants
contribute to this remote acidity in precipitation
remains to be seen. However, trajectory calculations to
estimate the transport on a global scale will be a useful
tool in such research.

CONCLUSIONS

Though the transport of materials in the atmosphere has
been studied for a number of years, there is still much

that can be learned in applying this knowledge to the acid deposition problem. By and large, the vigor of the atmosphere in both the horizontal and vertical rules where the final deposition of a given pollutant will be.

ACKNOWLEDGMENTS

The committee thanks N. Gillani, C. Patterson, and R. Husar for their help in preparing this appendix.

REFERENCES

Dittenhoefer, A.C. 1982. The effects of sulfate and non-sulfate particles on light scattering at the Mauna Loa Observatory. Water, Air and Soil Pollut. 18:129-154.

Draxler, R.R., and A.D. Taylor 1982. Horizontal dispersion parameters for long-range transport modeling. J. Atmos. Meteorol. 21:367-372.

Galloway, J.N., G.E. Likens, W.C. Keene, and J.M. Miller 1982. The composition of precipitation in remote areas of the world. J. Geophys. Res. 87:8771-8786.

Holzworth, G.C. 1972. Mixing heights, wind speeds, and potential for urban air pollution throughout the contiguous United States. U.S. EPA AP-101.

Huschke, R.E. (ed.) 1959. Glossary of Meteorology. Boston, Mass.: American Meteorological Society, p. 638.

Jickells, T., A. Knap, T. Church, J. Galloway, and J. Miller 1982. Acid rain in Bermuda. Nature 297:55-57.

Korshover, J. 1976. Climatology of stagnating anticyclones east of the Rocky Mountains, 1936-75. NOAA Technical Memorandum ERL ARL-55, 26 pp.

Miller, J.M., and A.M. Yoshigana 1981. The pH of Hawaiian precipitation. A preliminary report. Geophys. Res. Lett. 8:779-782.

Portelli, R.V. 1977. Mixing heights, wind speeds and ventilation coefficients for Canada. Environment Canada, Atmospheric Environment Service, Climatological Studies Number 31, UDC: 551.554.

Rahn, K.A., and R.J. McCaffrey 1980. On the origin and transport of the winter Arctic aerosol. Ann. N.Y. Acad. Sci. 308:486-503.

Appendix **C** Atmospheric Deposition Processes

1. INTRODUCTION

In this appendix we present an overview of current
scientific understanding about deposition phenomena, with
the objectives of identifying key literature sources on
this subject and providing the reader with the technical
basis necessary for effective evaluation of the available
literature. There are several important features of this
subject, which should be noted at the outset. First, the
ultimate deposition processes of interest are the end
products of a complex sequence of atmospheric phenomena
(cf. Figure 2.1). Deposition processes tend to reflect
these preceding events strongly. Much of the material
presented in this appendix therefore necessarily deals
with the predeposition processes, which may act as
important rate-influencing steps in the overall
source-deposition sequence.

A second important feature of initial interest is the
relative difference in states of our current understanding
of wet- and dry-deposition phenomena. Wet deposition is
comparatively simple to measure. As a consequence there
exists a substantial and growing base of data on wet
deposition from a variety of networks and field studies.
Precipitation processes tend to be rather complicated,
however, and currently a high level of uncertainty exists
regarding their mathematical characterization.

Dry deposition, on the other hand, tends to be
extremely difficult to measure, and the corresponding
data set is relatively meager. Partly because of this
fact most mathematical characterizations of dry-
deposition processes have been quite simple in form. The
tendency toward simplicity in most mathematical characterizations of dry deposition should not be taken to

imply that the physical processes themelves are simple.
As a consequence of these differences the following sections on dry and wet deposition have somewhat different
formats, with emphasis in each placed on areas of current
major activity.

Finally, it should be noted that very little of the
material presented in this appendix is new. A number of
reviews of both wet and dry deposition have been presented
during recent years, and the current treatment is merely
an attempt to consolidate these efforts.* In view of this
tendency toward redundancy, it is strongly recommended
that the reader proceed directly to the indicated journal
literature if more detailed pursuit of this subject is
desired.

2. DRY-DEPOSITION PROCESSES

2.1 MECHANISMS OF DRY DEPOSITION

2.1.1 Introduction

The rate of transfer of pollutants between the air and
exposed surfaces is controlled by a wide range of
chemical, physical, and biological factors, which vary in
their relative importance according to the nature and
state of the surface, the characteristics of the pollutant, and the state of the atmosphere. The complexity
of the individual processes involved and the variety of
possible interactions among them combine to prohibit easy
generalization; nevertheless, a "deposition velocity,"
v_d, analogous to a gravitational falling speed, is of
considerable use. In practice, knowledge of v_d enables
fluxes, F, to be estimated from airborne concentrations,
C, as the simple product, $v_d \cdot C$.

*Much of the material presented in this appendix was
prepared by Drs. B.B. Hicks and J.M. Hales as a
contribution to the Critical Assessment Document on
Acidic Deposition being prepared by North Carolina State
University under a cooperative agreement with the U.S.
Environmental Protection Agency. These contributions are
published here with permission of the authors and the
concurrence of the editors of the Critical Assessment
Document, Drs. A.P. Altschuller and R.A. Linthurst.

Particles larger than about 20-μm diameter will be deposited at a rate that is controlled by Stokes law, although with some enhancement due to inertial impaction of particles transported to near the surface in turbulent eddies. The settling of submicrometer-sized particles in air is sufficiently slow that turbulent transfer tends to dominate, but the net flux is often limited by the presence of a quasi-laminar layer adjacent to the surface, which presents a considerable barrier to all mass fluxes and especially to gases with very low molecular diffusivity. The concept of a gravitational settling velocity is inappropriate in the case of gases, but transfer is still often limited by diffusive properties very near the receptor surface.

Sehmel (1980b) presents a tabulation of factors known to influence the rate of pollutant deposition upon exposed surfaces. Figure C.2-1 has been constructed on the basis of Sehmel's list and has been organized to emphasize the greatly dissimilar processes affecting the fluxes of gases and large particles. Small, submicrometer-diameter particles are affected by all the factors indicated in the diagram; thus, simplification is especially difficult for deposition of such particles. In reality, Figure C.2-1 already represents a considerable simplification, since many potentially important factors are omitted. In particular, the emphasis of the diagram is on properties of the medium containing the pollutants in question; a similarly complicated diagram could be constructed to illustrate the effects of pollutant characteristics. For particles, critical factors include size, shape, mass, and wettability; for gases, concern is with molecular weight and polarization, solubility, and chemical reactivity. In this context, the acidity of a pollutant that is being transferred to some receptor surface by dry processes is a quality of special importance that may have strong impact on the efficiency of the deposition process itself.

Figure C.2-2 summarizes particle size distributions on a number, surface area, and volume basis. In this way, the three major modes are brought clearly to attention. The number distribution emphasizes the transient (or Aitken) nuclei range, 0.005-0.05-μm diameter, for which diffusion plays a role in controlling deposition. The area distribution draws attention to the so-called accumulation size range formed largely from gaseous precursors (0.05-2-μm diameter, affected by both diffusion and gravity). The remaining mode (2-50-μm

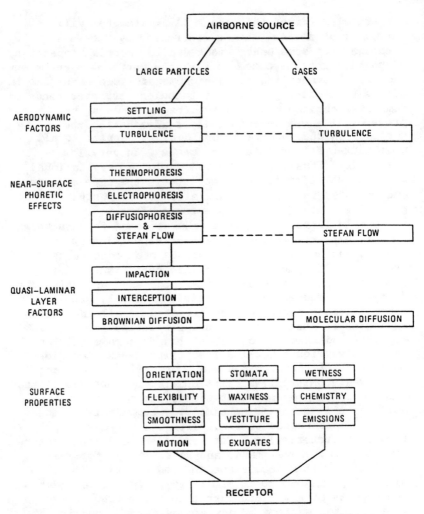

FIGURE C.2-1 A schematic representation of processes likely to influence the rate of dry deposition of airborne gases and particles. Note that some factors affect both gaseous and particulate transfer, whereas others do not. However, submicrometer particles are affected by all the factors that influence gases and large particles, and hence it is these "accumulation-size-range" aerosols that present the greatest challenge for deposition research.

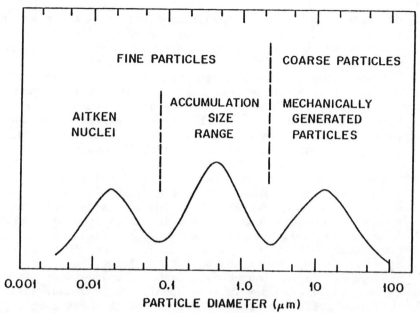

FIGURE C.2-2   A hypothetical particle-size spectrum, such as might be found down-wind of an industrial complex. The smaller aerosols have gaseous precursors and are formed by condensation of exhaust gases and by atmospheric chemical reactions (typically oxidation), followed by growth due to particle coagulation. The larger particles are partly soil-derived, suspended by natural erosion and agricultural practices, and partly the direct result of the combustion of fossil fuels. Acidic aerosols are primarily in the smaller mode of the particle-size spectrum, whereas the larger mode contains material that might tend to neutralize the acidic deposition of the smaller particles. In evaluating the net input of acidity to a surface, it is critical that both size fractions and gaseous contributions be included.

diameter, most evident in the volume distribution) is the mechanically generated particle range for which gravity causes most of the deposition. In most literature, 2-µm diameter is used as a convenient boundary between "fine" and "coarse" particles.

Atmospheric sulfates, nitrates, and ammonium compounds are primarily associated with the accumulation size range. Figure C.2-2 demonstrates that very little acidic or acidifying material is likely to be associated with the coarse particle fraction in background conditions. However, the larger particles include soil-derived minerals, some of which can react chemically with airborne and deposited acids. Moreover, it has been

suggested that some of these larger particles may provide
sites for the catalytic oxidation of sulfur dioxide (for
example when the particles are carbon; Chang et al. 1981,
Cofer et al. 1981). Little is known about the detailed
chemical composition of large particle agglomerates.
However it is accepted that their residence time is quite
short (i.e., they are deposited relatively rapidly), that
there are substantial spatial and temporal variations in
both their concentrations and their composition, and that
their contribution to acid dry deposition should not be
ignored.

To evaluate deposition rates, several different
approaches are possible. Field experiments can be
conducted to monitor changes in some system of receptors
from which average deposition rates can be deduced. More
intensive experiments can measure the deposition of
particular pollutants in some circumstances. Neither
approach is capable of monitoring the long-term, spatial-
average dry deposition of pollutants. To understand why,
we must first consider in some detail the processes that
influence pollutant fluxes and then relate these considerations to measurement and modeling techniques that are
currently being advocated. The logical sequence illustrated in Figure C.2-1 will be used to guide this
discussion.

2.1.2 Aerodynamic Factors

Except for the obvious difference that particles will
settle slowly under the influence of gravity, small
particles and trace gases behave similarly in the air.
Trace gases are an integral part of the gas mixture that
constitutes air and thus will be moved with all the
turbulent motions that normally transport heat, momentum,
and water vapor. However, particles have finite inertia
and can fail to respond to rapid turbulent fluctuations.
Table C.2-1 lists some relevant characteristics of
spherical particles in air (based on data tabulated by
Davies 1966, Friedlander 1977, and Fuchs 1964). The time
scales of most turbulent motions in the air are considerably greater than the inertial relaxation (or
stopping) times listed in the table. These time scales
vary with height, but even as close as 1 cm from a smooth,
flat surface, most turbulence energy will be associated
with time scales longer that 0.01 sec, so that even
100-μm-diameter particles would follow most turbulent

TABLE C.2-1 Dynamic Characteristics of Unit Density Aerosol
Particles at STP, Corrected for Stokes-Cunningham Effects[a]

Particle Radius (μm)	Diffusivity (cm^2/s)	Stopping Time (s)	Settling Speed (cm/s)
0.001	1.28×10^{-2}	1.33×10^{-9}	1.30×10^{-6}
0.007	3.23×10^{-3}	2.67×10^{-9}	2.62×10^{-6}
0.005	5.24×10^{-4}	6.76×10^{-9}	6.62×1^{-6}
0.01	1.35×10^{-4}	1.40×10^{-8}	1.37×10^{-5}
0.02	3.59×10^{-5}	2.97×10^{-8}	2.91×10^{-5}
0.05	6.82×10^{-6}	8.81×10^{-8}	8.63×10^{-5}
0.1	2.21×10^{-6}	2.28×10^{-7}	2.23×10^{-4}
0.2	8.32×10^{-7}	6.87×10^{-7}	6.73×10^{-4}
0.5	2.74×10^{-7}	3.54×10^{-6}	3.47×10^{-3}
1.0	1.27×10^{-7}	1.31×10^{-5}	1.28×10^{-2}
2.0	6.10×10^{-8}	5.03×10^{-5}	4.93×10^{-2}
5.0	2.38×10^{-8}	3.08×10^{-4}	3.02×10^{-1}
10.0	1.38×10^{-8}	1.23×10^{-3}	1.2×10^{0}

[a]Data are from Fuchs (1964), Davies (1966), and Friedlander (1977).

fluctuations. However, natural surfaces are normally
neither smooth nor flat, and it is clear that in many
circumstances the flux of particles will be limited by
their inability to respond to rapid air motions.

Naturally occurring aerosol particles are not always
spherical, although it seems reasonable to assume so in
the case of hygroscopic particles in the submicrometer
size range. Chamberlain (1975) documents the ratio of
the terminal velocity of nonspherical particles to that
of spherical particles with the same volume. In all
cases, the nonspherical particles have a lower terminal
settling speed than equivalent spheres. The settling
speed differential is indicated by a "dynamical shape
factor," α, as listed in Table C.2-2.

Thus, trace gases and small particles are carried by
atmospheric turbulence as if they were integral components of the air itself, whereas large particles are
also affected by gravitational settling, which causes
them to fall through the turbulent eddies. In general,
however, the distribution of pollutants in the lower
atmosphere is governed by the dynamic structure of the
atmosphere as much as by pollutant properties.

TABLE C.2-2 Dynamic Shape Factors as by which Nonspherical
Particles Fall More Slowly than Spherical (from
Chamberlain, 1975)

Shape	Ratio of axes	α
Ellipsoid	4	1.28
Cylinder	1	1.06
Cylinder	2	1.14
Cylinder	3	1.24
Cylinder	4	1.32
Two spheres touching, vertically	2	1.10
Two spheres touching, horizontally	2	1.17
Three spheres touching, as triangle	–	1.20
Three spheres touching, in line	3	1.34-1.40
Four spheres touching, in line	4	1.56-1.58

In daytime, the lower atmosphere is usually well mixed
up to a height typically in the range 1 to 2 km, as a
consequence of convection associated with surface heating
by insolation. Pollutants residing anywhere within this
mixed layer are effectively available for deposition
through the many possible mechanisms. However, at night,
the lower atmosphere becomes stably stratified and
vertical transfer of nonsedimenting material is so slow
that, at times, pollutants at heights as low as 50 to 100
m are isolated from surface deposition processes. Thus,
in daytime, atmospheric transfer does not usually limit
the rate of delivery of pollutants to the surface boundary layer in which direct deposition processes are active.
    The fine details of turbulent transport of pollutants
remain somewhat contentious. Notable among the areas of
disagreement is the question of flux-gradient relationships in the surface boundary layer. It is now well
accepted that the eddy diffusivity of sensible heat and
water vapor exceeds that for momentum in unstable (i.e.,
daytime) but not in stable conditions over fairly smooth
surfaces (see Dyer 1974, for example). However, it is
not clear that the well-accepted relations governing
either heat or momentum transfer are fully applicable to
the case of particles or trace gases; some disagreement
exists even in the case of water vapor. The situation is

even more uncertain in circumstances other than over large expanses of horizontally uniform pasture. When vegetation is tall, pollutant sinks are distributed throughout the canopy so that close similarity with the transfer of more familiar quantities such as heat or momentum is effectively lost. There is even considerable uncertainty about how to interpret profiles of temperature, humidity, and velocity above forests (see Garratt 1978, Hicks et al. 1979, Raupach et al. 1979).

2.1.3 The Quasi-laminar Layer

In the immediate vicinity of any receptor surface, a number of factors associated with the molecular diffusivity and the inertia of pollutants become important. Large particles carried by turbulence can be impacted on the surface as they fail to respond to rapid velocity changes. The physics of this process is similar to the physics of sampling by inertial collection.

Inertial impaction is a process that augments gravitational settling for particles that fall into a size range typically between 2- and 20-μm diameter (q.v. Slinn 1976b). Larger-sized particles tend to bounce, and capture is therefore less efficient, while smaller-sized particles experience difficulty in penetrating the quasi-laminar layer that envelops receptor surfaces. From the viewpoint of acidic particles, inertial impaction is a process of questionable relevance since most acidic species are associated with smaller particles (see Figure C.2-2), which are not strongly affected by this process. However, Figures C.2-2 and C.2-3 show that many airborne materials exist in the size range likely to be affected by inertial impaction. Since many of the chemical constituents of soil-derived particles are capable of neutralizing deposited acids, inertial impaction may have important indirect effects on acidic deposition.

To illustrate the role of molecular or Brownian diffusivity, it is informative to consider the simple case of a knife-edged thin plate, mounted horizontally and with edge normal to the wind sector. As air passes over (and under) the plate, a laminar layer develops, of thickness $\delta = c(vx/u)^{1/2}$, where v is kinematic viscocity, x is the downwind distance from the edge of the plate, and u is wind speed. According to Batchelor (1967), the value of the numerical constant c is 1.72. Thus, for a plate of dimensions 5 cm in a wind speed of

222

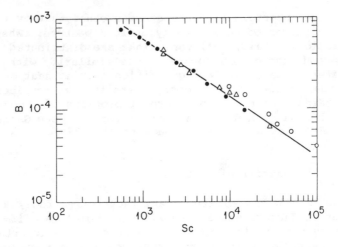

FIGURE C.2-3   Laboratory verification of Schmidt-number scaling for particle transfer to a smooth surface. The quantity plotted is $B \cong v_d/u_*$, evaluated for transfer
across a quasi-laminar layer of molecular control immediately adjacent to a smooth
surface. Data are from Harriott and Hamilton (1965; open circles), Hubbard and Light-
food (1966; triangles), and Muzushinz et al. (1971; solid circles), as reported by
Lewellen and Sheng (1980). The line drawn through the data is Equation (C.2-1), with
exponent $\alpha = -2/3$ and constant of proportionality $A \cong 0.06$.

1 m/s, we should imagine a boundary-layer thickness
reaching about 1.5-mm thick at the trailing edge.

Over nonideal surfaces, the internal viscous boundary
layer is frequently neither laminar nor constant with
time. The layer generates slowly as a consequence of
viscosity and surface drag as air moves across a
surface. The Reynolds number Re ($\equiv ux/v$, where u is
the wind speed, x is the downwind dimension of the
obstacle, and v is kinematic viscosity) is an index of
the likelihood that a truly laminar layer will occur.
For large Re, air adjacent to the surface remains
turbulent: viscosity is then incapable of exerting its
influence. In many cases, it seems that the surface
layer is intermittently turbulent. For these reasons,
and because close similarlity between ideal surfaces
studied in wind tunnels and natural surfaces is rather
difficult to swallow, the term "quasi-laminar layer" is
preferred.

Wind-tunnel studies of the transfer of particles to
the walls of pipes tend to support the concept of a
limiting diffusive layer adjacent to smooth receptor

surfaces. Transfer across such a laminar layer is conveniently formulated in terms of the Schmidt number, $Sc = v/D$, where v is viscosity and D is the pollutant diffusivity. The conductance, or transfer velocity v_1, across the quasi-laminar layer is proportional to the friction velocity u_*:

$$v_1 = Au_* \, Sc^\alpha, \qquad\qquad\qquad (C.2-1)$$

where A and α are determined experimentally. Most studies agree that the exponent α is about $-2/3$, as is evident in the experimental data represented in Figure C.2-3. However, a survey by Brutsaert (1975a) indicates exponents ranging from -0.4 to -0.8. The value of the constant A is also uncertain. The line drawn through the data of Figure C.2-3 corresponds to $A \simeq 0.06$, yet the wind-water tunnel results of Moller and Schumann (1970) appears to require $A \simeq 0.6$. These values span the value of $A \simeq 0.2$ recommended for the case of sulfur dioxide flux to fibrous, vegetated surfaces (Shepherd 1974, Wesely and Hicks 1977).

Laminar boundary-layer theory imposes the expectation that particle deposition to exposed surfaces will be strongly influenced by the size of the particle, with smaller particles being more readily deposited by diffusion than larger. It is clear that many artificial surfaces or structures made of mineral material will have characteristics for which the laminar-layer theories might be quite appropriate. However, the relevance to vegetation can be questioned. Microscale surface roughness elements can penetrate the barrier presented by this quasi-laminar layer and should be suspected as sites for enhanced deposition of both particles and gases (see Chamberlain 1980).

2.1.4 Phoretic Effects and Stefan Flow

Particles near a hot surface experience a force that tends to drive them away from the surface. For very small particles (<0.03-μm diameter, according to Davies 1967), this "thermophoresis" can be visualized as the consequence of hotter, more energetic air molecules impacting the side of the particle facing the hot surface. For larger particles, radiometric forces become important (Cadle 1966). In theory, thermal radiation can cause temperature gradients across particles that are not good thermal conductors, resulting in a mean motion of the particle away from a hot surface. In summary, the thermophoresis depends on the local temperature gradient in the air, on the thermal properties of the particle, on the Krudsen number $Kn \equiv \lambda/r$ (where λ is the mean free path of air molecules and r is the radius of the particle), and on the nature of the interaction between the particle and air molecules (see Derjaguin and Yalamov 1972). As a rule of thumb, the thermophoretic velocity of very small particles (<0.03-μm diameter) is likely to be about 0.03 cm/s (estimated from values quoted by Davies 1967). For particles exceeding 1-μm diameter, the velocity will be about four times less.

The process of diffusiophoresis results when particles reside in a mixture of intermixing gases. In most natural circumstances, the principal concern is with water vapor. Close to an evaporating surface, a particle will be impacted by more water molecules on the nearer side. Since these water molecules are lighter than air molecules, there will be a net "diffusiophoresis" toward the evaporating surface. In essence, these "phoretic" forces result from the flow of molecules of some special kind through the gas mixture and the "drag" exerted on particles. Since diffusiophoresis and thermophoresis depend on the size and shape of the particle of interest, neither can be predicted with precision, nor can safe generalizations be made. These subjects are sufficiently complicated that they constitute specialities in their own right. Excellent discussions have been given by Friedlander (1977) and Twomey (1977). These phoretic forces vary with particle size but are generally small, and their influence on dry deposition can usually be disregarded.

Many workers include Stefan flow in general discussion of diffusiophoresis, but because of the conceptual difference between the mechanisms involved it seems better to consider them separately. Stefan flow results from injection into the gaseous medium of new gas molecules at an evaporating or subliming surface. Every gram-molecule of substrate material that becomes a gas displaces 22.41 liters of air, at STP. Thus, for example, a Stefan flow velocity of 22.41 mm/s will result when 18 g of water evaporates from a 1-m^2 area every second. Generalization to other temperatures and pressures is straightforward. Daytime evaporation rates from natural

vegetation often exceed 0.2 g/m^{-2} s for considerable
times during the midday period, resulting in Stefan flow
of more than 0.2 mm/s away from the surface. Detailed
calculation for specific circumstances is quite simple.
For the present, it is sufficient to note that Stefan
flow is capable of modifying surface deposition rates by
an amount that is larger than the deposition velocity
appropriate for many small particles to aero- dynamically
smooth surfaces.

Electrical forces have often been mentioned as
possible mechanisms for promoting deposition (as well as
the retention, see Section 2.1.5) of small particles,
particularly through the "viscous" quasi-laminar layer
immediately above receptor surfaces. Wason et al. (1973)
report exceedingly high rates of deposition of particles
in the size range 0.6 to 6 μm to the walls of pipes
whenever a space charge is present. Chamberlain (1960)
demonstrated the importance of electrostatic forces in
modifying deposition velocities of small particles, when
fields are sufficiently high. Plates charged to produce
local field strengths of more than 2000 V cm^{-1} experienced considerably more deposition of small particles
than uncharged plates, by factors between 2 and 15.
However in fair-weather conditions, field strengths are
typically less than 10 V cm^{-1} so that the net effect on
particle transfer is likely to be small. Further studies
of the ability of electrostatic forces to assist the
transfer of particulate pollutants to vegetative surfaces
were conducted by Langer (1965) and Rosinski and Nagomoto
(1965). According to Hidy (1973), a series of experiments
was conducted using single conifer needles and conifer
trees. "For single needles or leaves, electrical charges
on 2-μm-diameter ZnS dust with up to eight units of
charge had no detectable effect at wind speeds of 1.2 to
1.6 m/s. The average collection efficiency was found to
be 6% for edgewise cedar or fir needles, with broadside
values an order of magnitude lower. Bounce-off after
striking the collector was not detected, but re-
entrainment could take place above 2 m/s wind speed.
Tests on branches of cedar and fir by Rosinski and
Nagomoto (1965) suggested similar results as for single
needles." It should be noted, however, that the
electrical mobility of a particle is a strong negative
function of particle size, ranging from 2 cm/s per V/cm
of field strength for 0.001-μm-diameter particles, to
0.0003 cm/s per V/cm for 0.1-μm particles (Davies 1967).

2.1.5 Surface Adhesion

Most workers assume that pollutants that contact a surface will be captured by it. For some gases, this assumption is clearly adequate. For example, nitric acid vapor is sufficiently reactive that most surfaces should act as nearly perfect sinks. Less reactive chemicals will be less efficiently captured. The case of particles is of special interest, however, because of the possibility of bounce and resuspension.

The role of electrostatic attraction in binding deposited particles to substrate surfaces remains something of a mystery. The process by which particles become charged and set up mirror-charges on the underlying surface is fairly well accepted. The resulting van der Waals forces are often mentioned as the major mechanism for binding particles once deposited. For large, nonspherical particles, dipole moments can be set up in natural electric fields, and these can help promote the adhesion at surfaces. These matters have been conveniently summarized by Billings and Gussman (1976), who provide mathematical relationships for evaluating the electrical energy of a particle on the basis of its size, shape, dielectric constant, and the strength of the surrounding electrical field. For smaller particles, the principal charging mechanism is thermal diffusion, leading to a Boltzmann charge distribution.

Condensation of water reduces the effectiveness of electrostatic adhesion forces, since leakage paths are then set up and charge differentials are diminished. However, the presence of liquid films at the interfaces between particles and surfaces causes a capillary adhesive force that compensates for the loss of electrostatic attraction. These "liquid-bridge" forces are most effective in high humidities and for coarse particles (<20 μm, according to Corn 1961).

Billings and Gussman (1976) draw attention to the effect of microscale surface roughness in promoting adhesion of particles to surfaces. Much of the experimental evidence is for particle diameters much greater than the height of surface irregularities (e.g., Bowden and Tabor 1950). It is the opposite case that is likely to be of greater interest in the present context, as will be discussed later.

2.1.6 Surface Biological Effects

The efficiency with which natural surfaces "capture" impacting particles or molecules will be influenced considerably by the chemical composition of the surface as well as its physical structure. The "lead candle" technique for detection of atmospheric sulfur dioxide is a historically interesting example of how chemical substrates can be selected to affect the deposition rates of particular pollutants.

Uptake rates of many trace gases by vegetation are controlled by biological factors such as stomatal resistance. In daytime, this is known to be the case for sulfur dioxide (Shepherd 1974, Spedding 1969, Wesely and Hicks 1977) and usually for ozone in most situations (Wesely et al. 1978). The similarity between sulfur and ozone is not complete, however, because the presence of liquid water on the foliage will tend to promote SO_2 deposition and to impede uptake of ozone; the former gas is quite soluble until the solution becomes too acidic, whereas the latter is essentially insoluble (q.v. Brimblecombe 1978).

Pubescence of leaves has received considerable attention. Chamberlain (1967) tested the roles of leaf stickiness and hairiness in a series of wind-tunnel tests. He concludes that "with the large particles (32 and 19 μm) the velocity of deposition to the sticky artificial grass was greater than to the real grass, but with those of 5 μm and less, it was the other way round, thus confirming . . . that hairiness is more important than stickiness for the capture of the smaller particles." The importance of leaf hairs appears to be verified by studies of the uptake of ^{210}Pb and [210]Po particles by tobacco leaves (Fleischer and Parungo 1974, Martell 1974) and by the wind-tunnel work of Wedding et al. (1975), who report increases by a factor of 10 in deposition rates for particles to pubescent leaves, compared with smooth, waxy leaves. It remains to be seen how greatly biological factors of this kind influence the rates of deposition of airborne particles to other kinds of vegetation.

2.1.7 Deposition to Liquid-Water Surfaces

Trace-gas and aerosol deposition on open water surfaces is of significant practical interest, especially consid-

ering the acidification of poorly buffered inland waters.
Air blowing from land across a coastline will slowly
equilibrate with the new surface at a rate that is
strongly dependent on the stability regime involved. If
the water is much warmer than upwind land, dynamic
instability over the water will cause relatively rapid
adjustment of the air to its new lower boundary, but if
the water is cooler, stratified flow will occur and
adjustment will be very slow. In the former (unstable)
case, dry-deposition rates of all soluble or chemically
reactive pollutants are likely to be much higher than in
the latter. Clearly, air blowing over small lakes will
be less likely to adjust to the water surface than when
blowing over larger water bodies. Thus, during much of
the summer, inland water surfaces will tend to be cooler
than the air, and hence protected from dry deposition,
because of the strongly stable stratification that will
then prevail. This phenomenon will occur more frequently
over small water bodies than larger ones (see Hess and
Hicks 1975).

Following the guidance of chemical engineering
gas-transfer studies, workers such as Kanwisher (1963),
Liss (1973), and Liss and Slater (1974) have considered
the role of Henry's law constant and chemical reactivity
in controlling the rate of exchange of trace gases between
the atmosphere and the ocean. In general, acidic and
acidifying species like SO_2 are readily removed on
contact with a water surface. Thus Hicks and Liss (1976)
neglected liquid-phase resistance and derived net
deposition velocities appropriate for the exchange of
reactive gases across the air-sea interface. The work of
Hicks and Liss is intended to apply to water bodies of
sufficient size that the bulk exchange relationships of
air-sea interaction research are applicable. Their
considerations indicate that deposition velocities for
highly soluble and chemically reactive gases such as
NH_3, HCl, and SO_2 are likely to be between 0.10 and 0.15
percent of the wind speed measured at 10-m height. The
analysis leading to this conclusion assumes that the
molecular and eddy diffusivities can be combined by
simple addition. This assumption has been shown to
approximate the transfer of water vapor and sensible heat
from water surfaces. However, for fluxes of trace gases
the validity of this assumption is questionable. Slinn
et al. (1978) argue that it is better to introduce
molecular diffusivity through a term analogous to the
Schmidt number of Equation (2-1), with the exponent

$\alpha \simeq -2/3$. (In contrast, the linear assumption used by Hicks and Liss implies $\alpha = -1.0$.) In view of the uncertainties mentioned in discussion of Equation (C.2-1), further comment on the implications and ramifications of these alternative assumptions is not warranted.

In the limiting case of a trace gas of low solubility, the deposition velocity is determined by the large liquid-phase resistance, which is essentially proportional to the Henry's law constant.

It is probable that breaking waves will modify the simple gas-transfer formulations derived from chemical engineering pipe-flow and wind-tunnel work. It is not clear to what extent such features account for the apparent discrepancy between the various Schmidt number dependencies of the kind expressed by Equation (C.2-1). However, the fractional power laws are basically extensions of laboratory work, whereas the unit-power, additive-diffusivities result is an approximation to field data. It is to be hoped that the two approaches produce results that will converge in due course.

Figure C.2-4 previews the discussion of wind-tunnel particle deposition results that will be given later. Such wind-tunnel work indicates exceedingly low deposition velocities for particles in the size range of most acidic pollutants. As in the case of gas exchange, there are conceptual difficulties in extending these results to the open ocean. The role of waves in the transfer of small particles between the atmosphere and water surfaces remains essentially unknown. Not only does engulfment by breaking waves provide an alternative path across the quasi-laminar sublayer where molecular (or Brownian) diffusion normally controls the transfer, but also waves are a source of droplets that can scavenge particulate material from the air (see, however, the study of Alexander 1967, which indicates otherwise). Hicks and Williams (1979) have proposed a simple model of air-sea particle exchange that extends smooth-surface, wind- and water-tunnel results (as in Figure C.2-4) to natural circumstances by permitting rapid transfer to occur whenever waves break. This results in very low deposition velocities in light winds, but rapidly increasing when winds increase above about 5 m/s. Slinn and Slinn (1980) also suggest that particle transfer is more rapid than the wind-tunnel studies of Figure C.2-4 might indicate but present an alternative hypothesis for this more rapid transfer: that hygroscopic particles grow rapidly when exposed to high humidities such as are

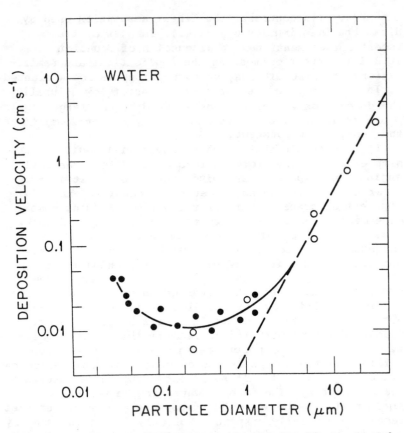

FIGURE C.2-4   Results of wind-tunnel studies of particle deposition to water surfaces. Solid circles are due to Moller and Schumann (1979), open circles to Sehmel and Sutter (1974). The dashed line at the right represents the terminal settling speed for 1.5 g cm^{-3} particles.

found in air adjacent to a water surface, resulting in increased gravitational settling and impaction to the water surface.

2.1.8 Deposition to Mineral Surfaces

Acid deposition is an obvious source of worry to architects, historians, and others concerned with the potentially accelerated deterioration of structures. Many popular building materials react chemically with acidic air pollutants, generating new chemical species

that can contribute directly to the decay process even if
they are rapidly and efficiently washed off by precipitation. Furthermore, in some cases the chemical product
causes a visual degradation that cannot easily be
rectified, such as the blackening of metal work exposed
to hydrogen sulfide.

The presence of water at the surface is known to be a
key factor in promoting the fracturing and erosion of
stone. Water penetrates pores and cracks and causes
mechanical stresses both by freezing and by hydration and
subsequent crystallization of salts (see Fassina 1978,
Gauri 1978, Winkler and Wilhelm 1970). The earlier
discussion of surface effects that influence dry
deposition indicated that surface scratches and fractures
will cause accelerated dry-deposition rates in localized
areas. Moreover, phoretic effects are likely to be more
important than in the case of foliage (because dry
surfaces exhibit wider temperature extremes than moist
vegetation). Stefan flow associated with dewfall is also
probably more important than for vegetation. Hicks
(1981) has summarized a number of relevant points as
follows:

1. In daytime, particle fluxes will be greatest to
the coolest parts of exposed surfaces.
2. Both particle and gas fluxes will be increased
when condensation is taking place at the surface, and
decreased when evaporation occurs.
3. If the surface is wet, impinging particles will
have a better chance of adhering, and soluble trace gases
will be more readily "captured."
4. The chemical nature of the surface is important;
if reaction rates with deposited pollutants are rapid,
then surfaces can act as nearly perfect sinks.
5. Biological factors can influence uptake rates, by
modifying the ability of the surface to capture and bind
pollutants.
6. The texture of the surface is important. Rough
surfaces will provide better deposition substrates than
smoother surfaces and will permit easier transport of
pollutants across the near-surface quasi-laminar layer.

2.1.9. Fog and Dewfall

The processes that cause aerosol particles to nucleate,
coalesce, and grow into cloud droplets are precisely the
same as those that assist in the generation of fog.

Whenever surface air supersaturates, fog droplets form on whatever hygroscopic nuclei are available. These small droplets slowly settle onto exposed surfaces or are deposited by interception and impaction. The characteristics of the liquid that is deposited are much the same as those of cloud liquid water (see Section 3).

The conditions under which low-altitude surface fogs form are the cases of strong stratification in which vertical turbulent transport is minimized. The frequency of occurrence of fogs varies widely with location and with time of year. The depth is also highly variable. However, it must be assumed that fogs constitute a mechanism whereby the lower atmosphere (say the bottom hundred meters or so) can be cleansed of particulate and some gaseous pollutants.

At higher elevations, fog droplets are precisely the same as the cloud droplets that in other circumstances would grow and finally precipitate in substantially diluted form. The importance of cloud droplet interception has been demonstrated recently by Lovett et al. (1982) at an altitude of 1200 m in New Hampshire. Most of the net deposition of acidic species is by cloud droplet interception.

The presence of liquid water on exposed surfaces will obviously help promote the deposition of soluble gases and wettable particles. This surface water arises through the action of three separate mechanisms. Some plants expel fluid from foliage, usually at the tips of leaves, by a process known as guttation. Moisture can evaporate from the ground and recondense on other exposed surfaces, a mechanism known as distillation. However, these mechanisms are frequently confused with dewfall, which is properly the process by which water vapor condenses on surfaces directly from the air aloft. In practice, the origin of the surface moisture is immaterial to pollutants that come in contact with it. However, dewfall and distillation are processes that assist pollutant deposition through Stefan flow, whereas guttation does not. According to Monteith (1963), the maximum rate of dewfall is of the order of 0.07 mm/h, so that the maximum Stefan flow enhancement of the nocturnal deposition velocity is about 8 cm/h (see Section 2.1.4).

2.1.10   Resuspension and Surface Emission

Deposited particles can be resuspended into the air and
subsequently redeposited. The mechanisms involved are
much the same as those that cause saltation of particles
from the beds of streams and from eroding croplands.
These subjects are of great practical importance in their
own right and have been studied at length. Concern about
resuspension of radioactive particles near sites of
accidents or weapons tests injected a note of some
urgency into related studies during the 1950s and 1960s,
as evidenced in the large number of papers on the subject
included in the volume "Atmosphere–Surface Exchange of
Particulate and Gaseous Pollutants" (Engelmann and Sehmel
1976).

The momentum transfer between the atmosphere and the
surface is the driving force that causes surface particles
to creep, bounce, and eventually saltate. There is a
minimum frictional force that will cause particles of any
particular size to rise from the surface. Bagnold (1954)
identifies u_*^2 as a controlling parameter, so that it
is the few occurrences of strongest winds that are the
most important. While most thinking seems to center on
widespread phenomena like dust storms, Sinclair (1976)
points out that dust devils provide a highly efficient
light-wind mechanism for resuspending surface particles
and carrying them to considerable altitudes. Clearly,
very large particles will not be moved frequently, or
far. Very small particles are bound to the surface by
adhesive forces that have already been discussed and tend
to be protected in crevices or between larger particles.

Chamberlain (1982) has provided a theoretical basis
for linking saltation of sand particles and snowflakes
and for relating these phenomena to the generation of
salt spray at sea.

It is not clear how saltation and related phenomena
affect acid deposition. Surface particles that are
injected into the air by the action of the wind do not
normally move far, nor do they offer much opportunity for
interaction with other air pollutants (firstly because
they are confined in a fairly shallow layer near the
suface and secondly because they have a very short
residence time). Their effects are largely local.

Much smaller particles (in the submicrometer size
range) are generated by reactions between atmospheric
oxidants and organic trace gases emitted by some
vegetation, especially conifers (see Arnts et al. 1978).

Once again, it is not obvious how these should best be
considered in the present context of acid deposition.
This is but one of many natural surface sources that
provide a conceptual mechanism for injecting particles
and trace gases into the lower atmosphere.

2.1.11 The Resistance Analog

Discussion of the relative importance of the various
factors that contribute to the net flux of a particular
atmospheric pollutant and determination of which process
might be limiting in specific circumstances is simplified
by considering a resistance model analogous to Ohm's
law. Figure C.2-5 illustrates the way in which the
concept is usually applied. An atmospheric resistance,
r_a, is identified with the transfer of material through
the air to the vicinity of the final receptor surfaces.
This resistance is defined as that associated with the
transfer of momentum; it is dependent on the roughness of
the surface, the wind speed, and the prevailing atmospheric stability. The aerodynamic resistance can be
written as

$$r_a = C_{fn}^{-1} - \psi_C/k)/u_*, \qquad \qquad (C.2-2)$$

where C_{fn} is the appropriate friction coefficient (the
square root of the familiar drag coefficient) in neutral
stability, u_* is the friction velocity (a scaling
quantity defined as the root mean covariance between
vertical and longitudinal wind fluctuations), k is the
von Kármán constant, and ψ_C is a stability correction
function that is positive in unstable, negative in
stable, and zero in neutral stratifications (see Wesely
and Hicks 1977). Equation (C.2-2) is obtained by
straightforward manipulation of standard micrometeorological relations, as given by Wesely and Hicks, for
example. The value of k is usually taken to be about
0.4. Table C.2-3 lists typical values of the friction
coefficient for a range of surfaces.

The surface boundary resistance, r_b, is that which
accounts for the difference between momentum transfer
(i.e., frictional drag) at the surface and the passage of
some particular pollutant through the near-surface
quasi-laminar layer. In the agricultural meteorology
literature, a quantity B^{-1} is frequently employed for
this purpose (Brutsaert 1975a). The relationship between

235

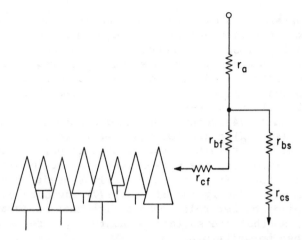

FIGURE C.2-5   A diagrammatic illustration of the resistance model frequently used to help formulate the roles of processes like those given in Figure C.2-1. Here, r_a is an aerodynamic resistance controlled by turbulence and strongly affected by atmospheric stability, r_{bf} and r_{bs} represent surface boundary-layer resistances that are determined by molecular diffusivity and surface roughness, and r_{cf} and r_{cs} are the net residual resistances required to quantify the overall deposition process, to the eventual sink. The second subscripts f and s are intended to indicate pathways to foliage and to soil, respectively. There are many other pathways that might be important; the diagram is not intended to be more than a simple visualization of some of the important factors.

these quantities can be clarified by relating both to the micrometeorological concept of a roughness length, z_0 (the height of apparent origin of the neutral logarithmic wind profile). Then the total atmospheric resistance, R, between the surface in question and the height of measurement, z, can be written as

$$R = (ku_*)^{-1}[\ln(z/z_{oc}) - \psi_c]$$
$$= (ku_*)^{-1}[\ln(z/z_0) + \ln(z_0/z_{oc}) - \psi_c]$$
$$= r_a + (ku_*)^{-1} \cdot \ln(z_0/z_{oc}) , \qquad (C.2-3)$$

where z_{oc} is a roughness length scale appropriate for the transfer of the pollutant. The residual boundary-layer resistance, $r_b = R - r_a$, is then

$$r_b = (ku_* ^{-1} \cdot \ln(z_0/z_{oc}) , \qquad (C.2-4)$$

which alternatively is written as

$$r_b = (u_* B)^{-1} . \qquad (C.2-5)$$

B is, therefore, a measure of the nondimensionalized limiting deposition velocity for concentrations measured sufficiently close to a receptor surface that the resistance to momentum transfer can be disregarded.

It should be noted that some workers refer to r_b as the aerodynamic resistance and use the symbol r_a for it (e.g., O'Dell et al. 1977).

Shepherd (1974) recommends using a constant value $kB^{-1} = \ln(z_0/z_{0c}) = 2.0$ for transfer to vegetation, on the basis of results obtained over rough, vegetated surfaces. However, the role of the Schmidt number in accounting for diffusion near a surface needs to be taken into account. Wesely and Hicks (1977) advocate the use of a Schmidt number relationship like that of Equation (C.2-1), so that the surface boundary-layer resistance would then be written as

$$r_b \simeq 5 \ Sc^{2/3} u_*. \qquad (C.2-6)$$

Equation (C.2-6) implies a value of 0.2 for A in the boundary-layer relationship given by Equation (C.2-1), as was mentioned earlier.

The final resistances in the conceptual chain of processes represented diagrammatically by Figure C.2-5

TABLE C.2-3   Estimates of Roughness Characteristics Typical of Natural Surfaces[a]

Surface	Approx. Canopy Height (m)	Roughness Length (cm)	Neutral Friction Coefficient, C_{fn}
Smooth ice	0	0.003	0.042
Ocean	0	0.005	0.045
Sandy desert	0	0.03	0.055
Tilled soil	0	0.10	0.066
Thin grass	0.1	0.70	0.095
Tall thin grass	0.5	5.	0.16
Tall thick grass	0.5	10.	0.21
Shrubs	1.5	20.	0.25
Corn	2.3	30.	0.29
Forest	10.	50.	0.23
Forest	20.	100.	0.24

[a]Values of the friction coefficient C_{fn} ($\equiv u_*/u$) are evaluated for neutral conditions, at a height 50 cm above the surface or top of the canopy.

are those that permit material to be transferred to the surface itself. For many pollutants, it is necessary only to consider the canopy foliage resistance r_{cf}, but for some it is also necessary to consider uptake at the ground by invoking a resistance to transfer to soil (or a forest floor), r_{cs}. In concept, it is also appropriate to differentiate between boundary-layer resistances r_{bf} and r_{bs}, for transfer to foliage and soil, respectively, as is shown in the diagram. Many other resistances can be identified and might often need to be considered, but further complication of Figure C.2-5 is unnecessary. Its main purpose is illustrative.

Transfer of many trace gases to foliage occurs by way of stomatal uptake, which, because of stomatal physiology, imposes a strong diurnal cycle on the overall deposition behavior. Following initial work by Spedding (1969), studies of foliar uptake of sulfur dioxide have repeatedly confirmed the controlling role of stomatal resistance. Chamberlain (1980) summarizes results of experiments by Belot (1975) and Garland and Branson (1977), who compared surface conductances of sulfur dioxide with those for water vapor over a broad range of stomatal openings (which largely govern stomatal resistance). The conclusion that stomatal resistance is the controlling factor when stomates are open appears to be well founded. However, once again, it is necessary to apply corrections to account for the diffusivity of the trace gas in question; the higher the molecular diffusivity of the gas, the lower the stomatal resistance.

Fowler and Unsworth (1979) point out that SO_2 deposition to wheat continues, even when stomates are closed, at a rate that suggests significant deposition at the leaf cuticle. Thus, it is not always sufficient to compute the canopy-foliage resistance r_{cf} on the assumption that SO_2 uptake is via stomates alone (although this may indeed be a sufficient approximation in most circumstances). Instead, it is more realistic to estimate r_{cf} from its component parts via

$$r_{cf} \approx (r_{st}^{-1} + r_{cut}^{-1})^{-1}/(\text{LAI}) \qquad (C.2-7)$$

(following Chamberlain 1980), where r_{st} is the stomatal resistance, and r_{cut} is cuticular resistance. LAI is the leaf area index (the total area of foliage per unit horizontal surface area). Note that in most literature the LAI is assumed to be the single-sided leaf area index. However, sometimes both sides of the leaves are counted.

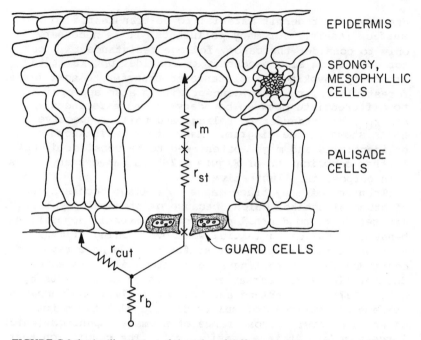

EPIDERMIS

SPONGY, MESOPHYLLIC CELLS

PALISADE CELLS

GUARD CELLS

FIGURE C.2-6   An illustration of the roles of different resistances associated with trace gases uptake by a leaf. Material is transferred along several possible pathways, of which two are shown. These involve cuticular uptake via a resistance, r_{cut}, and transfer through stomatal pores (via r_{st}) into substomatal cavities, with subsequent transfer to mesophyllic tissue (via r_m). The way in which the various resistances are combined to provide the best visualization of the overall transfer process is not clear.

The resistance analogy permits a closer look at the mechanisms that transfer gaseous material to leaves. Figure C.2-6 illustrates the pathways involved: via stomatal openings into the interior of the leaf (involving stomatal and mesophyll resistances, r_{st} and r_m) or through the epidermis (involving a cuticular resistance, r_{cut}).

The resistance model is somewhat limited by the manner in which it structures the chain of relevant processes, each being represented by a resistance to transfer that occupies a prescribed location in a conceptual network. The structure of this network is sometimes not clear, and furthermore, there are important processes that do not conveniently fit into the resistance model. Mean drift velocities (e.g., gravitational settling of particles) are not easily accommodated in the simple resistance

picture, and it is doubtful whether some of the biological factors are relevant to the question of particle
transfer. Studies of leaves show that stomates are
typically slits of the order of 2-20 μm long. For
stomatal uptake of particles to be a controlling factor
of deposition, we would need to hypothesize spectacularly
good aim by the particles.

2.2 METHODS FOR MEASURING DRY DEPOSITION

2.2.1 Direct Measurement

There is little question that the deposition of large
particles is accurately measured by collection devices
exposed carefully above a surface of interest. Deposit
gauges and dust buckets have been important weapons in
the geochemical armory for a long time. They are intended
to measure the rate of deposition of particles that are
sufficiently large that deposition is controlled by
gravity. In studies of radioactive fallout conducted in
the 1950s and 1960s, these same devices were used. In
the case of debris from weapons tests, the major local
fallout was of so-called hot particles, originating with
the fragmentation of the casing of the weapon and its
supporting structures, and the suspension of soil in the
vicinity of the explosion. These large particles fall
over an area of rather limited extent downwind of the
explosion. This area of greatest fallout was the major
focus of the work on fallout dry deposition. It was
largely in this context that dustfall buckets were used
to obtain an estimate of how much radioactive deposition
occurred. It was recognized that collection vessels
failed to reproduce the microscale roughness features of
natural surfaces. However, this was not seen as a major
problem, since the emphasis was on evaluating the maximum
rate of deposition that was likely to occur, so that
upper limits could be placed on the extent of possible
hazards. Nevertheless, efforts were made to "calibrate"
collection vessels in terms of fluxes to specific types
of vegetation, soils, etc. (see Hardy and Harley 1958).
Much further downwind, most of the deposition was
shown to be associated with precipitation, since the
effective source of the radioactive fallout being
deposited was typically in the upper troposphere or the
lower stratosphere. The acknowledged inadequacies of
collection buckets for dry deposition were then of only

little concern, since dry fallout composed a small
fraction of the total surface flux.

In the context of present concerns about acid deposition, we must worry not only about large, gravitationally settling particles but also about small
"accumulation-size-range" particles that are formed in
the air from gaseous precursors and about trace gases
themselves. All of these materials contribute to the net
flux of acidic and acidifying substances by dry processes.
It is known that collection vessels do indeed provide a
measure of the flux of large particles. However,
accumulation-size-range particles, typically of less than
1-μm diameter, do not deposit by gravitational settling
at a significant rate. These small particles are
transported by turbulence through the lower atmosphere
and are deposited by impaction and interception on
surface roughness elements, with the assistance of a wide
range of surface-related effects (e.g., electrophoresis,
Stefan flow) many of which will be influenced by the
detailed structure of the surface involved.

Early work on the deposition of radioactive fallout
made use of collection vessels and surrogate surface
techniques that were frequently "calibrated" in terms of
fluxes to specific types of vegetation, soils, etc.
Studies of this kind were relatively easy, especially in
the case of radioactive pollutants, since very small
quantities of many important species could be measured
accurately by straightforward techniques. Most of the
radioactive materials that were of interest do not exist
in nature, and so experimental studies benefitted from a
zero background against which to compare observed data.
Moreover, major emphasis was on the dose of radioactivity
to specific receptors, a quantity that is strongly
influenced by contributions of large, "hot" particles in
situations of practical interest. Such circumstances
included deposition of bomb debris, fission products, and
soil particles from the radioactive cloud downwind of
nuclear explosions. In such cases, highest doses were
incurred near the source and were due to these larger
particles.

The applicability of collection vessels and surrogate
surfaces in studies of the dry deposition of acidic
pollutants is in dispute. Principal among the conceptual
difficulties concerning their use is their inability to
reproduce the detailed physical, chemical, and biological
characteristics of natural surfaces, which are known to
control (or at least strongly influence) pollutant uptake

in most instances. Furthermore, the continued exposure
of already-deposited materials to airborne trace gases
and aerosol particles undoubtedly causes some changes to
occur, but of unpredictable magnitude and unknown
significance. A recent intercomparison between different
kinds of surrogate surfaces and collection vessels has
indicated that fluxes derived from exposing dry buckets
are more than those obtained using small dishes, which in
turn exceed values obtained using rimless flat plates
(Dolske and Gatz 1982). This provides a tantalizing
tidbit of evidence for an ordering of performance
characteristics according to the total exposed surface
area per unit horizontal projection. In this context,
the similarity with arguments concerning leaf area index
seems especially attractive. Micrometeorological data
obtained during the same experiment fall between the
extremes represented by the buckets and the flat plates.

Dasch (1982) reports on a comparison between many
different configurations of flat-plate collection
surfaces, pans, and buckets. The results indicate that
glass surfaces provide the greatest flux estimates for
almost all chemical species considered and Teflon the
lowest. Bucket data generally fall midway in the range.

Tracer techniques that were developed in the
radioecology era for investigating fluxes to natural
surfaces offer some promise. A β-emitting isotope of
sulfur, ^{35}S, lends itself to use in studies of SO_2
uptake by crops since measurements of low rates of sulfur
accumulation are then possible. Garland (1977), Garland
and Branson (1977), Garland et al. (1973), and Owers and
Powell (1974) report the results of a number of studies
of $^{35}SO_2$ uptake by various vegetated surfaces ranging
from pasture to pine plantation and by nonvegetated
surfaces such as water.

In concept, it is feasible to extend studies of this
kind to the deposition of sulfurous particles, but as yet
no such experiment has been reported. However, analogous
studies of particle deposition using nonradioactive
aerosol tracers have been carried out. In wind-tunnel
experiments, Wedding et al. (1975) employed uranine dye
particles in conjunction with lead chloride particles to
study the influence of leaf microscale roughness on
particle capture characteristics; uranine particles are
relatively easily measured by fluorimetry, whereas
measurements of lead deposition require far more pain-
staking chemical analysis of the deposition surface. The
particle sizes used by Wedding et al. were in the range
of 3- to 7-μm diameter.

Considerably larger particles have been used in many studies. In detailed wind-tunnel studies, Chamberlain (1967) used lycopodium spores (~30-μm aerodynamic diameter). Workers at Brookhaven National Laboratory extended these wind-tunnel techniques to real-world circumstances by conducting a series of experiments employing pollen grains in the same general size range (Raynor et al. 1970, 1971, 1972, 1974).

In general, these methods of tracer measurement have not been applied to natural circumstances for the particle sizes of major interest in the present context of acid deposition. An important exception concerns studies of deposition on snow surfaces. The retention of deposited material at the top of or within a snowpack has been studied in some detail and continues to be an intriguing area of research. Particulate materials such as sulfate were considered by Dovland and Eliassen (1976), who studied the accumulation upon snow surfaces during periods of no precipitation and found average deposition velocities in the range 0.1 to 0.7 cm/s depending on the assumption made regarding the contribution by gaseous SO_2 deposition. Similar work by Barrie and Walmsley (1978) yielded average sulfur dioxide deposition velocities to snow in the range 0.3 to 0.4 cm/s, with standard error equivalent to about a factor or 2.

Dillon et al. (1982) and Eaton et al. (1978) present examples of the use of calibrated watersheds to estimate atmospheric deposition. Dry-deposition fluxes are estimated as a residual between measured fluxes out of a conceptually closed system and measured wet deposition into it. Considerable effort is required to document annual chemical mass balances for specific watersheds. Once the effort is made, it appears possible to draw fairly well-founded conclusions regarding dry deposition, although obviously such estimates might result as the difference between fairly large numbers. According to Eaton et al., the annual dry-deposition flux estimate obtained at the Hubbard Brook Experimental Forest in New Hampshire is accurate to about ±35 percent (one standard error). The data do not permit apportionment between gaseous and particulate sulfur inputs, but the total sulfur flux corresponded to a deposition velocity of about 0.6 cm/s.

2.2.2 Laboratory Studies

Figure C.2-1 illustrates the overall complexity of the
problem of dry deposition. While it is indisputable that
no indoor experiment can provide a comprehensive evaluation of pollutant deposition that would be applicable to
the natural countryside, laboratory studies provide the
unique attraction of controllable conditions. It is
feasible to study the relative importance of various
factors thought to be of importance, as in Figure C.2-1
and especially as in Figure C.2-6, and to formulate these
processes in a logical manner. In this general category,
we must include the extensive wind-tunnel work referred
to earlier, the pipe-flow and flat-plate studies conducted in experiments more aligned to problems of chemical
engineering, and the chamber experiments favored by
ecologists and plant physiologists. Distinction between
these kinds of experiment is often difficult. Many
exposure chambers and pipe-flow studies have features of
wind tunnels.

The utility of chamber studies is well illustrated by
the series of results reported by Hill (1971). By
comparing the rates of deposition of various trace gases
to oat and alfalfa canopies exposed in large chambers,
Hill concluded that solubility was a critical parameter
in determining uptake rates of trace gases by vegetation.
The ordering of deposition velocities was: hydrogen
fluoride > sulfur dioxide > chlorine > nitrogen
dioxide > ozone > carbon dioxide > nitric oxide >
carbon monoxide. Furthermore, the chamber studies
indicated a wind-speed dependence of the kind predicted
by turbulent transfer theory and demonstrated a physio-
logical effect of chlorine and ozone on stomatal
opening: exposure to high concentrations of either
quantity caused partial stomatal closure, thus limiting
the fluxes of all trace gases that are stomatally
controlled.

Experiments conducted by Judeikis and Wren (1977,
1978) yielded valuable information on the deposition of
hydrogen sulfide, dimethyl sulfide, sulfur dioxide,
nitric oxide, and nitrogen dioxide to nonvegetated
surfaces (Table C.2-4). The values listed were derived
from initial deposition rates obtained before surface
accumulation limited uptake rates. For comparison,
surface resistances derived from Hill's (1971) studies of
trace-gas uptake by alfalfa are also listed. On the
whole, the ordering of deposition velocities suggested by

Hill's work appears to be supported, providing some justification for extending the ordering to CO, H_2S, and $(CH_3)_2S$ in the manner indicated in the table. Residual surface resistance to uptake of soluble gases by solid, dry surfaces appears to be substantially greater than for vegetation, which is as would be expected.

The values listed in Table C.2-4 represent resistances to transport very near the surface, to which atmospheric resistances must be added to obtain values representative of natural, outdoor conditions. The reciprocals of the tabulated numbers provide upper limits of the appropriate deposition velocities.

Similarly informative data have been obtained about particle deposition on surfaces that can be contained in wind tunnels. Studies of this kind are an obvious extension of pipe-flow investigations by workers such as Friedlander and Johnstone (1957) and Liu and Agarwal (1974), which provide strong support for theories involving particle inertia and Schmidt number scaling. Wind tunnels provide a means to extend chamber and

TABLE C.2-4 Resistances to Deposition of Selected Trace Gases, Measured for Solid Surfaces in a Cylindrical Flow Reactor (Judeikis and Stewart, 1976) and for Alfalfa in a Growth Chamber (Hill, 1971); Solid-Surface Data are Derived from Table 2 of Judeikis and Wren (1978); the Alfalfa Values are Obtained from Table 1 of Hill (1971)

| Pollutant | Substrate Surface | | |
	Adobe Clay	Sandy Loam	Alfalfa
CO	-	-	∞
H_2S	62	67	-
$(CH_3)_2S$	3.6	16	-
NO	7.7	5.3	10.
CO_2	-	-	3.3
O_3	-	-	0.7
NO_2	1.3	1.7	0.5
Cl_2	-	-	0.5
SO_2	1.1	1.7	0.4
HF	-	-	0.3

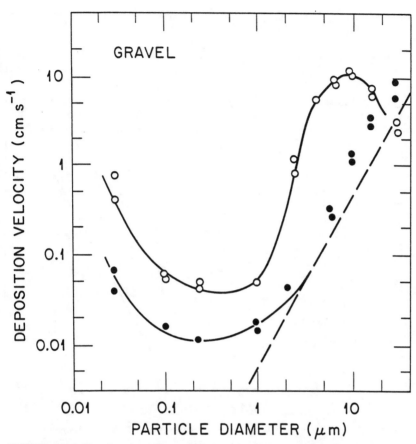

FIGURE C.2-7   Results of wind-tunnel studies of particle deposition to 1.6-cm-diameter dry gravel (after Sehmel et al. 1973a). Solid circles were obtained at about 2.4 m/s, open circles at about 16 m/s.

pipe-flow investigations to situations more closely approximating natural conditions.

Results obtained in studies of particle deposition to dry gravel (Sehmel et al. 1973a) are shown in Figure C.2-7. The wind-speed effect evident in these data is fairly typical and applies also in the case of vegetation (Figure C.2-8). Experiments on deposition to wet gravel were also conducted. These indicated deposition velocities some 30 percent less than the values evident in Figure C.2-7 (for particles in the 0.2- to 1.0-µm

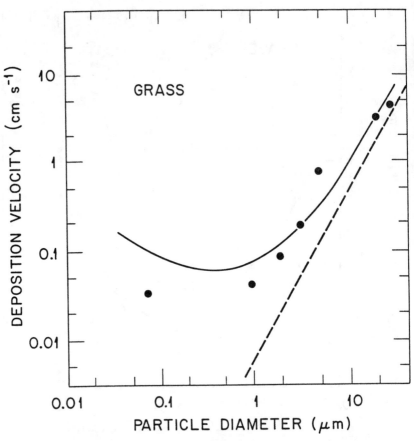

FIGURE C.2-8 Results of wind-tunnel studies of particle deposition to grass as reported by Chamberlain (1967; dots; natural grass; $u_* \cong 70$ cm/s) and by Sehmel et al. (1973; the curve; artificial grass; $u_* \cong 16$ cm/s).

size range), as might be expected from considerations of Stefan flow and diffusiophoresis. When surface roughness was increased, deposition velocities also increased.

Chamberlain (1967) extended his earlier (1966) wind-tunnel studies of gas transfer to "grass and grass-like surfaces" by considering particle deposition to rough surfaces. Sehmel (1970) conducted similar wind-tunnel experiments, employing monodisperse particles ranging from about 0.5- to 20-μm diameter. Figure C.2-8 combines results from Chamberlain (1967), and Sehmel et al. (1973b). The Chamberlain data refer to live grass, but the Sehmel et al. data were obtained

using 0.7-cm-high artificial grass. Moreover, the two
sets of data were obtained at different wind speeds
(Chamberlain, $u_* \simeq 70$ cm/s; Sehmel et al., $u_* \simeq 19$ cm/s).
Further tests conducted by Chamberlain (1967) indicated
that deposition velocities to natural grass exceeded
those to artificial grass by a factor of about 2 for
particles smaller than about 5 μm. This appears to be
contrary to the indication of Figure C.2-7, where v_d
(natural) of Chamberlain is seen to be about half the
v_d (artificial) of Sehmel et al. for sizes less than a
few micrometers. The difference in u_* between their
experiments amplifies this discrepancy, rather than
resolving it. However, both data sets provide evidence
for the deposition velocity particle-size dependency that
is predicted by theory and supported by all such
laboratory investigations.

2.2.3 Micrometeorological Measurement Methods

The factors that control pollutant fluxes are frequently
surface properties such as stomatal resistance and soil
moisture (for soluble gases), cuticular resistance and
the available leaf area (for strongly surface-reactive
gases like HF and HNO_3), and microscale roughness (for
submicrometer particles). Any measurement technique that
interferes with a controlling property is likely to yield
erroneous results, and hence there has been considerable
effort expended to develop and apply methods of measurement that impose no surface or environmental modification.
In concept, if an area is sufficiently homogeneous, flat,
and contains no areas of strong sources or sinks, pollutant fluxes can be assumed to be constant with height.
Therefore questions regarding dry deposition can be
addressed by measuring the flux of material through a
horizontal layer of air at some more convenient height
above the surface. The intent of any such study is to
investigate dry-deposition fluxes in carefully documented
natural situations in order to identify and quantify
controlling properties. The results of these investigations are formulations of surface mechanisms, surface
boundary-layer resistances, stomatal resistances, etc.
The demanding site criteria are required to enable these
results to be obtained from the experiments; the surface
parameterizations that are derived are far more widely
applicable.

Several micrometeorological methods are suitable for measuring dry-deposition fluxes in intensive case studies. The flux can be measured directly by eddy correlation, a process that evaluates instantaneous products of the the vertical wind speed, w, and pollutant concentration, C, in order to derive the time-average vertical flux F_C as

$$F_C = \overline{\rho w'C'}, \qquad\qquad (C.2-8)$$

where ρ is air density and the primes denote deviations from mean values. The over-bar indicates a time average. This is an extremely demanding task and constitutes a specialized field of micrometeorology in its own right. Details of experimental procedures are given by, for example, Dyer and Maher (1965), Kaimal (1975), and Kanemasu et al. (1979).

Figure C.2-9 shows some examples of sensor output signals that are fundamental to the eddy-correlation technique. Fast-response sensors of any pollutant concentration can be used; the trace shown for CO_2 in the diagram is an interesting example of considerable agricultural relevance. As a basic requirement, sensors suitable for eddy-correlation applications should have response times shorter than 1 sec for operation at convenient heights on towers. For application aboard aircraft (Bean et al. 1972, Lenschow et al. 1980), considerably faster response is required.

Eddy-correlation methods have been used in field experiments addressing the fluxes of ozone (Eastman and Stedman 1977), sulfur (Galbally et al. 1979, Hicks and Wesely 1980), nitrogen oxides (Wesely et al. 1982b), carbon dioxide (Desjardins and Lemon 1974, Jones and Smith 1977), and small particles (Wesely et al. 1977).

Rates of transfer through the lower atmosphere are governed by intensities of turbulence generated by both mechanical mixing and convection. In this context, there are three atmospheric quantities that cannot be separated: the vertical flux of a material, the local gradient of its concentration ($\partial C/\partial z$), and its corresponding eddy diffusivity (K). Knowledge of any two of these quantities will permit the third to be evaluated. Often, when sensors suitable for direct measurement of pollutant fluxes are not available, assumptions regarding the eddy diffusivity are made to

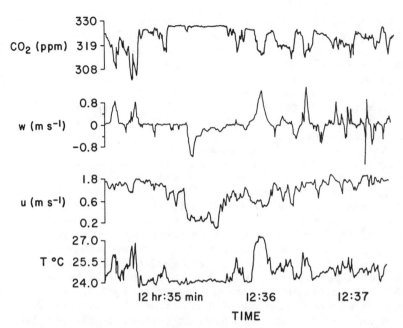

FIGURE C.2-9   An example of atmospheric turbulence near the surface. These traces of CO_2 concentration, vertical velocity (w), wind speed (u), and temperature (T) were obtained over a corn canopy by workers at Cornell University.

provide a method for estimating fluxes from measurements of vertical concentration gradients:

$$F_c = \rho K (\partial C / \partial z).$$
(C.2-9)

Droppo (1980) and Hicks and Wesely (1978) have summarized a number of critical considerations. In particular, with a typical value of u_* = 40 cm/s and neutral stability, the concentration difference between adjacent levels differing in height by a factor of 2 is about 9 percent, for a 1 cm/s deposition velocity. In unstable (daytime) conditions, smaller gradients would be expected for the same v_d; in stable conditions, they would be greater.

The demands for high resolution by the concentration measurement technique are obvious. Nevertheless, a substantial quantity of excellent information has been obtained, especially concerning fluxes of SO_2 (Fowler 1978, Garland 1977, Whelpdale and Shaw 1974).

It should be emphasized that the stringent site uniformity requirements mentioned above for use of eddy-correlation approaches are also relevant for gradient studies. The detection of a statistically significant difference between concentrations at two heights is not necessarily evidence of a vertical flux and can only be interpreted as such after the appropriate siting criteria have been satisfied.

Gradients of particle concentration present special problems since it is often not possible to derive internally consistent results from alternative measurements. Droppo (1980) concludes that "The particulate source and sink processes over natural surfaces cannot be considered as a simple unidirectional single-rate flux." Thus, the proper interpretation of gradient data in terms of fluxes might not be possible for airborne particles, even in the best of siting circumstances, because of the role of the surface in emitting and resuspending particles. In this case, eddy-correlation methods will still provide an accurate determination of the flux through a particular level, but this flux will be made up of a downward flux of airborne material and an upward flux of similar material of surface origin. Disentangling the two is likely to present a considerable problem.

None of the various micrometeorological methods has yet been developed to the extent necessary for routine application. Rather, they are research methods that can be used in specific circumstances, requiring considerable experimental care and the use of sensitive equipment and fairly complicated data analysis. They are more suitable for investigating the processes that control dry deposition than for monitoring the flux itself. Nevertheless, some new techniques that might be appropriate for dry-deposition monitoring are currently under development. A "modified Bowen ratio" method is being developed in the hope that it might permit an accurate determination of vertical fluxes without the need for rapid response or great resolution (Hicks et al. 1981). High-frequency variance methods are also being advocated but have yet to be fully investigated; for these, sensors having rapid response are required. An eddy-accumulation method that bypasses the need for rapid response of the pollutant sensor is of long-standing interest (e.g., Desjardins 1977) but has yet to be applied to the pollutant flux problem with significant success.

2.3 FIELD INVESTIGATIONS OF DRY DEPOSITION

2.3.1 Gaseous Pollutants

Table C.2-5 summarizes a number of recent field experiments on trace gas deposition. The listing is drawn from a variety of sources (especially Chamberlain 1980, Garland 1979, Sehmel 1979, 1980a); it is not meant to be exhaustive but is intended to demonstrate that much of the available data on surface fluxes of trace gases refers to daytime conditions, when "canopy" resistances are usually the controlling factors. Extrapolation of these values to nighttime conditions is dangerous on two grounds; first, because of the large changes that might accompany stomatal closure and, second, because of the much greater influence of aerodynamic resistance in nighttime, stable conditions.

Figure C.2-10 illustrates the large diurnal cycle that is typical of the dry-deposition rates of most pollutants. These observations were made over a pine plantation in North Carolina, using eddy correlation to measure each quantity (Hicks and Wesely 1980). The diurnal cycle of sensible heat flux meshes well with expectations based on the available heat energy (i.e., on net radiation), and the friction velocities determined by direct measurement conform to expectations based on the known roughness characteristics of the site. The eddy fluxes of total sulfur demonstrate a diurnal cycle that appears to be as strong as for the meteorological properties, a result that is not surprising when it is remembered that many of the causative factors are common (e.g., vertical turbulent exchange). The quality of the data appears to be quite good--sensible heat fluxes and friction velocities are difficult to measure, and the ability to measure each with the run-to-run smoothness evident in the diagram instills considerable confidence in the sulfur data, measured with the same apparatus. Nevertheless, it would be unwise to place too much confidence in the fine details of the sulfur-flux data. While the strong (downward) fluxes of sulfur evident during the midday periods are probably accurately represented, it is possible that the indicated nightime fluxes are contaminated by a water-vapor interference or some other factor that is especially significant in the calmer, more humid nighttime situation. Thus, some caution must be associated with interpreting the negative (upward) fluxes of sulfur evident on two periods as evidence of emission or resuspension from the

TABLE C.2-5   Recent Experiments on Trace-Gas Deposition to Natural Surfaces

Worker	Method	Results and Comments
SO_2		
Hill (1971)	$^{35}SO_2$ with stable SO_2 carrier over alfalfa	$v_d \approx 2.3$ cm/s in daytime; Implies $r_c \approx 0.4$ s/cm
Garland et al. (1973)	$^{35}SO_2$ over pasture	$v_d \approx 1.2$ cm s^{-1} in daytime $r_c \approx 0.6$ s cm^{-1}
Owers and Powell (1974)	$^{35}SO_2$ over pasture	$v_d \approx 1.3$ cm/s, daytime
Shepherd (1974)	SO_2 gradients over grass	$v_d \approx 1.3$ cm/s in summer ≈ 0.3 cm/s in autumn
Whelpdale and Shaw (1974)	SO_2 gradients over snow, water, and grass	$v_d \approx 1$ cm/s in daytime for grass and water snow
Garland (1977)	SO_2 gradients, calcareous soils	$v_d \approx 1.2$ cm/s $r_c \approx 0.01$ s/cm
Fowler (1978)	SO_2 gradients, over —wheat —soybean	$v_d \approx 0.4$ cm/s $v_d \approx 1.3$ cm/s

Reference	Method	Deposition velocity
Dannevik et al. (1976)	SO_2 gradients over wheat	$v_d \approx 0.4$ cm/s
Garland and Branson (1977)	$^{35}SO_2$ to pine	$v_d \approx 0.1 - 0.6$ cm/s
Belot (as summarized by Chamberlain 1980)	$^{34}SO_2$ to pine	$v_d < 1$ cm/s
Galbally et al. (1979)	Eddy correlation over pine forest	$v_d \approx 0.2$ cm/s
Dovland and Eliassen (1976)	Accumulation to snow	$v_d \approx 0.1$ cm/s
Barrie and Walmsley (1978)	Accumulation to snow	$v_d \approx 0.2$ cm/s

NO_2

Reference	Method	Deposition velocity
Wesely et al. (1982)	Eddy correlation – soybeans	$v_d \approx 0.6$ cm/s in daytime $r_c \approx 1.3$ s/cm in daytime; ≈ 15 s/cm at night

O_3

Reference	Method	Deposition velocity
Galbally and Roy (1980)	Gradients over wheat	$v_d \approx 0.7$ cm/s Implies $r_c \approx 1.4$ s/cm
Wesely et al. (1978, 1982b)	Eddy correlation over a range of natural surfaces	$r_c \approx 0.8$ s/cm in daytime ≈ 1.8 s/cm at night

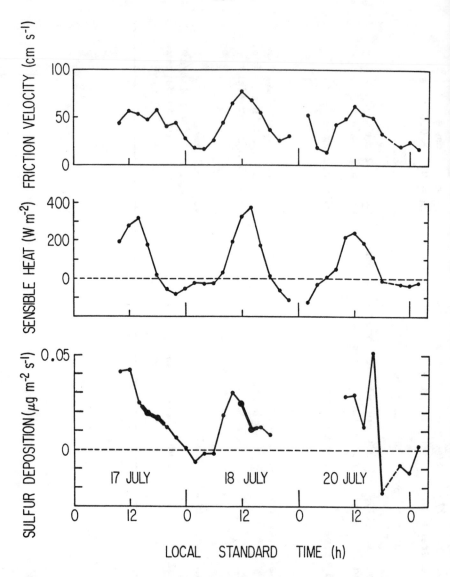

FIGURE C.2-10 Records of sulfur flux, sensible heat flux, and friction velocity through 3 days of an intensive study of dry deposition to a pine plantation (Hicks and Wesely 1980). The darker portions of the sulfur data indicate periods when gaseous sulfur could not be detected. At all times, the data refer to total sulfur, usually made up of gaseous and particulate contributions.

canopy. However, attempts to explain the phenomenon in terms of some interference have so far failed.

Similarly, large diurnal cycles of SO_2 deposition are reported by Fowler (1978), who introduces the further complexity of enhanced SO_2 deposition to wheat covered with dewfall. Using the notation of Figures C.2-5 and C.2-6, Fowler finds typical daytime values to be

$$r_a = 0.25 \text{ s/cm,}$$
$$r_b = 0.25 \text{ s/cm,}$$
$$r_{st} = 1.0 \text{ s/cm,}$$
$$r_{cut} = 2.5 \text{ s/cm.}$$

For deposition to dry soil, Fowler suggests the use of $r_{cs} = 10.0$ s/cm, and $r_{cs} = 0$ when the soil is wet.

It might be noted, in passing, that the aerodynamic resistance r_a influences the deposition of all nonsedimenting pollutants, and thus it is not possible for any trace gas to have a deposition velocity greater than $1/r_a$, i.e., about 4 cm/s in the daytime conditions of Fowler's experiment. Because of stability effects, the maximum possible deposition velocity at night would be considerably lower. Many of the exceedingly large deposition velocities reported in the open literature appear to exceed the limits imposed by our knowledge of the aerodynamic resistance. Thus, several of the results included in the exhaustive tabulation presented by Sehmel (1980b) should be viewed more as indications of experimental error than as determinations of a physical quantity.

Figure C.2-11 addresses the question of the time variation of the aerodynamic resistance, r_a. Values plotted are the maximum deposition velocity permitted by the prevailing aerodynamic resistance, evaluated directly from eddy fluxes of heat and momentum determined during the pine plantation experiment of Figure C.2-10. The reciprocals of the plotted velocities provide evaluations of r_a. It is seen that in daytime, deposition velocities could be as much as 20 cm/s if the surface resistance is zero, implying $r_a \approx 0.05$ s/cm during midday periods. At night, however, r_a can increase to 10 s/cm on infrequent occasions, but often exceeds 0.5 s/cm. The recommendations of Fowler are probably representative of the long-term average.

The importance of diurnal cycles in pollutant deposition and the close relationship with other meteorological quantities is further illustrated by Figure

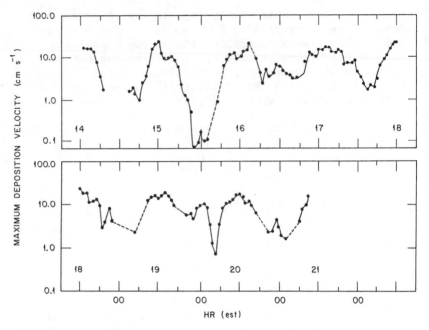

FIGURE C.2-11   Values of the maximum possible deposition velocity of trace gases, determined as the inverse of the aerodynamic resistance r_a for the pine plantation experiment of Hicks and Wesely (1980).

C.2-12, which provides examples of the trend from nighttime, through dawn, and into the afternoon of the residual canopy resistance r_c for ozone and water vapor determined using eddy correlation (Wesely et al. 1978) These data were obtained over corn (<u>Zea mays</u>) in July 1976. The upper sequence shows good matching between r_c for ozone and water vapor, with the former exceeding the latter by a small amount, on the average. As the day progresses, r_c increases gradually, presumably as a consequence of increasing water stress and eventual stomatal closure. The lower data sequence has two features of considerable interest. First, the gradual initial decrease of r_c for O_3 corresponded to a period of evaporation of dewfall (note the relatively low value of r_c for H_2O during the same period), suggesting that the presence of liquid water on the leaf surfaces might inhibit ozone deposition (much as might be expected on the basis of ozone insolubility in H_2O). This would not be the case for SO_2 deposition (Fowler 1978). Second, the peak in both evaluations of r_c at about 10 a.m. is associated with the passage of clouds, which caused a rapid and strong decrease in incoming radiation and lasted for about an hour. The peak is seen as further evidence for stomatal control, since some stomatal closure would be expected with reduced insolation.

The above discussion of both SO_2 and O_3 deposition confirms the generalization made by Chamberlain (1980) that the deposition of such quantities might be modeled after the case of water-vapor transfer with considerable confidence.

Recently, Wesely et al. (1982b) have reported a field

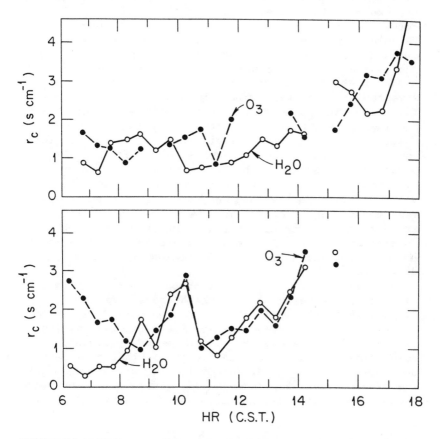

FIGURE C.2-12 Evaluations of the residual "canopy resistance," r_c, to the transfer of ozone and water vapor, based on eddy fluxes measured above mature corn in central Illinois on 29 July 1976 (upper sequence) and 30 July 1976 (lower sequence). Data are from Wesely et al. (1978).

study in which both O_3 and NO_2 fluxes were measured.
Bulk canopy resistances to ozone uptake for a soybean
canopy exceeded water-vapor values by about 0.5 cm/s
during daytime, with r_c for NO_2 still greater by a
similar amount.

2.3.2 Particulate Pollutants

No technique for measuring particle fluxes has been
developed to the extent necessary to provide universally
accepted data. In every case, research and development
are continuing at this time. Use of gradient methods,
for example, is limited by the inability to resolve
concentration differences of the order of 1 percent.
Turbulence methods require rapid response, yet sensitive
chemical sensors that are not often available. In both
cases, practical application is hindered by the need for
a site meeting stringent micrometeorological criteria.
Nevertheless, several applications of micrometeorological
flux-measuring methods have been published. Table C.2-6
provides a list that illustrates the narrow range of
available information. The evidence points to a difference between the deposition characteristics of small
particles and sulfate; the latter seems to be transferred
with deposition velocities somewhat greater than the
value of 0.1 cm/s that has been assumed in most assessment studies and greater than the values appropriate for
small particles, on the average. At this time, the
possibility that sulfate fluxes are promoted by the
strong effect of a few large particles cannot be
dismissed.

As must be expected, taller canopies are associated
with higher values of v_d, on the average. Figure
C.2-13 shows how small particle fluxes varied with time
of day over a pine plantation in North Carolina during
1977 (Wesely and Hicks 1979). These eddy-correlation
results display a run-to-run smoothness that engenders
considerable confidence; moreover, they are supported by
the finding that simultaneous eddy fluxes of momentum and
heat closely satisfied the usual surface roughness and
energy balance constraints. There is little doubt that
the surface under scrutiny (or at least the air below the
sensor) did indeed represent a source rather than a sink
for substantial periods (Arnts et al. 1978). A basic question then arises about the significance of the measured
deposition rates, since these probably represent a net

TABLE C.2-6 Field Experimental Evaluations of the Deposition Velocity of Submicrometer Diameter Particles

Surface	Size and Method	Results and Comments
Snow		
Dovland and Eliassen (1976)	Lead aerosol, surface sampling	0.16 cm/s in stable stratification, greater values in neutral. All light-wind data
Wesely and Hicks (1979)	0.05-0.1 μm particles eddy correlation	Net fluxes small but upwards; v_d too small to be determined
Open Water		
Sievering et al. (1979)	0.2-1.0 μm particles, gradients	Gradients highly variable. Range of v_d typically 0.2-1.0 cm/s in magnitude. Including reversed gradients in long-term average reduces average v_d to near zero (see Hicks and Williams 1979)
Williams et al. (1978)	0.05-0.1 μm particles, eddy correlation	Preliminary indications only: v_d very small, 95% certainty < 0.05 cm/s
Bare Soil		
Wesely and Hicks (1979)	0.05-0.1 μm particles, eddy correlation	Surface frequently a source v_d very low, on the average but often large for short periods
Grass		
Sehmel et al. (1973b)	Polydispersed rhodamine-B particles with mass median diameter 0.7 μm, deposited to artificial grass exposed outdoors	Average $v_d \simeq 0.2$ cm/s

TABLE C.2-6 (Continued)

Surface	Size and Method	Results and Comments
Chamberlain (1960)	Radon daughters deposited to natural grass. Work attributed to Megaw and Chadwick	$v_d \approx 0.20$ cm/s
Hudson and Squires (1978)	Cloud condensation nuclei fluxes measured by gradient methods over sagebrush and grass. Particle size probably 0.002–0.04 μm	$v_d \approx 0.04$ cm/s
Davidson and Friedlander (1978)	~0.03-μm particles, gradients over wild oats	Average $v_d \approx 0.9$ cm/s
Wesely et al. (1977)	0.05–0.1 μm particles, eddy correlation	Direction of flux sometimes changes. During deposition periods, $v_d \approx 0.8$ cm/s, but much lower on the average
Everett et al. (1979)	Particulate lead and sulfur, gradients	v_d greater for sulfur (~1 cm/s) than for lead from more local sources
Sievering (1982)	0.15–0.3 μm particle gradients over mature rye and wheat	v_d averaged 0.4 ± 0.3 cm/s in light winds, unstable stratification
Hicks et al.(1982)	Sulfate by eddy correlation	v_d as high as 0.7 cm/s in daytime, about 0.2 cm/s as a long-term average
Wesely et al. (1982)	Sulfate by eddy correlation	v_d largest for daytime lush grass (~0.5 cm/s), much less for short dry grass (~0.2 cm/s), strongly

Crops

Reference	Measurement	Results
Droppo (1980)	Particulate trace metals, gradients: senescent maize	v_d varying widely with element, ranging up to about 1 cm/s
Wesely and Hicks (1979)	0.05–0.1 μm particles, eddy correlation: senescent maize	Strong diurnal variation in the direction of the flux. Long-term average $v_d \approx 0.1$ cm s^{-1}

Trees

Reference	Measurement	Results
Hicks and Wesely (1978, 1980)	Sulfate particles, eddy correlation, Loblolly pine	Strong diurnal variability but less marked than for small particles; average $v_d \approx 0.7$ cm/s
Wesely and Hicks (1979)	0.05–0.1 μm particles, eddy correlation	Very strong diurnal variation with the canopy a net source. During deposition periods, v_d probably greater than 0.6 cm s^{-1}
Lindberg et al. (1979)	Pb, Cd, S, etc. particles, foliar washing	$v_d > 0.1$ cm/s for all quantities, on the average
Wesely et al. (1982)	Sulfate particles, eddy correlation	v_d not significantly different from zero for a winter deciduous forest

result of continuing but varying surface emission and a
deposition flux that is also varying with time. The
relevance of the answers obtained is then unclear. In
particular, it is not obvious how to relate such results
to the common situation in which we wish to evaluate the
atmospheric deposition rate of some particulate pollutant
that is not emitted or resuspended from the surface.

Figure C.2-13 identifies periods of the 1977 pine
plantation study during which no gaseous sulfur was
detectable. These occasions were used by Hicks and
Wesely (1978) to evaluate residual canopy resistances for
particulate sulfur that averaged about 1.5 cm/s (with a
standard error margin of about ±15 percent) for July 17
and about 1.1 cm/s (±25 percent) for July 18.

Results of two tests of sulfate gradient equipment
over arid grassland, reported by Droppo (1980), yield
values of 0.10 and 0.27 cm/s for v_d, in very light
winds (≈1 m/s). The residual surface resistances
evaluated from his data are 7.7 and 3.3 cm/s, respectively. These values are considerably higher than the
pine plantation results quoted above but might not be
wholly discordant when the nature of the surface present
in the gradient studies is taken into account.

Results of an extensive series of eddy-correlation
measurements of particulate sulfur fluxes to a variety of
vegetated surfaces have been summarized by Wesely et al.
(1982a). In daytime conditions, deposition velocities to
grass range from about 0.2 to 0.5 cm/s. Values for a
deciduous forest in winter (few leaves) are not
significantly different from zero. In general, somewhat
lower values are appropriate at night. In almost all of
the cases summarized by Wesely et al., normalization of
surface transfer conductances by u_* appears to reduce
the residual variance. Hicks et al. (1982) present
supporting data from another study of the same series,
also over grassland.

Considerable controversy remains concerning the value
of v_d appropriate for formulating the deposition of
sulfate aerosol (and presumably all similar particles).
Garland (1978) advocates the continued use of values of
0.1 cm/s or less, since experiments conducted over grass
in England failed to detect a significant gradient.
However, some of the experiments listed in Table C.2-6
indicate quite high deposition velocities for sulfate
particles. A possible explanation in terms of a strong
contribution by particles much larger than the usual
accumulation size mode has been discussed (Garland 1978),
and different deposition velocities (0.025 and 0.56 cm/s)

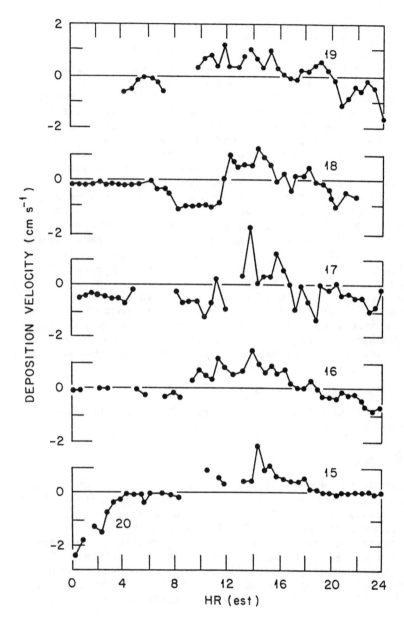

FIGURE C.2-13   Deposition velocity of small particles (~0.1 μm) measured by eddy correlation above a pine plantation in North Carolina in 1977 (Hicks and Wesely 1978, Wesely and Hicks 1979). Note the strong diurnal cycle, with frequent extended periods of emission rather than deposition (as indicated by the negative "deposition velocities").

have been postulated for the submicrometer and larger particles, respectively.

There are great uncertainties about results obtained by deposition plates or other surrogate collection surfaces. All such methods assume that the collection characteristics of some artificial surface are the same as those of the natural surface of interest. Clearly, this assumption will be valid when particles are sufficiently large that gravity is the controlling factor. However, small particles are transferred predominantly by turbulence, with subsequent impaction on the surface of microscale surface roughness elements; these latter factors are not easily reproduced by commonly used artificial collecting devices. The use of collection vessels to monitor the accumulation of particles in them continues to be a widespread practice; however, relating the data obtained to natural circumstances is difficult (q.v. Hicks et al. 1981). In a special category of its own, however, is the method of foliar washing, as used by Lindberg et al. (1979). As applied in careful studies of particle dry deposition at the Walker Branch Watershed in Tennessee, this method of removing and analyzing material deposited on vegetation has succeeded in demonstrating long-term average values of v_d larger than the usually accepted values for several elements.

2.4 MICROMETEOROLOGICAL MODELS OF THE DRY-DEPOSITION PROCESS

2.4.1 Gases

Almost all models of dry deposition of trace gases have as their foundation either the resistance analogy illustrated in Figures C.2-5 and C.2-6 or some equivalent to it. The convenience of this approach is obvious: it permits separate processes to be formulated and combined in a manner that mimics nature, while providing a clear-cut mechanism for determining which processes can be omitted from consideration in specific circumstances. The relevance of the resistance approach to the matter of particle deposition is not so obvious, especially when gravitational settling must be considered.

It is useful to start by identifying the gaseous properties of interest for gases and to identify possible controlling processes.

SO_2: Uptake by plants is largely via stomates during daytime, with about 25 percent apparently via the epidermis of leaves (Fowler 1978). At night, stomatal resistance will increase substantially, but cuticular resistance should be unchanged. When moisture condenses on the depositing surface, associated resistances to transfer should be allowed to decrease to near zero (Fowler 1978, Murphy 1976). To a liquid-water surface, water-vapor appears to provide an acceptable analogy to SO_2 flux.

O_3: Behavior is like SO_2, but with significant cuticular uptake at night ($r_{cut} \sim 2-2.5$ cm/s at night, see r_c quoted by Wesely et al. 1982) and with surface moisture effectively minimizing uptake. Deposition to water surfaces, in general, is very slow.

NO_2: Similar to O_3 in overall deposition characteristics, but with a significant additional resistance (possibly mesophyllic, see Wesely et al. 1982b) of about 0.5 cm/s. Even though NO_2 is insoluble in water in low concentrations, deposition to water surfaces might be quite efficient. Chamber studies (Table C.2-4) indicate similar overall surface resistances for SO_2 and NO_2.

NO: Typical canopy resistances are in the range 5-20 cm/s, as indicated by chamber studies (Table C.2-4) and field experiments (Wesely et al. 1982b). NO appears to be emitted by surfaces at times, possibly as a consequence of NO_2 deposition and of the intimate linkage with ozone concentrations (Galbally and Roy 1980).

HNO_3: No direct information is available; however, on the basis of its high solubility and chemical reactivity, substantial similarity to HF should be expected. Consequently, the use of $r_c = 0$ appears to be a reasonable first approximation.

NH_3: Again, no direct measurements are available but in this case similarity with SO_2 appears likely. Natural surfaces may be emitters of NH_3 because of a number of biological processes occurring in and on soil.

Variations in aerodynamic resistance must be expected
to modulate all the behavior patterns summarized above.
In many circumstances, deposition rates at night will be
nearly zero solely because atmospheric stability is so
great that material cannot be transferred through the
lower atmosphere. The evaluations given in Figure C.2-12
are especially informative, since even over a pine forest
whose surface roughness operates to maximize v_d, an
occasion was encountered on one evening out of eight in
which atmospheric stability was sufficient to constrain
the deposition velocity of all airborne material to less
than 0.1 cm/s (with the exception of gravitationally
settling particles).

Some models focus on micrometeorological aspects of
the pollutant transfer question, others on the biological.
Meteorological models tend to follow the lead of agricultural workers. Shreffler (1976) considers profiles of
pollutant concentration above and through a canopy, making
use of the familiar concepts of zero-plane displacement,
leaf area density, and aerodynamic roughness. Interfacial transfer is formulated as recommended by Brutsaert
(1975a). The results are shown to agree, in general
form, with the experimental data of Chamberlain (1966,
for Th-B).

Roth (1975) also uses micrometeorological relations to
investigate gaseous dry deposition, adopting the general
resistance format and emphasizing the time dependency of
the aerodynamic resistance and surface (i.e., canopy)
resistance.

Brutsaert (1975b) applies theory to cases of sensible
heat and water vapor and extends them to consider a
general scalar quantity. He emphasizes that the common
micrometeorological assumption that the roughness length
is the same for all quantities can lead to considerable
error.

O'Dell et al. (1977) consider details of leaf uptake
mechanisms, with emphasis on stomatal and mesophyllic
resistances. The model addresses questions of plant
physiology in detail and is intended to permit comparison
of uptake rates of different gaseous species, once air
concentrations near leaves are known.

None of these above models considers all the processes
that have been discussed above, nor is it likely that any
model will have sufficient generality to permit its use
for all pollutants in all circumstances. To circumvent
many of the difficulties involved, Sheih et al. (1979)
prepared a land-use map of North America and coupled this
with information regarding surface conditions in order to

derive a spatial distribution of surface resistances
appropriate for formulating the deposition of sulfur
dioxide. By coupling these values with aerodynamic
resistances characteristic of different stability
conditions and for different seasons, they produced a map
of SO_2 deposition velocities suitable for use in
numerical models and for interpreting concentration
data. This approach has not been extended to other
gaseous pollutants.

2.4.2 Particles

Modeling of particle deposition is complicated by three
major factors: (1) gravitational settling, which causes
particles to fall through the atmospheric turbulence that
provides the conceptual basis for conventional micrometeorological models (Yudinge 1959); (2) particle
inertia, which permits particles to be projected through
the near-surface laminar layer by turbulence but also
prohibits them from responding to the high-frequency
turbulent motions that transport material near receptor
surfaces; and (3) uncertainty regarding the processes
that control particle capture. These three factors are
interrelated in such a manner that clear-cut differentia-
tion of their separate consequences is not possible.
    The problem has attracted the attention of many
theoreticians, and many numerical models have been
developed. Each model represents a selected combination
of processes, chosen for consideration on the basis of
the modeler's understanding of the problem. Without
adequate consideration of all the mechanisms involved,
none of these models can be considered as a simulator of
natural behavior. This is not to question the worth of
such models, but rather to emphasize that each should be
applied with caution and only to those situations
commensurate with its own assumptions.
    The many numerical models can be classified in several
different ways. Some extend chemical engineering results
to surface geometries that are intended to represent plant
communities. Others extend agrometeorological air-canopy
interaction models by including critical aspects of
aerosol physics. Both approaches have benefits, and the
final solution will probably include aspects of each.
    An excellent review of model assumptions has been
given by Davidson and Friedlander (1978). They trace the
evolution of models from the 1957 work of Friedlander and
Johnstone, which concentrated on the mechanism of inertial

impaction and assumed that particles shared the eddy
diffusivity of momentum, to the canopy filtration models
of Slinn (1974) and Hidy and Heisler (1978). Early work
concerned deposition to flat surfaces and made various
assumptions about the surface collection process.
Friedlander and Johnstone (1957) permitted particles to
be carried by turbulence to within one free-flight
distance of the surface, upon which they were assumed to
be impacted by inertial penetration of the quasi-laminar
"viscous" sublayer. Beal (1970) introduced viscous
effects to limit the transfer of small particles, while
retaining inertial impaction of larger particles. Sehmel
(1970) assumed that all particles that contact the
surface will be captured and used empirical evidence
obtained in his wind-tunnel studies to determine the
overall resistance to transfer, assumed to apply at a
distance of one particle radius from the surface.
Sehmel's work has been updated recently to provide an
estimate of deposition velocities to canopies of a range
of geometries in different meteorological conditions (see
Sehmel, 1980a).

The above models are based largely on observations and
theory regarding the deposition of particles to smooth
surfaces, usually of pipes. More micrometeorologically
oriented models have been presented by workers such as
Chamberlain (1967), who extended the familiar meteorological concepts of roughness length and zero plane
displacement to the case of particle fluxes. Much of
this work was considered as an extension of models
developed for the case of gaseous deposition to
vegetation, which in turn were based on an extensive
background of agricultural and forest meteorology,
especially concerning evapotranspiration. A recent
development of this genre is the canopy model of Lewellen
and Sheng (1980), which utilizes recent techniques in
turbulence modeling to reproduce the main features of
subcanopy flow and combines these with particle-
deposition formulations like those represented in Figure
C.2-1. Lewellen and Sheng emphasize their model's
omission of several potentially critical mechanisms,
especially electrical migration, coagulation, evolution
of particle size distributions, diffusiophoresis, and
thermophoresis. To this list we can add a number of
other factors about which little is known at this time
such as subcanopy chemical reactions, interactions with
emissions, and the effect of microscale roughness
elements.

Although outwardly simpler than the case of particle deposition to a canopy, that to a water surface has given rise to a similar variety of models. Once again, however, different models focus on different mechanisms. That of Sehmel and Sutter (1974) is based on their wind-tunnel observations and lacks a component that can be identified with wave effects. Slinn and Slinn (1980) invoke the rapid growth of hygroscopic aerosol particles in very humid air to propose rather rapid deposition to open water; deposition velocities of the order of 0.5 cm/s appear possible in this case. On the other hand, Hicks and Williams (1979) propose negligible fluxes unless the surface quasi-laminar layer is interrupted by breaking waves. At present, none of these models has strong experimental evidence to support it. However, experimental and theoretical studies are proceeding, and a resolution of the matter can certainly be expected.

2.5 CONCLUSIONS TO SECTION 2

All the many processes that combine to permit airborne materials to be deposited at the surface have aspects that are strongly surface dependent. While broad generalities can be made about the velocities of deposition of specific chemical species in particular circumstances, there will be wide temporal and spatial variabilities of most of the controlling properties. The detailed nature of the vegetation covering the surface is often a critical consideration. If depositional inputs to some special sensitive area need to be estimated, then this can only be accomplished if characteristics specific to the vegetation cover of the area in question are taken into account in an adequate manner.

Recent field studies investigating the fluxes of small particles have confirmed wind-tunnel results that point to a surface limitation. Studies of the rate of deposition of particles to the internal walls of pipes and investigations of fluxes to surfaces more characteristic of nature, exposed in wind tunnels, tend to confirm theoretical expectations that surface uptake is controlled by the ability of particles to penetrate a quasi-laminar layer adjacent to the surface in question. The mechanisms that limit the rate of transfer of particles involve their finite mass. Particles fail to respond to the high-frequency turbulent fluctuations that cause transfer to take place in the immediate vicinity of a surface. However, the momentum of particles also causes

an inertial deposition phenomenon that serves to enhance
the rate of deposition of particles in the 10- to 20-µm
size range.

The general features of particle deposition to smooth
surfaces are fairly well understood. Studies conducted
so far support the theoretical expectation that particles
smaller than about 0.1 µm in diameter will be deposited
at a rate that is largely determined by Brownian diffusivity. In this instance, the limiting factor is the
transfer by Brownian motion across the quasi-laminar
layer referred to above. On the other hand, particles
larger than about 20 µm in diameter are effectively
transferred via gravitational settling, at rates determined by the familiar Stokes-Cunningham formulation.
Particles in the intermediate size ranges are transferred
very slowly. The minimum value of the "well" of the
deposition velocity versus particle size curve is
approximately 0.001 cm/s.

However, natural surfaces are rarely aerodynamically
smooth. Wind-tunnel studies have shown that the "well"
in the deposition velocity curve is filled in as the
surface becomes rougher. Although studies have been
conducted, in wind tunnels, of deposition fluxes to
surfaces such as gravel, grass, and foliage, the
situation involving natural vegetation such as corn, or
even pasture, remains uncertain. It is well known that
many plant species have foliage with exceedingly complicated microscale surface roughness features. In
particular, leaf hairs increase the rate of particle
deposition; studies of other factors, such as electrical
charges associated with foliage and stickiness of the
surface, indicate that a natural canopy might be considerably different from the simplified surfaces suitable
for investigation in laboratory and wind-tunnel
investigations.

Caution should be exercised in extending laboratory
studies using artificially produced aerosol particles to
the situation of the deposition of acidic quantities.
Special concern is associated with the hygroscopic nature
of many acidic species. Their growth as they enter a
region of high humidity and their liquid nature when they
strike the surface are both potentially important factors
that might work to increase otherwise small deposition
velocities. Moreover, there is evidence that acidic
particles, especially sulfates, might be carried by
larger particles; the rates of deposition of such
complicated particle structures are essentially unknown.
However, the shape of particles can have a considerable

influence on their gravitational settling speed and probably on their impaction characteristics as well.

It is not clear to what extent special considerations appropriate for acidic species, such as those mentioned above, contribute to the finding of unexpectedly high deposition velocities for atmospheric sulfate particles, as reported in some recent North American studies. European work has been fairly uniform in producing deposition velocities close to 0.1 cm/s, while North American experience has generated larger values.

It is informative to consider the flux of any airborne quantity to the surface underneath in terms of an electrical analog, the so-called resistance model developed initially in studies of agrometeorology. In this model, the flux of the atmospheric property in question is identified with the flow of current in an electrical circuit; individual resistances can then be associated with readily identifiable atmospheric and surface properties. While the electrical analogy has obvious shortcomings, it permits an easy visualization of many contributing processes and enables a comparison of their relative importance. Micrometeorological studies of the fluxes of atmospheric heat and momentum show that the aerodynamic resistance to transfer (i.e., the resistance to transfer between some convenient level in the air and a level immediately above the quasi-laminar layer) ranges from between 0.1 s/cm in strongly unstable, daytime conditions, to more than 10 s/cm in many nocturnal cases. There are several resistance paths that permit gaseous pollutants to be transferred into the interior of leaves. An obvious pathway is directly through the epidermis of leaves, involving a cuticular resistance. An alternative route, known to be of significantly greater importance in many cases, is via the pores of leaves, involving a stomatal resistance that controls transfer to within stomatal cavities and a subsequent mesophyllic resistance that parameterizes transfer from substomatal cavities to leaf tissue. Comparison between resistances to transfer for water vapor, ozone, sulfur dioxide, and gases that are similarly soluble and/or chemically reactive shows that in general such quantities are transferred via the stomatal route, whenever stomates are open. Otherwise, cuticular resistance appears to play a significant role. Cuticular uptake of ozone and of quantities like NO and NO_2 appears to be quite significant, whereas for SO_2 this pathway appears to be less important. When leaves are wet, such as after heavy dewfall, uptake of sulfur dioxide is exceedingly efficient until the pH of the surface water becomes sufficiently acidic to impose a chemical limit on the rate of absorption of gaseous SO_2. However, the insolubility of ozone causes dewfall to inhibit ozone dry deposition.

The same conceptual model can be applied to the case of particle transfer with considerable utility. While the roles of factors such as stomatal opening become less clear when particles are being considered, the concept of a residual surface resistance to particle uptake appears to be rather useful. Studies of the transfer of sulfate particles to a pine forest have shown that this residual surface resistance is of the order of 1 to 2 s/cm. It appears probable that substantially larger values for residual surface resistance will be appropriate for nonvegetated sufaces, especially to snow, for which the values are more likely to be approximately 15 s/cm. At this time, an exceedingly limited quantity of field information is available; however, it appears that in North American conditions the surface resistance to uptake of sulfate particles will be in the range 1.5 to 15 s/cm.

While sulfate particles have received most of the recent emphasis, the general question of acid deposition requires that equal attention be paid to nitrate and ammonium particles. There is little information regarding the deposition velocities of these particles. However, there is no strong indication that they are different from the case of sulfate.

Regarding trace-gas uptake, sulfur dioxide has received the majority of recent attention. Chamber studies and some recent field work indicate that highly reactive materials such as hydrogen fluoride (and presumably iodine vapor, nitric acid vapor, etc.) are readily taken up by a vegetative surface, whereas a second set of pollutants including SO_2, NO_2, and O_3 seem to be easily transferred via stomates, and a third category of relatively unreactive trace gases are poorly taken up.

Transfer to water surfaces presents special problems, especially when the surface concerned is snow. As mentioned above, surface resistances to particle uptake by snow appear to be of the order 15 s/cm. Soluble gases will be readily absorbed by all water surfaces, and so equivalence to transfer of water vapor might be expected. An important exception occurs in the case of SO_2, in which case absorbed SO_2 can increase the acidity of the surface moisture layer to the extent that further SO_2 transfer is cut off. Trace-gas transfer to liquid-water surfaces is influenced by the Henry's law constant.

Wind-tunnel studies of particle transfer to water surfaces all show exceedingly small deposition velocities for particles in the 0.1- to 1-μm size range. Several workers have suggested mechanisms by which larger deposition velocities might exist in natural circumstances; for example, the growth of hygroscopic particles in highly humid, near-surface air can cause accelerated deposition of such particles, and breaking waves might provide a route that bypasses the otherwise-limiting quasi-laminar layer in contact with a water surface. Once again field observations are lacking.

While larger deposition velocities of soluble trace gases to open water surfaces appear quite likely, water bodies are frequently sufficiently small that air-surface thermal equilibrium cannot be achieved. Air blowing from warm land across a small cool lake, for example, will not rapidly equilibrate with the smooth, cooler surface. Flow will then be stable and largely laminar, with the consequence that very small deposition velocities will apply for all atmospheric quantities. In many circumstances, especially in daytime summer occasions, deposition velocities are likely to be so small as to be disregarded for all practical purposes. On the other hand, during winter when the land surface is frequently cooler than the water, the resulting corrective activity over small water bodies will cause the air to come into fairly rapid equilibrium with the water, and rather high deposition velocities, in agreement with the open water surface expectations, will probably be attained.

An associated special case concerns the effect of dewfall, which can accelerate the net transfer of trace gases and particles in some circumstances. The velocities of deposition involved are small, however they permit an accumulation of material at the surface in conditions in which the atmospheric considerations are likely to predict minimal rates of exchange (i.e., limited by stability to an extreme extent). When surface fog exists, the highly humid conditions will permit airborne hygroscopic particles to nucleate and grow rapidly. This process provides a mechanism for cleansing the lower layers of the atmosphere of most acidic airborne particles. The small fog droplets that are formed around the hygroscopic acidic nuclei are transferred by the classical process of fog interception to foliage and other surface roughness elements.

Conclusions of recent workshops (e.g., Hicks et al. 1981) have indicated that it is not possible to measure the dry deposition of acidic atmospheric materials using exposed collection vessels, since they fail to collect trace gases and small particles in a manner that can be related in a direct fashion to natural circumstances. However, surrogate surface methods appear to be useful in indicating space and time variations of deposition in some cases. It is possible to measure the flux of some airborne quantities by micrometeorological means, without interfering with the natural processes involved. These studies, and laboratory and wind-tunnel investigations, provide evidence that the controlling properties in the deposition of many gaseous pollutants are associated with surface structure, rather than with atmospheric properties. The exception to this generalization is the nocturnal case, in which atmospheric stability may often be sufficient to impose a severe restriction on the rate of delivery of all airborne quantities to the surface below.

The conclusions presented above can be summarized as follows:

• Dry deposition of small aerosol particles and trace gases is a consequence of many atmospheric, surface, and pollutant-related processes, any one of which may dominate under some set of conditions. The complexity of each individual process makes it unlikely that a comprehensive simulation will be developed in the near future.

• The convenient simplicity afforded by the concept of a deposition velocity (or its inverse, the total resistance to transfer) makes it possible to incorporate dry-deposition processes in models in a manner that is adequate for modeling and assessment purposes. The simplicity of the deposition velocity approach imposes obvious limitations on its applications. For example, the use of average deposition velocities is inappropriate when it is desired to look at time- or space-resolved details of deposition fluxes.

• Sufficient is known about the processes that control the deposition of trace gases that in many instances deposition velocities can be considered to be known functions of properties such as wind speed, atmospheric stability, surface roughness, and biological factors such as stomatal aperture. Important exceptions are for the case of insoluble (or poorly soluble) gases and for deposition to nonsimple surfaces such as forests in rough terrain.

• The deposition of particles larger than about
20-μm diameter is controlled by gravity and can be
evaluated using the Stokes-Cunningham relationship.
These large particles might contribute to the deposition
of acidic and acidifying substances.

• The deposition of small particles remains an
issue of considerable disagreement. On the whole, model
predictions agree with the results of laboratory and
wind-tunnel studies, at least for test surfaces that are
usually smoother than pasture, but field experiments
provide data that indicate greater deposition velocities.
The reasons for the apparent disagreement are not yet
clear.

• Over water surfaces, there are almost no field
data on the deposition of small particles. Different
models have been put forward, predicting a wide range of
deposition velocities. At this time, there is little
evidence that would permit us to choose between them.
The situation for trace gases like sulfur dioxide and
ammonia is much better. On the whole, models agree with
the available field data, although there is disagreement
between the models on how factors such as molecular
diffusivity should be handled.

• Dry deposition to the surfaces of materials used
in the construction of buildings and monuments, for
example, can be measured in many instances by taking
sequential samples of the surface over extended periods.

• Particulate material at the surface can creep,
bounce, and eventually resuspend under the influence of
wind gusts. The large particles entrained in this way
can cause a local modification of the acid deposition
phenomenon that is associated with accumulation-size
aerosol particles and trace gases of more distant origin.

• For both case-study measurement purposes and for
long-term monitoring, accurate measurements of pollutant
air concentrations are necessary. For monitoring purposes, measurement of airborne pollutant concentrations
in a manner carefully designed to permit evaluation of
dry-deposition rates by applying time-varying deposition
velocities specific to the pollutant and site in question
appears to be the most attractive option.

• Micrometeorological methods for measuring dry-
deposition fluxes have been developed from the techniques
conventionally used to determine fluxes of sensible heat,
moisture, and momentum. These methods are techno-
logically demanding, and their use in routine monitoring
applications is not yet possible.

2.6 REFERENCES

Alexander, L.T. 1967. Does salt water spray trap strontium-90 from the air? U.S.A.E.C. Health and Safety Laboratory Quarterly Summary Report HASL-181, I-21-I-24.

Arnts, R.R., R.L. Seila, R.L. Kuntz, F.L. Mowry, K.R. Knoerr, and A.C. Dudgeon. 1978. Measurement of α-pinene fluxes from a loblolly pine forest. Proceedings Fourth Joint Conference on Sensing of Environmental Pollutants. Washington, D.C.: American Chemical Society, pp. 829-833.

Bagnold, R.A. 1954. The Physics of Blown Sand and Desert Dunes. London: Methuen and Company.

Barrie, L.A., and J.L. Walmsley. 1978. A study of sulphur dioxide deposition velocities to snow in northern Canada. Atmos. Environ. 12:2321-2332.

Batchelor, G.K. 1967. An Introduction to Fluid Mechanics. New York: Cambridge University Press. 615 pp.

Beal, S.K. 1970. Deposition of particles in turbulent flow on channel or pipe walls, Nucl. Sci. Eng. 40:1-11.

Bean, B.R., R.F. Gilmer, R.L. Grossman, and R. McGavin. 1972. An analysis of airborne measurements of vertical water vapor flux during BOMEX. J. Atmos. Sci. 29:860-869.

Belot, Y. 1975. Etude de la captation des pollutants atmospheriques par les vegetaux, C.E.A., Fontenay-aux-Roses, France.

Billings, C.E., and R.A. Gussman. 1976. Dynamic behavior of aerosols, pp. 40-65 of Handbook on Aerosols. R.E. Dennis, ed. U.S. ERDA TIC-26608, 142 pp.

Bowden, F.P., and D.Tabor. 1950. The Friction and Lubrication of Solids. Oxford: Clarendon Press.

Brimblecombe, P. 1978. Dew as a sink for sulphur dioxide. Tellus 30:151-157.

Brutsaert, W.H. 1975a. The roughness length for water vapor, sensible heat, and other scalars. J. Atmos. Sci. 32:2028-2031.

Brutsaert, W.H. 1975b. A theory for local evaporation (or heat transfer) from rough and smooth surfaces at ground level. Water Resources Res. 11:543-550.

Cadle, R.D. 1966. Particles in the Atmosphere and Space. New York:Reinhold. 226 pp.

Chamberlain, A.C. 1960. Aspects of the deposition of radioactive and other gases and particles. Int. J. Air Pollut. 3:63-88.

Chamberlain, A.C. 1966. Transport of gases to and from grass and grass-like surfaces. Proc. Roy. Soc. London A 290:236-265.

Chamberlain, A.C. 1967. Transport of Lycopodium spores and other small particles to rough surfaces. Proc. Roy. Soc. London A 296:45-70.

Chamberlain, A.C. 1975. The movement of particles in plant communities. Pp. 155-203 in Vegetation and the Atmosphere, Volume 1, Principles. J.L. Monteith, ed. London: Academic Press.

Chamberlain, A.C. 1980. Dry deposition of sulfur dioxide. Pp. 185-197 in Atmospheric Sulfur Deposition. D.S. Shriner, C.R. Richmond, and S.E. Lindberg, eds. Ann Arbor, Mich.: Ann Arbor Science Publishers.

Chamberlain, A.C. 1982. Roughness length of sea, sand and snow. Boundary-Layer Meteorol.

Chang, S.G., R. Toossi, and T. Novakov. 1981. The importance of soot particles and nitric acid in oxidizing SO_2 in atmospheric aqueous droplets. Atmos. Environ. 15:1287-1292.

Cofer, W.R., D.R. Schryer, and R.S. Rogowski. 1981. The oxidation of SO_2 on carbon particles in the presence of O_2, NO_2 and N_2O. Atmos. Environ. 15:1281-1286.

Corn, M. 1961. The adhesion of solid particles to solid surfaces.

Dannevik, W.P., S. Frisella, L. Granat, and R.B. Husar. 1976. SO_2 deposition measurements in the St. Louis region, pp. 506-511 of Proceedings, Third Symposium on Atmospheric Turbulence, Diffusion, and Air Quality, Raleigh, N.C. Boston, Mass.: American Meteorological Society.

Dasch, J.M. 1982. A comparison of surrogate surfaces for dry deposition collection, Proceedings, Fourth International Conference on Precipitation Scavenging, Dry Deposition, and Resuspension, Santa Monica, California, 29 November-3 December.

Davidson, C.I., and S.K. Friedlander. 1978. A filtration model for aerosol dry deposition: application to trace metal deposition from the atmosphere. J. Geophy. Res. 83:2343-2352.

Davies, C.N. 1966. Deposition from moving aerosols. Pp. 393-446 in Aerosol Science. C. N. Davies, eds. New York: Academic Press. 468 pp.

Davies, C.N. 1967. Aerosol properties related to surface contamination. Pp. 1-5 of Surface Contamination. B.R. Fish, ed. New York: Pergamon Press. 415 pp.

Derjaguin, B.V., and Yu.I. Yalamov. 1972. The theory of thermophoresis and diffusiophoresis of aerosol particles and their experimental testing. Pp. 1-200 of Topics in Current Aerosol Research, Part 2. G.M. Hidy and J.R. Brock, eds. New York: Pergamon Press. 384 pp.

Desjardins, R.L. 1977. Energy budget by an eddy correlation method. J. Appl. Meteorol. 16:248-250.

Desjardins, R.L., and E.R. Lemon. 1974. Limitations of an eddy-correlation technique for the determination of the carbon dioxide and sensible heat fluxes. Boundary-Layer Meteorol. 5:475-488.

Dillon, P.J., D.S. Jeffries, and W.A. Scheider. 1982. The use of calibrated lakes and watersheds for estimating atmospheric deposition near a large point source, Water, Air Soil Pollut.

Dolske, D.A., and D.F. Gatz. 1982. A field intercomparison of sulfate dry deposition monitoring and measurement methods: preliminary results, Proceedings, American Chemical Society Acid Rain Symposium, Las Vegas, Nevada, 30 March 1982.

Dovland, H., and A. Eliassen, 1976. Dry deposition on a snow surface. Atmos. Environ. 10:783-785.

Droppo, J.G. 1980. Experimental techniques for dry deposition measurements. Pp. 209-221 of Atmospheric Sulfur Deposition. D.S. Shriner, C.R. Richmond, and S.E. Lindberg, eds. Ann Arbor, Mich.: Ann Arbor Science Publishers. 568 pp.

Dyer, A.J. 1974. A review of flux-profile relationships. Boundary-Layer Meteorol. 7:363-372.

Dyer, A.J., and F.J. Maher. 1965. The "Evapotron". An Instrument for the Measurement of Eddy Fluxes in the Lower Atmosphere, CSIRO (Australia) Division of Meteorological Physics Technical Paper Number 15. 31 pp.

Eastman, J.A., and D.H. Stedman. 1977. A fast response sensor for ozone eddy-correlation measurements. Atmos. Environ. 11:1209-1212.

Eaton, J.S., G.E. Likens, and F.H. Borman. 1978. The input of gaseous and particulate sulfur to a forest ecosystem. Tellus 30:546-551.

Engelmann, R.J., and G.A. Sehmel, eds. 1976. Atmosphere-Surface Exchange of Particulate and Gaseous Pollutants, U.S. ERDA CONF-7409Z1. 988 pp.

Everett, R.G., B.B. Hicks, W.W. Berg, and J.W. Winchester. 1979. An analysis of particulate sulfur and lead gradient data collected at Argonne National Laboratory. Atmos. Environ. 13:931-943.

Fassina, V. 1978. A survey of air pollution and deterioration of stonework in Venice. Atmos. Environ. 121:2205-2211.

Fleischer, R.L., and F.P. Parungo. 1974. Aerosol particles on tobacco trichomes. Nature 250:158-159.

279

Fowler, D. 1978. Dry deposition of SO_2 on agricultural crops. Atmos. Environ. 12:369-373.

Fowler, D., and M. H. Unsworth. 1979. Turbulent transfer of sulphur dioxide to a wheat crop. Q. J. Roy. Meteorol. Soc. 105:767-784.

Friedlander, S.K. 1977. Smoke, Dust, and Haze. New York: John Wiley and Sons. 317 pp.

Friedlander, S.K., and H.F. Johnstone. 1957. Deposition of suspended particles from turbulent gas streams. Ind. Eng. Chem. 49:1151-1156.

Fuchs, N.A. 1964. The Mechanics of Aerosols. New York: Pergamon Press. 408 pp.

Galbally, I.E., and C.R. Roy. 1980. Ozone and nitrogen oxides in the southern hemisphere troposphere. Pp. 431-438 in Proceedings, Quadrennial International Ozone Symposium. 4-9 August 1980, Boulder, Colo. (available from IAMAP).

Galbally, I.E., J.A. Garland, and M.J.G. Wilson. 1979. Sulfur uptake from the atmosphere by forest and farmland. Nature 280:49-50.

Garland, J.A. 1977. The dry deposition of sulphur dioxide to land and water surfaces. Proc. Roy. Soc. London A 354:245-268.

Garland, J.A. 1978. Dry and wet removal of sulphur from the atmosphere. Atmos. Environ. 12:349-362.

Garland, J.A. 1979. Dry deposition of gaseous pollutants. Pp. 95-103 in WMO No. 538. Papers presented at the WMO Symposium on the Long-Range Transport of Pollutants and Its Relation to General Circulation Including Stratospheric/Tropospheric Exchange Processes, Sofia, 1-5 October 1979.

Garland, J.A., W.S. Clough, and D. Fowler. 1973. Deposition of sulphur dioxide on grass. Nature 242:256-257.

Garland, J.A., and J.R. Branson. 1977. The deposition of sulphur dioxide to a pine forest assessed by a radioactive tracer method. Tellus 29:445-454.

Garratt, J.R. 1978. Flux profile relations above tall vegetation. Q. J. Roy. Meteorol. Soc. 104:199-212.

Gauri, K.L. 1978. The preservation of stone. Sci. Am. 238:196-202.

Hanna, S.R., and R.P. Hosker. 1980. Atmospheric removal processes for toxic chemicals, NOAA Technical Memorandum, ERL-ARL-102. 34 pp.

Hardy, E.P., Jr., and J.H. Harley, eds. 1958. Environmental contamination from weapons tests. U.S. AEC Health and Safety Laboratory Report, HASL-42A.

Harriott, P., and R.M. Hamilton. 1965. Solid-liquid mass transfer in turbulent pipe flow. Chem. Eng. Sci. 20:1073.

Hess, G.D., and B.B. Hicks. 1975. The influence of surface effects on pollutant deposition rates over the Great Lakes. Pp. 238-247 in the Proceedings of the Second Federal Conference on the Great Lakes, ICMSE.

Hicks, B.B. 1981. Wet and dry surface deposition of air pollutants and their modeling. Pp. 183-196 in Conservation of Historic Stone Buildings and Monuments. National Materials Advisory Board. Washington, D.C.: National Academy Press. 365 pp.

Hicks, B.B., and P.S. Liss. 1976. Transfer of SO_2 and other reactive gases across the air-sea interface. Tellus 28:348-354.

Hicks, B.B., and M.L. Wesely. 1978. An examination of some micrometeorological methods for measuring dry deposition. U.S. EPA report EPA-600/7-78-116. 27 pp.

Hicks, B.B., and M.L. Wesely. 1980. Turbulent transfer processes to a surface and interaction with vegetation. Pp. 199-207 in Atmospheric Sulfur Deposition. D.S. Shriner, C.R. Richmond, and S.E. Lindberg, eds. Ann Arbor, Mich.: Ann Arbor Science Publishers. 568 pp.

Hicks, B.B., and R.M. Williams. 1979. Transfer and deposition of particles to water surfaces, Chapter 26, pp. 237-244 in Atmospheric Sulfur Deposition. D.S. Shriner, C.R. Richmond, and S.E. Lindberg, eds. Ann Arbor, Mich.: Ann Arbor Science Publishers. 568 pp.

Hicks, B.B., G.D. Hess, and M.L. Wesely. 1979. Analysis of flux-profile relationships above tall vegetation, an alternative view, Q. J. Roy. Meteorol. Soc. 105:1074-1077.

Hicks, B.B., M.L. Wesely, and J.L. Durham. 1981. Critique of methods to measure dry deposition; concise summary of workshop, presented at the 1981 National ACS Meeting, Atlanta, Georgia. Ann Arbor, Mich.: Ann Arbor Science Publishers.

Hicks, B.B., M.L. Wesely, R.L. Coulter, R.L. Hart, J.L. Durham, R.E. Speer, and D.H. Stedman. 1982. An experimental study of sulfur deposition to grassland. Proceedings, Fourth International Conference on Precipitation Scavenging, Dry Deposition, and Resuspension, Santa Monica, California, 29 November-3 December.

Hidy, G.M. 1973. Removal processes of gaseous and particulate pollutants, Chapter 3, pp. 121-176 in Chemistry of the Lower Atmosphere. S.I. Rasool, ed. New York: Plenum Press. 335 pp.

Hidy, G.M., and S.L. Heisler. 1978. Transport and deposition of flowing aerosols, in Recent Developments in Applied Aerosol Technology. D. Shaw, ed. New York: John Wiley and Sons.

Hill, A.C. 1971. Vegetation: a sink for atmospheric pollutants. J. Air Pollut. Control Assoc. 21:341-346.

Hubbard, D.W., and E.N. Lightfoot. 1966. Correlation of heat and mass transfer data for high Schmidt and Reynolds numbers. I/Ec Fundam. 5:370.

Hudson, J.G., and P. Squires. 1978. Continental surface measurements of CCN flux. J. Atmos. Sci. 35:1289-1295.

Jones, E.P., and S.D. Smith. 1977. A first measurement of sea-air CO_2 flux by eddy correlation. J. Geophys. Res. 82:5990-5992.

Judeikis, H.S., and T.B. Stewart. 1976. Laboratory measurement of SO_2 deposition velocities on selected building materials and soils. Atmos. Environ. 10:769-776.

Judeikis, H.S., and A.G. Wren. 1977. Deposition of H_2S and dimethyl sulfide on selected soil materials. Atmos. Environ. 11:1221-1224.

Judeikis, H.S., and A.G. Wren. 1978. Laboratory measurements of SO_2 and NO_2 depositions onto soil and cement surfaces. Atmos. Environ. 12:2315-2319.

Kaimal, C.J. 1975. Sensors and techniques for the direct measurement of turbulent fluxes and profiles in the atmospheric surface layer, Atmos. Technol. Number 7, 7-14.

Kanemasu, E.T., M.L. Wesely, B.B. Hicks, and J.L. Heilman. 1979. Techniques for calculating energy and mass fluxes. Chapter 3.2 in Modification of the Aerial Environment of Plants. B.J. Barfield and J.F. Gerber, eds. ASAE Monograph No. 2, pp. 156-182.

Kanwisher, J. 1963. On the exchange of gases between the atmosphere and the sea. Deep-Sea Res. 10:195-207.

Langer, G. 1965. Particle deposition on and re-entrainment from coniferous trees, Part II. Kolloid Z. 204:119-124.

Lenschow, D.H., A.C. Delany, B.B. Stankov, and D.H. Stedman. 1980. Airborne measurements of the vertical flux of ozone in the boundary layer. Boundary-Layer Meteorol. 19:249-308.

Lewellen, W.S., and Y.P. Sheng. 1980. Modeling of Dry Deposition of SO_2 and Sulfate Aerosols, Electric Power Research Institute Report EA-1452, 46 pp. (Available from EPRI, Research Reports Center, Box 50490, Palo Alto, Calif. 94303.)

Lindberg, S.E., R.C. Harris, R.R. Turner, D.S. Shriner, and D.D. Huff. 1979. Mechanisms and Rates of Atmospheric Deposition of Selected Trace Elements and Sulfate to a Deciduous Forest Watershed. Oak Ridge National Laboratory Report ORNL/TM-6674, 514 pp.

Liss, P.S. 1973. Processes of gas exchange across an air-sea interface. Deep-Sea Res. 20:221-238.

Liss, P.S., and P.G. Slater. 1974. Flux of gases across the air-sea interface. Nature 247:181-184.

Liu, B.Y.H., and J.K. Agarwal. 1974. Experimental observation of aerosol deposition in turbulent flow. Aerosol Sci. 5:145-155.

Lovett, G.M., W.A. Reiners, and R.K. Olson. 1982. Cloud droplet deposition in subalpine balsam fir forests: hydrological and chemical inputs. Science 218:1303-1304.

Martell, E.A. 1974. Radioactivity of tobacco trichomes and insoluble cigarette smoke particles. Nature 249:215-217.

Meszaros, A., and K. Vissy. 1974. Concentrations, size distribution and chemical nature of atmospheric aerosol particles in remote oceanic areas. J. Aerosol Sci. 5:101-109.

Mizushina, T., F. Ogino, Y. Oka, and H. Fukuda. 1971. Turbulent heat and mass transfer between wall and fluid streams of large Prandtl and Schmidt numbers. Int. J. Heat and Mass Transfer 14:1705-1716.

Moller, U., and G. Schumann. 1970. Mechanisms of transport from the atmosphere to the earth's surface. J. Geophys. Res. 75:3013-3019.

Monteith, J.L. 1963. Dew, facts and fallacies. In The Water Relations of Plants. A.J. Rutter and F.H. Whitehead, eds. New York: John Wiley and Sons. Pp. 37-56.

Murphy, B.D. 1976. The influence of ground cover on the dry deposition rate of gaseous materials, Oak Ridge National Laboratory Report UCCCND/CSD-19. 28 pp.

O'Dell, R.A., M. Taheri, and R.L. Kabel. 1977. A model for uptake of pollutants by vegetation. J. Air Pollut. Contr. Assoc. 27:1104-1109.

Owers, M.J., and A.W. Powell. 1974. Deposition velocity of sulphur dioxide on land and water surfaces using a S^{35} method. Atmos. Environ. 8:63-67.

Raupach, M.R., J.B. Stewart, and A.S. Thom. 1979. Comments on "Analysis of flux-profile relationships above tall vegetation an alternative view" by Hicks et al. Q. J. Roy. Meteorol. Soc. 105:1077-1078.

Raynor, G.S., E.C. Odgen, and J.V. Hayes. 1970.
Dispersion and deposition of ragweed pollen from
experimental sources. J. Appl. Meteorol. 9:885-895.

Raynor, G.S., E.C. Ogden, and J.V. Hayes. 1971.
Dispersion and deposition of timothy pollen from
experimental sources. J. Appl. Meteorol. 9:347-366.

Raynor, G.S., E.C. Ogden, and J.V. Hayes. 1972.
Dispersion and deposition of corn pollen from
experimental sources. Agronomy J. 64:420-427.

Raynor, G.S., J.V. Haynes, and E.C. Ogden. 1974.
Particulate dispersion into and within a forest.
Boundary-Layer Meteorol. 7:429-456.

Rosinski, J., and C.T. Nagomoto. 1965. Particle
deposition on and reentrainment from coniferous trees,
Part I. Kolloid Z. 204:111-119.

Roth, R., 1975: Der vertikale Transport von
Luftbeimengungen in der Prandtl-Schmidt und die
Deposition-Velocity. Meteorol. Rundsch. 28:65-71.

Sehmel, G.A. 1970. Particle deposition from turbulent
airflow. J. Geophys. Res. 75:1966-1781.

Sehmel, G.A. 1979. Deposition and Resuspension Processes,
Battelle, Pacific Northwest Laboratory Publication
PNL-SA-6746.

Sehmel, G.A. 1980a. Model predictions and a summary of
dry deposition velocity data. Pp. 223-235 of
Atmospheric Sulfur Deposition. D.S. Shriner, C.R.
Richmond, and S.F. Lindberg, eds. Ann Arbor, Mich.:
Ann Arbor Science Publishers. 568 pp.

Sehmel, G.A. 1980b. Particle and gas dry deposition: a
review. Atmos. Environ. 14:983-1012.

Sehmel, G.A., and S.L. Sutter. 1974. Particle deposition
rates on a water surface as a function of particle
diameter and air velocity. J. Rech. Atmos. 3:911-920.

Sehmel, G.A., W.H. Hodgson, and S.L. Sutter. 1973a. Dry
Deposition of Particles, Battelle Pacific Northwest
Laboratory Report BNWL-1850. Part 3, pp. 157-162.

Sehmel, G.A., S.L. Sutter, and M.T. Dana. 1973b. Dry
Deposition Processes, Battelle Pacific Northwest
Laboratories Report BNWL-1751. Part 1, pp. 43-49.

Sheih, C.M., M.L. Wesely, and B.B. Hicks. 1979. Estimated
dry deposition velocities of sulfur over the eastern
United States and surrounding regions. Atmos. Environ.
13:1361-1368.

Shepherd, J.G. 1974. Measurements of the direct
deposition of sulphur dioxide onto grass and water by
the profile method. Atmos. Environ. 8:69-74.

Shreffler, J.H. 1976. A model for the transfer of gaseous
pollutants to a vegetational surface. J. Appl.
Meteorol. 15:744-746.

Sievering, H. 1982. Profile measurements of particle dry deposition velocity at an air-land interface. Atmos. Environ. 16:301-306.

Sievering, H., M. Dave, D.A. Dolske, R.L. Hughes, and P. McCoy. 1979. An experimental study of loading by aerosol transport and dry deposition in the southern Lake Michigan basin. U.S. Environmental Protection Agency Report EPA-905/4-79-016.

Sinclair, P.C. 1976. Vertical transport of desert particulates by dust devils and clear thermals. Pp. 497-527 of Atmosphere—Surface Exchange of Particulate and Gaseous Pollutants. R.J. Engelmann and G.A. Sehmel, eds. U.S. ERDA CONF-740921. 988 pp.

Slinn, W.G.N. 1974. Analytical Investigations of Inertial Deposition of Small Aerosol Particles from Laminar Flows into Large Obstacles--Parts A and B, PNL Ann. Report to the USAEC, DBER, 1973; BNWL-1850, Pt. 3, Battelle—Northwest, Richland, Wash. Available from NTIS, Springfield, Va.

Slinn, W.G.N. 1976a. Formulation and a solution of the diffusion, deposition, resuspension problem. Atmos. Environ. 10:763-768.

Slinn, W.G.N. 1976b. Dry deposition and resuspension of aerosol particles--a new look at some old problems. Pp. 1-40 of Atmospheric-Surface Exchange of Particulate and Gaseous Pollutants--1974. R.J. Englemann and G.A. Sehmel, Coords. Available as ERDA CONF-740921 from NTIS, Springfield, Va.

Slinn, S.A., and W.G.N. Slinn. 1980. Predictions for particle deposition and natural waters. Atmos. Environ. 14:1013-1016.

Slinn, W.G.N., L. Hasse, B.B. Hicks, A.W. Hogan, D. Lal, P.S. Liss, K.O. Munnich, G.A. Sehmel, and O. Vittori. 1978. Some aspects of the transfer of atmospheric trace constituents past the air-sea interface. Atmos. Environ. 12:2055-2087.

Spedding, D.J. 1969. Uptake of sulphur dioxide by barley leaves at low sulphur dioxide concentrations. Nature 224:1229-1234.

Twomey, S. 1977. Atmospheric Aerosols. Amsterdam, Netherlands: Elsevier Scientific Publishing Company. 302 pp.

Wason, D.T., S.K. Wood, R. Davies, and A. Lieberman. 1973. Aerosol transport: particle charges and re—entrainment effects. J. Colloid Interface Sci. 43:144-149.

Wedding, J.B., R.W. Carlson, J.J. Stiekel, and F.A. Bazzaz. 1975. Aerosol deposition on plant leaves, Environ. Sci. Technol. 9:151-153.

Wesely, M.L., and B.B. Hicks. 1977. Some factors that affect the deposition rates of sulfur dioxide and similar gases on vegetation. J. Air Pollut. Control Assoc. 27:1110-1116.

Wesely, M.L., and B.B. Hicks. 1979. Dry deposition and emission of small particles at the surface of the earth. Pp. 510-513 in Proceedings Fourth Symposium on Turbulence, Diffusion and Air Quality, Reno, Nev., 15-18 January 1979. Boston, Mass.: Am. Meteorol. Soc.

Wesely, M.L., B.B. Hicks, W.P. Dannevik, S. Frisella, and R.B. Husar. 1977. An eddy correlation measurement of particulate deposition from the atmosphere. Atmos. Environ. 11:561-563.

Wesely, M.L., J.A. Eastman, D.R. Cook, and B.B. Hicks. 1978. Daytime variation of ozone eddy fluxes to maize. Boundary-Layer Meteorol. 15:361-373.

Wesely, M.L., D.R. Cook, R.L. Hart, B.B. Hicks, J.L. Durham, R.E. Speer, D.H. Stedman, and R.J. Trapp. 1982a. Eddy-correlation measurements of dry deposition of particulate sulfur and submicron particles. In Proceedings, Fourth International Conference on Precipitation Scavenging, Dry Deposition, and Resuspension, Santa Monica, California, 29 November-3 December.

Wesely, M.L., J.A. Eastman, D.H. Stedman, and E.D. Yalvac. 1982b. An eddy-correlation measurement of NO_2 flux to vegetation and comparison to O_3 flux. Atmos. Environ. 16:815-820.

Whelpdale, D.M., and R.W. Shaw. 1974. Sulphur dioxide removal by turbulent transfer over grass, snow, and other surfaces. Tellus 26:196-204.

Whitby, K.T. 1978. The physical characteristics of sulfur aerosols. Atmos. Environ. 12:135-159.

Williams, R.M., M.L. Wesely, and B.B. Hicks. 1978. Preliminary eddy correlation measurements of momentum, heat, and particle fluxes to Lake Michigan. Argonne National Laboratory Radiological and Environmental Research Division Annual Report, January-December 1978, ANL-7865 Part III, pp. 82-87.

Winkler, E.M., and E.J. Wilhelm. 1970. Saltburst by hydration pressures in architectural stone in urban atmosphere. Geol. Soc. Am. Bull. 81:576-572.

Yudinge, M.E. 1959. Physical considerations on heavy-particle diffusion. Pp. 185-191 of Advances in Geophysics, Volume 6. H.E. Landsberg and J. van Mieghem, eds. New York: Academic Press.

3. PRECIPITATION SCAVENGING PROCESSES

3.1 STEPS IN THE SCAVENGING SEQUENCE

3.1.1 Introduction

Precipitation Scavenging is defined generally as the composite process by which airborne pollutant gases and particles attach to precipitation elements and thus deposit to the Earth's surface.* This process typically contains many parallel and consecutive steps, and as an introduction to this section it is appropriate to provide a brief overview of these intermeshing pathways. In a general sense there are four major events in which a natural or pollutant molecule† may participate, prior to its wet removal from the atmosphere; depicted pictorially in Figure C.3-1, these are as follows:

1-2.   The pollutant and the condensed atmospheric water (cloud, rain, snow, . . .) must intermix within the same airspace.

2-3.   The pollutant must attach to the condensed-water elements.

---

*One should note that this definition pertains to removal from the gaseous medium of the atmosphere combined with deposition to the ground. An alternative definition, employed often throughout the open literature, pertains to the simple attachment of airborne pollutants to liquid water elements, without regard to whether the material is subsequently conveyed to the Earth's surface. Which of these definitions is used is unimportant so long as the precise definition is understood. The definition of "scavenging" adopted here will be utilized consistently throughout this text. When specific reference to the alternative situation is made, the terms "attachment" and "capture" will be employed essentially interchangeably.

†Initial portions of this section will treat precipitation scavenging in a general sense, with limited reference to specific types of atmospheric material. The reader should continue to note, however, that the "natural or pollutant molecules" of primary concern in the present context are species associated with acid-base formation, such as SO_2, HNO_3, NH_3, sulfate, chloride, and metallic cations.

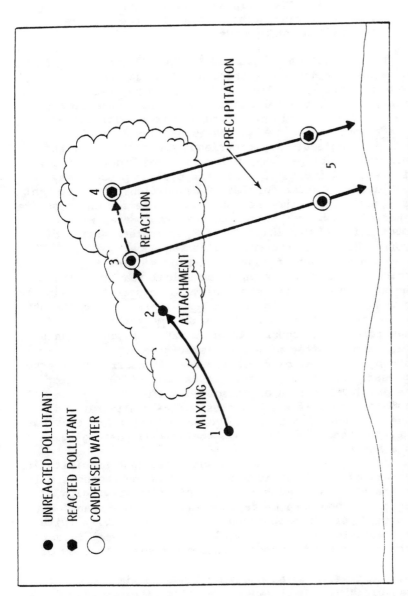

FIGURE C.3-1 Steps in the scavenging sequence: pictorial representation.

3-4.   The pollutant <u>may</u> react physically and/or
       chemically within the aqueous phase.

3-5 or (4-5). The pollutant-laden water elements <u>must</u>
       be delivered to the Earth's surface via the
       precipitation process.

The interaction diagram of Figure C.3-2 gives a
somewhat more detailed portrayal of these four major
events. Here the individual steps are represented as
transitions of the pollutant between various states in
the atmosphere, and one can note that a multitude of
<u>reverse</u> processes is also possible; thus a particular
pollutant molecule may experience numerous cycles through
this complex of pathways prior to deposition. Indeed,
Figure C.3-2 indicates that this cycling process may
continue even after "ultimate" deposition. By pollutant
off-gassing and other resuspension processes the deposited
material can be re-emitted to the atmosphere, with the
possibility of participating in yet another series of
cycles throughout the scavenging sequence.

Another important feature of Figure C.3-2 is its
indication that, while physicochemical reaction within
the aqueous phase is potentially an <u>important</u> step in the
scavenging process, it is not <u>essential</u>. This contrasts
to the remaining forward steps that <u>must</u> take place if
scavenging is to occur. Despite its nonessential nature,
this step is often of utmost importance in influencing
scavenging rates, owing to its role in modifying reverse
processes in the sequence. An example of this effect is
the devolatilization of dissolved sulfur dioxide via wet
oxidation to sulfate. This effectively eliminates
gaseous desorption from the condensed water and thus has
a strong tendency to enhance the overall scavenging rate
as a result.

From Figure C.3-2 one can note also that precipitation
scavenging of pollutant materials from the atmosphere is
intimately linked with the precipitation scavenging of
<u>water</u>. If one were to replace the word "pollutant" with
"water vapor" in each of the steps, Figure C.3-2 (with
the exception of box 4) would provide a general description of the natural precipitation process. In view of
this intimate relationship, it is not surprising that
pollutant wet-removal behavior tends to mimic that of
precipitation. Pollutant-scavenging efficiencies of
storms, for example, are often similar to water-extraction
efficiencies. This relationship is useful in the
practical estimation of scavenging rates and will

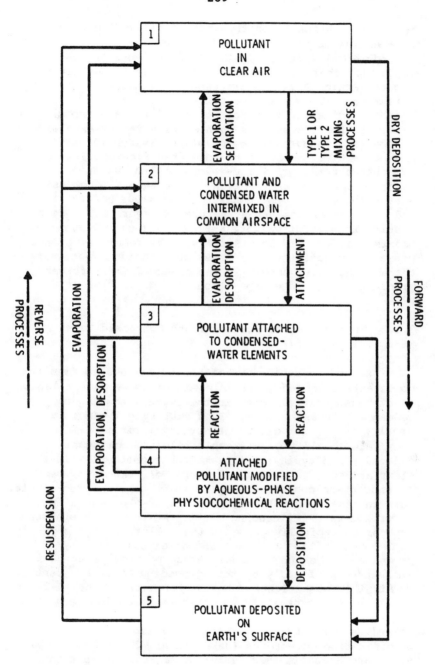

FIGURE C.3-2 Scavenging sequence: interaction diagram.

reappear continually in the ensuing discussion of wet-removal behavior.

Figure C.3-2 is interesting also because of its indication that, if some particular step in the diagram occurs particularly slowly compared with the others, then this step will dominate behavior of the overall process. This is similar to the "rate-controlling step" concept in chemical kinetics and has been applied rather extensively in practical scavenging calculations (Slinn 1974a). Finally, it is important to note that Figure C.3-2 presents a framework for development and evaluation of mathematical models of scavenging behavior. Successful scavenging models must emulate these steps effectively and tend to reflect the structure of Figure C.3-2 as a result. This point will re-emerge later when scavenging models are examined specifically. The remainder of the present subsection will address qualitative aspects of the scavenging sequence, in the order of their forward progress to ultimate deposition.

3.1.2 Intermixing of Pollutant and Condensed Water (Step 1-2)

On first consideration one often is inclined to dismiss pollutant/condensed-water intermixing as an unimportant, or at least trivial, step in the overall scavenging sequence. It is neither. In a statistical sense it usually is neither cloudy nor precipitating in the immediate locality of a freshly released pollutant molecule; and typically this molecule must exist in the clear atmosphere for several hours, or even days, before it encounters condensed water with which it may co-mingle. This in itself establishes step 1-2 as a potentially important rate-influencing event. Moreover, this extended dry period typically presents the pollutant with significant opportunities to react and/or deposit via dry processes; thus the chemical makeup of precipitation is influenced profoundly by this preceding chain of events.

Significant insights into the behavior of step 1-2 can be gained via past analyses of storm formation (e.g., Godske et al. 1957) and the atmospheric water cycle (Newell et al. 1972). Several statistical analyses of precipitation occurrence (Baker et al. 1969; Junge, 1974; Rodhe and Grandell 1972, 1981; Slinn, 1973b) have been applied as general interpretive descriptors of this step. These will not be examined in detail here; rather, we shall concentrate on the mechanisms by which step 1-2 can occur, from a more pictorial viewpoint.

Two types of mixing processes exist in which pollutant and condensed water can come to occupy common airspace; these are as follows:

1. Relative movement of the initially unmixed pollutant and condensed water, in a manner such that they merge into a common general volume;

and

2. In situ phase change of water vapor, thus producing condensed water in the immediate vicinity of pollutant molecules.

The relative importance of Type-1 and Type-2 mixing processes will depend to some extent on the pollutant. If a particular pollutant is easily scavengable, and if precipitation is occurring at the pollutant's release location, then Type-1 processes are likely to contribute significantly. If these two conditions are not met, the pollutant will usually mix intimately with makeup water vapor for some future cloud, and Type-2 processes will predominate. Based on in-cloud versus below-cloud scavenging estimates (Slinn 1983) it is not unreasonable to estimate that, as a global average, roughly 90 percent of all precipitation scavenging occurs as the consequence of a Type-2 process.

As indicated in Figure C.3-2, reverse processes exist that can serve to reseparate pollutant and condensed water. Evaporation, for example, can reinject pollutant from cloudy to clear air, and relative motion such as precipitation "fall-through" can remove hydrometeors from contact with elevated plumes. Cloud formation-re-evaporation cycles are particularly significant in this respect. Junge (1964), for example, estimates that a single cloud condensation nucleus is likely to experience of the order of ten or more evaporation-condensation cycles before it is ultimately delivered to the Earth's surface with precipitation. The rate-influencing effect of such cycling on precipitation scavenging is obvious. Additional types of cycles will be described below, in conjunction with succeeding steps of the scavenging sequence.

3.1.3 Attachment of Pollutant to Condensed-Water Elements (Step 2-3)

The microphysics of the pollutant-attachment process has been the subject of extensive research, and numerous reviews of this area have been prepared (e.g., Davies 1966, Dingle and Lee 1973, Hales 1983, Junge 1963, Pruppacher and Klett 1978, Slinn 1983, Slinn and Hales 1983). In the context of Figure C.3-1, this process is complicated somewhat in the sense that, depending on the particular attachment mechanism, Step 2-3 may occur either simultaneously or consecutively with Step 1-2.

Simultaneous comixing and attachment occur in the case of cloud-particle nucleation. This is a phase-transformation (Type-2) process wherein water molecules, thermodynamically inclined to condense from the vapor phase, migrate to some suitable surface for this purpose. Pollutant aerosol particles provide such surfaces within the air parcel, and the consequence is a cloud of droplets (or ice crystals)* containing attached pollutant material.

Different types of aerosol particles possess different capabilities to nucleate cloud elements and grow by the condensation process. As a consequence there is typically a competition for water molecules among the aerosol and associated cloud particles; some will capture water with high efficiency and grow substantially in size. Others will acquire only small amounts of water, and still others will remain essentially as "dry" elements. In addition, some particles may nucleate ice crystals, while others will be active only for the formation of liquid water. The nucleating capability of a particular aerosol particle is determined by its size, its morphological characteristics, and its chemical composition. Various

---

*At this point it is important to note that aerosols can participate in several types of phase transitions in cloud systems. These include vapor-liquid, vapor-solid, and liquid-solid transitions, in addition to a subset of interactions between numerous solid phases. Particles active as ice-formation nuclei are generally much less abundant than those active as droplet (or "cloud-condensation") nuclei. As will be demonstrated later, the relative abundance of ice nuclei can have a profound effect on precipitation-formation processes and related scavenging phenomena.

aspects of this subject are discussed at length in standard cloud-physics textbooks (e.g., Mason 1971, Pruppacher and Klett 1978) and in the periodical literature (e.g., Fitzgerald 1974).

An additional important aspect of the cloud-droplet nucleation and growth process is the fact that once initiated, cloud-droplet growth does not proceed instantaneously to some sort of thermodynamic equilibrium. Because of diffusional constraints on delivering water molecules from the surrounding atmosphere, the growth in droplet diameter slows appreciably as droplet size increases (cf. Slinn 1983). Superimposition of this lag on the continually fluctuating environment of a typical cloud results in a dynamic and complex physical system.

Finally, the competitive nature of the cloud-nucleation process results in significant impacts by the pollutant on the basic character of the cloud itself. If the local aerosol were populated solely by a relatively small number of large, hygroscopic particles, for example, one would expect any corresponding cloud to be composed chiefly of low populations of large droplets. If on the other hand the local aerosol were composed of large numbers of small, nonhygroscopic particles, the corresponding cloud should contain larger numbers of smaller droplets.

This is precisely what is observed in practice. Unpolluted marine atmospheres, for example, contain large sea-salt particles as a primary component of their aerosol burden. Warm marine clouds are noted for their wide drop spectra containing large drop sizes and their corresponding capability to form precipitation easily. Continental clouds, on the other hand, are typically composed of larger populations of smaller droplets. Figure C.3-3, which was prepared on the basis of results published by Squires and Twomey (1960), provides a good example of this point. Here measured convective-cloud droplet spectra are compared for two different cloud systems. The continental air-mass cloud exhibits a distinct tendency toward smaller drop sizes and larger populations, as compared with its maritime counterpart. It is interesting also in this context to note the estimates of Junge (1963) with regard to relative amounts of aerosol participating in the nucleation process. Junge suggests that while 50 to 80 percent of the mass of continental aerosols can be expected to participate as cloud nuclei, as much as 90 to 100 percent of maritime aerosols can become actively involved.

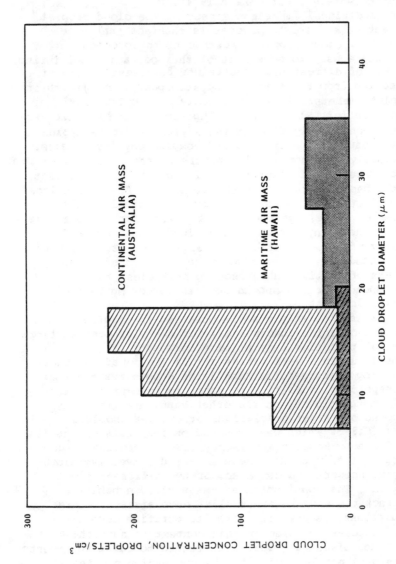

FIGURE C.3-3   Cloud droplet spectra in convective clouds formed in maritime and continental air masses. Adapted from Squires and Twomy (1960).

As a concluding note in the context of nucleating capability and water competition it should be pointed out that acid-forming particles, by their very nature, are chemically competitive for water vapor and thus tend to participate actively as cloud-condensation nuclei. This attribute tends to enhance their propensity to become scavenged early in storm systems and has a significant effect on the nature of the acid-rain formation process.

There are numerous mechanisms by which pollutants can attach to cloud and precipitation elements <u>after</u> the elements already exist, and thus in a manner that is <u>consecutive</u> with Step 1-2. These mechanisms are itemized in the following paragraphs; they are typically active for both aerosols and gases, although the relative importances and magnitudes vary widely with the state of the scavenged substance.

<u>Diffusional</u> attachment, as its name implies, results from diffusional migration of the pollutant through the air to the water surface. This process may be effective both in the case of suspended cloud elements and falling hydrometeors. It depends chiefly on the magnitude of the pollutant's molecular (or Brownian) diffusivity; and since diffusivity is inversely related to particle size, this mechanism becomes less important as pollutant elements become large. Diffusional attachment is of utmost importance for scavenging of gases and very small aerosol particles. For all practical purposes it can be ignored for aerosol particle sizes above a few tenths of a micrometer.

In concordance with Fick's law (Bird et al. 1960), diffusional transport to a water surface is dependent as well on the pollutant's concentration gradient in the vicinity of this surface. Thus if the cloud or precipitation element can accommodate the influx of pollutant readily, it will effectively depopulate the adjacent air, thus making a steep concentration gradient and encouraging further diffusion. If for some reason (e.g., particle "bounce off" or approach to solute saturation) the element cannot accommodate the pollutant supply, then further diffusion will be discouraged. If the cloud or precipitation element, through some sort of outgassing mechanism, <u>supplies</u> pollutant to the local air, then the concentration gradient will be reversed and diffusion will carry the pollutant <u>away</u> from the element.

Mixing processes inside of cloud or precipitation elements play an important role in determining the accommodation of gaseous species. If mixing is slow, for example, it is likely that the element's outer layer will saturate with pollutant and thus inhibit further attachment processes. This is quite often a limiting factor in cases involving gas scavenging by ice crystals. Internal mixing occurs as a consequence of diffusion and fluid circulation and has been analyzed at length by Pruppacher and his co-workers (cf. Pruppacher and Klett 1978).

In general diffusional attachment processes are sufficiently well understood to allow their mathematical description with reasonable accuracy, and numerous references are available as guides for this purpose (e.g., Hales 1983, Pruppacher and Klett 1978, Slinn 1983).

Inertial attachment processes are directly dependent on the size of the scavenged particle and thus are unimportant for gaseous pollutants. In a somewhat general sense this class of processes depends on motions of pollution particles and scavenging elements relative to the surrounding air, which arise because both have finite volume and mass. The most important example of inertial attachment is the impaction of aerosols on falling hydrometeors. Here the hydrometeor (because of its mass and volume) falls by gravity, sweeping out a volume of space. Some of the aerosol particles (because of their mass) cannot move sufficiently rapidly with the flow field to avoid the hydrometeor and thus are impacted. In principle impaction could occur, even if the aerosol particles were point masses with zero volume. Assigning a volume to a particle further increases its chance of collision, simply on the basis of geometric effects. The inclusion of aerosol volume has been generally referred to in the past literature as accretion.

A second example of inertial attachment is turbulent collision. In this case the particles and scavenging elements, subjected to a turbulent field, collide because of dissimilar dynamic responses to velocity fluctuations in the local air. This scavenging mechanism is thought to be of secondary importance and has received comparatively little attention in the past literature, although some recent theoretical analyses have suggested it to be significant for specific drop-size/particle-size ranges.

While the mechanisms of diffusional and inertial attachment are efficient for capturing very fine and very coarse particles, respectively, a region of low efficiency should exist in the 0.1-5 μm range, where the mechanism is effective. This effect is shown schematically in Figure C.3-3. Because its importance to scavenging was first recognized by Greenfield (1957) it has become known generally as the "Greenfield gap." Depending on circumstances there are several additional attachment mechanisms (including the two-stage nucleation-impaction mechanism mentioned earlier) that can serve to "fill" the Greenfield gap. Some of the more important of these are itemized in the following paragraphs.

Diffusiophoretic attachment to a capturing element can occur whenever the element grows via the condensation of water vapor. In effect the flux of condensing water vapor "sweeps" the surrounding aerosol particles to the element's surface. In a competitive cloud-element system where some droplets grow while others evaporate, diffusiophoresis can be a rather important secondary attachment mechanism. This is particularly true when the cloud contains mixtures of ice and liquid. Under such conditions the ice crystals have a pronounced tendency, owing to their lower equilibrium vapor pressure, to gain water at the expense of the droplets. Known as the Bergeron-Findeisen effect, this process is important in precipitation formation as well as in diffusiophoretic enhancement.

Thermophoretic attachment results from a temperature gradient in the direction of the capturing element. Here the element acts essentially as a miniature thermal precipitator. Warmer gas molecules on the outward side of the aerosol particle impart a proportionately larger amount of momentum, resulting in a driving force toward the capturing element.*

Thermophoresis depends directly on the temperature gradient in the vicinity of the capturing element. In cloud and precipitation systems local temperature gradients are caused most often by evaporation/condensation effects; thus thermophoresis is usually strongly associated with diffusiophoresis,† and in fact these two processes often tend to counteract each other.

---

*One should note that the precise mechanisms of thermal transport differ radically, depending on particle size (cf. Cadle 1965).

†As noted by Slinn and Hales (1983) inappropriate treatment of this relationship has caused erroneous conclusions to be drawn in some of the past literature. The reader should be cognizant of this if more detailed pursuit is intended.

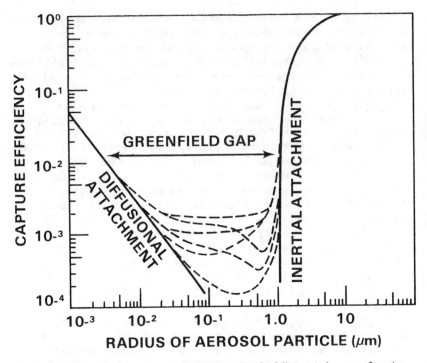

FIGURE C.3-4 Theoretical scavenging efficiency of a falling raindrop as a function of aerosol particle size. Adapted from Pruppacher and Klett (1978). Dashed lines correspond to contributions by electrical and phoretic effects under chosen humidity and raindrop-charge conditions (see original reference for details).

Phoretic processes are unimportant in the case of gaseous pollutants, owing to the overwhelming contributions of molecular diffusion. At present the theory of diffusiophoretic/thermophoretic particle attachment is at a state where reasonably quantitative assessments can be made for simple systems such as isolated droplets (Pruppacher and Klett 1978, Slinn and Hales 1971) (cf. Figure C.3.4). Rough estimates are possible also for more complex and interactive cloud/precipitation systems, but much remains to be done to bring our knowledge of this area to a really satisfactory state.

Electrical attachment of aerosol particles to cloud and precipitation elements has been the subject of continuing study over the past three decades. Understanding of this process is currently at a state where relationships between aerosols and isolated drops can be quantified with reasonable accuracy (cf. Wang and Pruppacher 1977). In general, electrical charging of cloud and/or precipitation elements must be moderately high for electrical effects to become competitive with other capture phenomena, although such charging is certainly possible in the atmosphere--particularly in convective-storm situations. Understanding of electrical deposition in clouds of interacting drops is still at a relatively unsatisfactory stage of development.

As a conclusion to this discussion of attachment processes it is appropriate to note that, while the mechanisms have been presented here on an individual basis, they tend in actuality to proceed in a simultaneous and competitive manner. Insofar as atmospheric cleansing is concerned, this is a fortunate circumstance, because some mechanisms tend to be operative in physical situations where others are ineffective. Figure C.3-4 gives an excellent illustration of this point. Here theoretical attachment efficiencies appropriate to a 0.31-mm radius raindrop are presented for various electrical and relative-humidity conditions, demonstrating the capability of phoretic and electrical mechanisms to "bridge" the Greenfield gap. This simultaneous and competitive interaction of mechanisms serves to complicate profoundly the mathematics of the scavenging process and lends an additional degree of difficulty to the problem of scavenging calculations. This aspect will continue to emerge throughout this section, especially during the discussion of scavenging models.

3.1.4 Aqueous-Phase Reactions (Step 3-4)

Aqueous-phase conversion phenomena have been discussed in some detail in Appendix A and will not be examined further here except to note their general importance within the framework of the overall scavenging sequence. As observed previously in the context of Figure C.3-2, aqueous-phase reactions are not essential to the scavenging process. Depending on the pollutant material, however, these reactions often can have the effect of stabilizing the captured material within the condensed phase and thus enhancing the scavenging efficiency appreciably. There is much to be learned before this important aspect is brought to a satisfactory stage of understanding.

3.1.5 Deposition of Pollutant with Precipitation (Steps 3-5 and 4-5)

Although a variety of mechanisms exist (e.g., impaction of fog on vegetation), the predominant means for depositing pollutant-laden condensed water to the Earth's surface is simply gravitational <u>sedimentation</u>. Sedimentation rates depend on hydrometeor fall velocities, which depend in turn on hydrometeor size; thus the processes by which the pollutant-laden cloud droplets grow to precipitation elements emerge as major determining factors in this final stage of the scavenging sequence.

Once attached to condensed water, a pollutant molecule has several alternative pathways for action (Figure C.3-2). If the captured pollutant possesses some degree of volatility, it may desorb back into the gas phase. Reverse chemical reactions may occur. Evaporation of the condensed water may, in effect, "free" the pollutant to the surrounding gaseous atmosphere. This multitude of pathways results in an active <u>competition</u> for pollutant. If the precipitation stage of the scavenging sequence is to be effective, it must interact successfully within this competitive framework.

Besides competing actively for pollutants, the above interactions produce a vigorous competition for <u>water</u>. This parallel relationship between pollutant scavenging and water scavenging, apparent in some of the preceding discussion regarding attachment processes, can be drawn even more emphatically when considering precipitation processes. The following paragraphs provide a brief overview of some of the more important mechanisms in this regard.

Once initial nucleation has occurred, cloud particles may grow further by <u>condensation</u> of additional water vapor. Net condensation will occur to the surface of a cloud element whenever water vapor molecules can find a more favorable thermodynamic state in association with it; and because clouds contain varieties of makeup elements having different thermodynamic characteristics, a competition for water vapor usually exists. Such interactions are discussed at length in standard textbooks (Mason 1971, Pruppacher and Klett 1978). Slinn (1983) has developed a conceptual scavenging model in which condensational growth is an important rate-limiting step.

Thermodynamic affinity for water-vapor molecules depends on the cloud-element's size, its pollutant burden, and its physical structure. These latter two factors often influence precipitation characteristics profoundly. In particular, the favored thermodynamic state of a water molecule in association with an ice crystal (as compared with a supercooled water droplet) results in rapid competitive growth of ice particles in mixed-phase clouds. This Bergeron-Findeisen process has been mentioned already in the context of diffusiophoretic and thermophoretic transport. Growth of large cloud elements via this process is the primary reason that ice-containing clouds tend to be so strongly effective as generators of precipitation water.

A further mechanism by which suspended cloud droplets can grow to form precipitation elements is <u>coagulation</u>. This process occurs via the collision of two or more cloud elements to form a new element containing the total mass (and pollutant burden)* of its predecessors. Coagulation occurs over size-distributed systems of cloud elements by a variety of physical mechanisms, and because of this it is a rather poorly understood and mathematically complex process. Comprehensive analyses of coagulation processes have been performed by Berry and Reinhardt (1974). Coagulation can be considered to be an important initiator of precipitation in single-phase clouds (water or ice). In mixed-phase clouds the Bergeron-Findeisen process can be expected to enhance the coagulation process by widening the droplet size distribution, as well as contributing to precipitation growth in a direct sense.

Once a moderate number of precipitation-sized elements have been generated, the process of <u>accretion</u> rapidly begins to dominate as a means for generating precipitation water. As noted previously, this process occurs by the "sweeping" action of large hydrometeors falling through the field of smaller elements, attaching them on the way. As was the case with coagulation, the accretion process tends to accumulate the pollutant burden of all collected elements.

---

*Coagulation is often referred to as <u>autoconversion</u> in the cloud-physics literature. It is interesting to notice in this context that while coagulation tends to accumulate nucleated pollutants, the Bergeron-Findeisen process tends to re-liberate nucleated pollutants to the air.

Accretion can occur via drop-drop, drop-crystal, and crystal-crystal interactions. Drop-crystal interactions are particularly important in mixed-phase clouds; when supercooled droplets are accreted by falling ice crystals, the process is usually referred to as <u>riming</u>.

Although the above discussion has been confined primarily to deposition in conjunction with rain and snow, it should be emphasized that <u>fog deposition</u>* often is an important secondary process for conveying pollutants to the Earth's surface. Classification of fog-bound pollutant deposition is problematic for two major reasons. The first of these is that no sharp demarkation exists between "fog droplets" and "water-containing aerosols"; thus the choice of considering fog deposition as simply the dry deposition of wet particles, or the wet deposition of contaminated water, depends primarily on personal preference. Secondly, there is no real distinction between fog droplets and precipitation. Cloud physicists often find it convenient to categorize condensed atmospheric water into "precipitation" and "cloud" classifications, with the presumption that cloud water has a negligible sedimentation velocity. Such a classification is of limited use when considering fog deposition, however, owing to the fact that fog droplets do have significant gravitational fall speeds. A 50-μm-diameter fog droplet, for example, will fall at a rate of about 10 cm/s. This, combined with the fact that typical fogs and clouds contain droplet-size distributions ranging between 0 to 100 μm (cf. Pruppacher and Klett 1978), suggests that gravitational transport of fog droplets will indeed be a significant pollution-deposition pathway under appropriate circumstances.

In addition to purely gravitational transport, fog droplets have a strong tendency to <u>impact</u> on projected surfaces. The rates of fog impaction depend in a complex fashion on drop size, wind velocity, and geometry of the projected object. The common observations of rime-ice accumulation on alpine forests and on power-transmission lines give direct testimony to the effectiveness of this process.

Chemical deposition by fogs is directly proportional to fog-bound pollutant concentration, and this fact often acts to enhance substantially the pathway's overall

---

*A "fog" is (rather pragmatically) defined here as any cloud that is in the proximity of the Earth's surface.

effectiveness. Owing to their proximity to the Earth's surface, fogs typically form in conjunction with high pollutant concentrations. Attaching particles and gases via the variety of mechanisms described in Section 3.1.3, the droplets typically accumulate extremely high burdens of material. It is not difficult to find evidence in support of this point. Scott and Laulainen (1979), for example, reported sulfate and nitrate concentrations approaching 500 μm/liter in water obtained near the bases of clouds over Michigan, while the SUNY group has reported (e.g., Falconer and Falconer 1980) numerous similar concentrations (as well as extremely low pH measurements) in clouds sampled at the Whiteface Mountain, New York, observatory.

Recently Waldman et al. (1982) have reported nitrate and sulfate concentrations in Los Angeles fogs ranging up to and beyond 5000 μm/liter. This compares with typical precipitation-borne concentrations of about 35 μm/liter for the northeastern United States.

Recently Lovett et al. (1982) have applied a simple impaction model to estimate fog-bound pollutant deposition to subalpine balsam fir forests and have concluded that chemical inputs via this mechanism exceed those by ordinary precipitation by 50-300 percent. This is undoubtedly an extreme case, and it would be more meaningful to possess a regional assessment indicating the general importance of fog deposition on an areal basis. This requires substantial effort however, involving climatological fogging analysis (cf. Court 1966) as well as numerous additional factors, and no really satisfactory evaluation of this type is currently available. Regardless of this it is appropriate to conclude that fog-deposition processes probably play an important, if secondary, role in pollutant delivery on a regional basis. In the future more effort should be addressed to this important research area.

3.1.6 Combined Processes and the Problem of Scavenging Calculations

The preceding discussion of individual steps in the scavenging sequence has been presented intentionally on a highly visual and nonmathematical basis, with appropriate references given for the reader interested in more detailed pursuit. Despite the qualitative nature of this presentation, however, it should be obvious that the most

direct and expedient approach to model development is
first to formulate mathematical expressions corresponding
to each of these steps and then to combine them in some
sort of a model framework that describes the composite
process. This subject will be examined in greater detail
in Section 3.4, which is addressed specifically to
scavenging models.

3.2 STORM SYSTEMS AND STORM CLIMATOLOGY*

3.2.1 Introduction

From the preceding discussion it is not difficult to
imagine that scavenging rates and pathways will be
dictated to a large extent by the basic nature of the
particular storm causing the wet removal to occur.
Storms containing water that is predominantly in the ice
phase, for example, will provide little opportunity for
attachment mechanisms associated with droplet nucleation,
accretion, or phoretic processes. The abundance of
liquid water and the temperature distribution in a given
storm will have a direct bearing on the degree to which
aqueous-phase chemistry can occur. Storms containing no
ice phase whatsoever will be generally ineffective as
generators of precipitation and thus will tend to inhibit
the scavenging process. An interesting indication of the
importance of storm type in this regard is presented in
Figure C.3-23, which presents estimated scavenging
efficiencies that vary extensively with storm classification. Different storm types differ profoundly with
regard to inflow, internal mixing, vertical development,
water-extraction efficiency, and cloud physics; and
consequently it is appropriate at this point to consider
briefly the major classes and climatologies of storm
systems occurring over the continental United States.

Two major points should be stressed at the outset to
this discussion. The first is the essential fact that
all storms are initiated by a <u>cooling</u> of air, which leads
to a condensation process. Such cooling may occur by the

---

*In the present text the term "storm" is intended to
denote any system in which precipitation occurs. This
definition thus encompasses all occurrences, ranging from
mild precipitation conditions up to and through the major
and cataclysmic events.

transport of sensible heat, such as when a comparatively warm, moist air parcel flows over a cold land surface. The dominant cooling mode for most storm systems, however, is expansion, which occurs via vertical motion of the air parcel to elevations of lower pressure. The second note-worthy point in this context is that the overwhelming majority of storm systems is strongly associated with fronts between one or more air masses. The primary reason for this fact, of course, is that thermodynamic per-turbations and discontinuities associated with the frontal surfaces provide the opportunity for vertical motion (and thus expansion processes) to occur. This relationship is an essential component of storm-classification systems and will emerge repeatedly in the following discussion.

Overlaps in the characteristics of different storm types render a strict classification largely impossible. For practical purposes, however, it is convenient to segregate mid-latitude continental storms into two classes, which are usually described as being "convective" and "frontal" in nature. These two major categories then can be subdivided further as deemed expedient for the purpose at hand, although it should be noted that sig-nificant overlap among storm types occurs even at this major level of classification. Frontal storms, for example, often possess significant convective character in their basic composition, and true convective storms often occur as the consequence of fronts. Because of this the following discussion will utilize storm classification primarily as a descriptive aid and will not belabor taxonomical detail beyond this rather pragmatic end.

3.2.2   Frontal-Storm Systems

Much of what is understood today regarding mid-latitude frontal-storm systems stems from the pioneering work of the Norwegian meteorologist Bjerknes, who conducted a systematic survey of large numbers of storm systems and from this developed a conceptual model of frontal-storm development and behavior. Characterized schematically in Figure C.3-5, the Bjerknes model can be understood most easily by considering a cool northern airmass, separated from a warm southern air mass by an east-west front, as indicated in Figure C.3-5A. The progression of figures represents a typical result of the atmosphere's natural

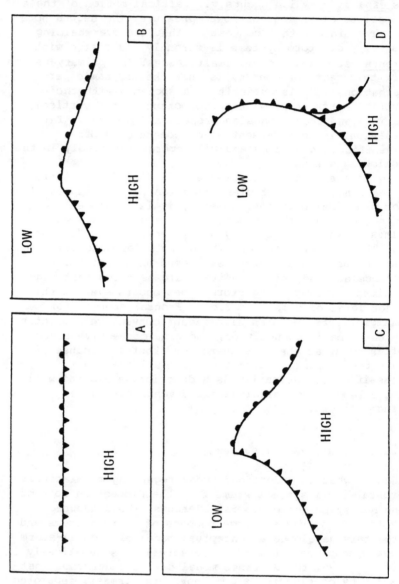

FIGURE C.3-5 Cyclonic storm development according to Bjerknes's conceptual model.

tendency to exchange heat from southern to northern
latitudes. This is often referred to as a "tongue" of
warm air intruding into the cold air mass. In the
northern hemisphere this wave will tend to propagate in
an easterly direction; thus the intrusion is bounded by
two moving fronts--a warm front followed by a cold front,
as shown in Figure C.3-5C.

Flows associated with the wave system occur in a
manner such that a depression in atmospheric pressure
occurs at the vertex of the warm-air intrusion, and as a
consequence a general counterclockwise or "cyclonic"
circulation pattern emerges. Because of this feature
Bjerknes's conceptual model is often referred to as the
"Bjerknes cyclone theory," and frontal storms associated
with this pattern are termed "cyclonic" storms. A typical
feature of storms of this type is the tendency for the
cold front to overtake the warm front and, ultimately, to
annihilate the wave. The "occluded" front created as a
consequence of this behavior is shown schematically in
Figure C.3-5D. In view of this birth-death sequence of
the Bjerknes cyclone model, the progression depicted in
Figure C.3-5 often has been termed the "life history" of
a cyclone. Some idea of spatial scale and the general
cyclonic flow pattern of a mature cyclone are given in
Figure C.3-6. In viewing these indicated flow patterns,
however, the reader should note carefully that considerable vertical structure exists in such systems, and
marked deviations of the wind field with elevation are
typical. In particular, one should take care not to
confuse the indicated general circulation patterns with
corresponding surface winds.

Although created from the limited observational base
available during the early twentieth century, the
fundamental precepts of the Bjerknes theory have proven
valid even as more sophisticated observational and
analytical facilities have become available. Certainly
nonidealities and deviations from this model occur; but
its general concepts have proven to be immensely valuable
as a conceptual basis and as an idealized standard for
the assessment of actual storm systems. Comprehensive
descriptive and theoretical material pertaining to such
systems is available in the classic text by Godske et al.
(1957), and more elaborate and modern extentions are
given in the periodical literature (e.g., Browning et al.
1973, Hobbs 1978).

308

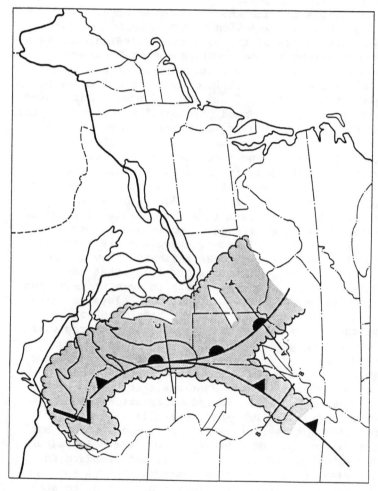

FIGURE C.3-6 General flow patterns in the vicinity of an idealized cyclonic storm system. Arrows denote general circulation patterns and should not be interpreted as surface winds.

Warm-Front Storms

It is important to note that the plan views exhibited by
Figure C.3-6 are gross simplications, since they do
nothing to characterize the three-dimensional nature of
the cyclonic system. If one were to construct a vertical
cross section of the warm front (A-A' in Figure C.3-6),
then typically one would observe an inclined frontal
surface as shown in Figure C.3-7.* In this situation the
presence of warm air aloft creates a relatively stable
environment, which inhibits vertical mixing of air between
the two air masses. The warm, moist air moves up over
the cold air wedge, expanding, cooling, and ultimately
forming clouds and precipitation. Typically the warm air
supplying moisture for this purpose has been advected
from deep within the southern air mass, carrying water
vapor and pollutant over extensive distances. This transport trajectory has been aptly compared to a "conveyor
belt" for moisture by Browning and his co-workers
(1973). It is appropriate to note that this moisture
conveyor belt is a conveyor belt for pollution as well.

Warm-front storms often are associated with long
periods of continuous precipitation, although significant
structure can exist within such systems. An important
structural element in this regard is the occurrence of
prefrontal rain bonds, which take the form of concentrated areas of precipitation imbedded within the major
storm system. At present the factors contributing to
rain-band formation are not totally understood, although
mechanisms such as seeding from aloft by ice crystals and
nonlinearities of the associated thermodynamic and flow
processes undoubtedly contribute to a major extent.

Warm-front storms usually can be expected to be rather
effective as scavengers of pollution originating from
within the warm air mass, especially if temperatures in
the feeder region are sufficiently high to allow the
presence of liquid water and the nucleation-accretion
process. Scavenging of pollutants from the underlying
cold air mass usually will be less effective, owing to
the relative scarcity of clouds and generally less
definitive flows in this sector. Scavenging in both
regions will of course depend on the physiochemical
nature of the pollutant of interest and the microphysical
attributes of the cloud system in general. Methods for

---

*See Table C.3-1 for definition of cloud abbreviations.

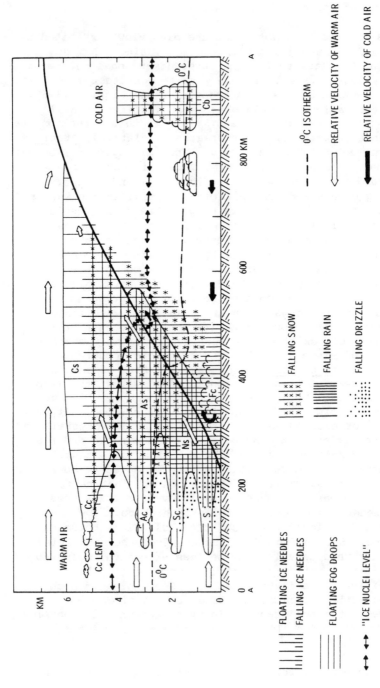

FIGURE C.3-7 Vertical cross section of a typical warm front (Section A-A' on Figure C.3-6). Adapted from Godske et al. (1957).

TABLE C.3-1 Summary of Cloud Types Appearing
in Figures C.3-7-C.3-9

Type	Abbreviation
Cirrus	Ci
Cirrostratus	Cs
Cirrocumulus	Cc
Altostratus	As
Atlocumulus	Ac
Stratus	St
Stratocumulus	Sc
Nimbostratus	Ns
Cumulus	Cu
Cumulonimbus	Cb

estimating scavenging rates in such circumstances are
discussed in Section 3.4.

Cold-Front Storms

A typical vertical cross section (B-B' in Figure C.3-6)
of a cold-front storm is shown in Figure C.3-8. This
differs substantially from the warm-front situation in
the sense that, instead of flowing over the frontal
surface, the warm air is forced ahead by the moving cold
air mass. This action produces a more steeply inclined
frontal surface, which, combined with the presence of
low-elevation warm air, creates a relatively unstable
situation leading to convective uplifting and the
formation of clouds and precipitation.

Although discussed here in a frontal-storm context,
this precold-front situation composes an important class
of convective storms, which will be discussed in some
detail later. Scavenging rates and efficiencies
associated with such storm systems will again depend on
the pollutant and the physical attributes of the
particular cloud system involved.

312

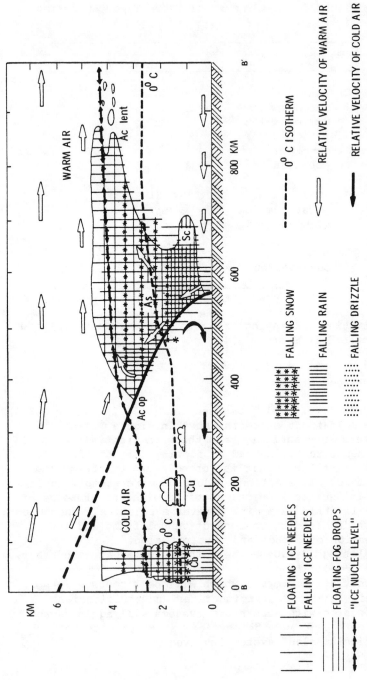

FIGURE C.3-8 Schematic vertical cross section of a typical cold front (Section B-B' in Figure C.3-6). Adapted grom Godske et al. (1957).

Occluded-Front Storms

Owing to the fact that occluded fronts are formed via
merger of warm and cold fronts, it seems reasonable to
expect that storms associated with occlusions should
share characteristics of the respective elementary
systems. Figure C.3-9, which shows a typical vertical
cross section (Section C-C' on Figure C.3-6) of an
occluded system demonstrates this point. Typically the
easterly flow of warm air aloft maintains a relatively
stable environment to the east of the occlusion, and
clouds and precipitation occur in this region largely as
a consequence of ascending flow from the south. Much
more detailed accounts of occluded systems can be found
in standard references such as the book by Godske et al.
(1957).

3.2.3 Convective-Storm Systems

An idealized cross section of a typical convective storm
is shown in Figure C.3-10. Such storms depend on
atmospheric instabilities to induce the necessary
vertical motions and concurrent cooling and condensation
processes; and as such they are most likely to occur
under warm, moist conditions where the energetics are
most conducive to this process. Often convective storm
systems occur as "clusters" of cells such as that shown
in Figure C.3-10 and exhibit a marked tendency to
exchange moisture and pollutant between cells; thus the
flow dynamics and scavenging characteristics of such
systems tend to be extremely complex.

Typically the moisture and pollutant input to a
convective cell occurs primarily through the storm's
updraft region (cf. Figure C.3-10), although entrainment
from upper regions is possible as well. Dynamics of this
process are such that violent updraft velocities often
occur; these are capable of lifting entrained air, water
vapor, and pollution to extremely high elevations
(sometimes breaching the stratosphere). Along this
course entrained pollutant is subjected to a large
variety of environments and scavenging mechanisms; as
will be noted in Section 3.4, convective storms tend to
be highly effective scavengers of air pollution.

As was stated earlier, convective storms often are
associated with frontal systems, although frontal
influence is not absolutely necessary for their presence.

314

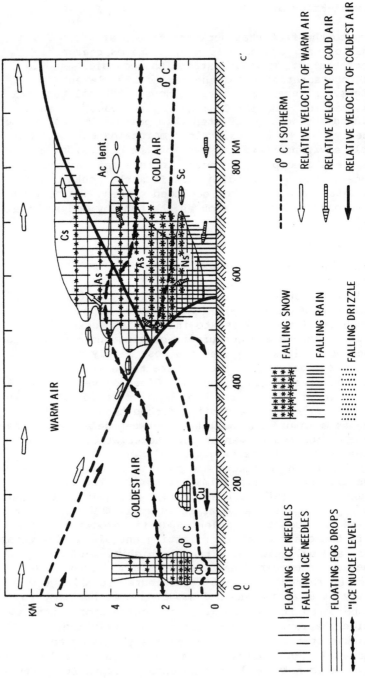

FIGURE C.3-9 Schematic vertical cross section of a typical occluded front (Section C-C' in Figure C.3-6). Adapted from Godske et al. (1957).

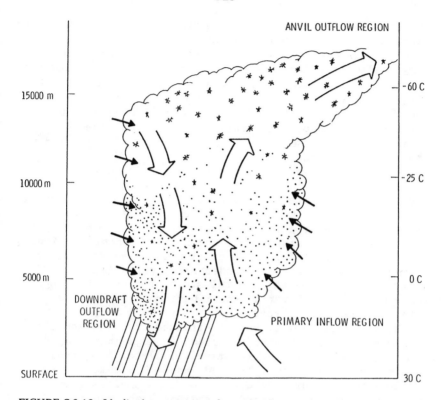

FIGURE C.3-10 Idealized cross section of an isolated convective storm.

An isolated air mass, for example, is totally capable of acquiring sufficient energy and water vapor to induce a convective disturbance on its own accord. Perturbations arising from fronts, however, often contribute to the creation of convective activity--if for no other reason than supplying a "trigger" to initiate convection in a conditionally unstable atmosphere.

3.2.4 Additional Storm Types: Nonideal Frontal Storms, Orographic Storms, and Lake-Effect Storms

As noted previously, the Bjerknes cyclone model represents something of an idealized concept, and numerous features can contribute to deviations from this "textbook" behavior. Orographic effects are highly important in this regard. Considering a cyclonic disturbance approaching the North American continent from across the

Pacific Ocean, for example, the frontal patterns typically lose much of their original identity after impacting with the mountainous regions of the west. In addition to the physical distortion of flow patterns, the lifting induced by the terrain encourages further precipitation, resulting in large spatial variability in rainfall patterns and pronounced local phenomena such as "rain shadows" and chinooks. Precipitation-formation and precipitation-scavenging processes associated with such systems tend to be highly complex.

Frontal systems often tend to reconstitute their structure after crossing the Rocky Mountains; but continental effects still impart a marked impact on their basic makeup. In the midwest-northeast region, for example, there is a tendency for the fronts to orient themselves in an east-west direction and become stationary for extended periods, often punctuated by several minor low-pressure areas. Even under relatively ideal conditions continental frontal storms tend to possess more convective flavor in their basic makeup than do their oceanic counterparts.

As indicated above, terrain-induced or "orographic" effects are usually most important in augmenting major storm systems, although relatively isolated orographic storms (such as oceanic "island-induced" storms) certainly do occur. Orographic effects obviously will tend to be most pronounced in regions where radical terrain changes occur; but even the small elevation changes typical of the Midwest can contribute significantly at times. Orographic effects also are suspected to influence storm behavior over substantial downwind distances. Lee waves from the Rocky Mountains, for example, have been suggested to trigger thunderstorm formation at extended distances.

Lake-effect storms are yet another example of a somewhat nonideal phenomenon that often is superimposed with more major meteorological patterns. Typically such storms occur during fall and early winter periods when land surfaces tend to be cooler than their adjoining water bodies. Considering an air parcel moving on an easterly course across Lake Michigan, for example, the warm lake surface tends to supply both heat and water vapor as it proceeds. As this parcel is advected across the downwind shore, however, two important things will occur. Firstly, the cold land mass will act to extract the heat from the air, and secondly the orographic lifting (of the order of a few tens of meters) will result in ascent, expansion, and further cooling. The

net result is a lake-effect storm. Such storms are
capable of inducing highly variable precipitation patterns
in specific areas around the Great Lakes region. Although
confined largely to this portion of the United States,
these storms are accountable for a majority of the snowfall accumulated in specific cities, such as Muskegon,
Michigan, and Buffalo, New York. Some appreciation for
the magnitude of this effect can be gained by looking at
the climatological precipitation map given in Figure
C.3-11.

3.2.5 Storm and Precipitation Climatology

The subject of storm climatology is exceedingly complex
and will be discussed here only to the point necessary to
describe some key attributes and indicate references for
more detailed pursuit. Factors especially important in
the context of precipitation scavenging are temporal and
spatial precipitation patterns, storm-trajectory behavior,
and storm-duration statistics. These will be discussed
in order in the following paragraphs.

Precipitation Climatology

Figure C.3-12 provides climatological averages of monthly
precipitation amounts at various stations throughout the
United States. This figure was taken directly from the
Climate Atlas of the United States (1968) and requires
little elaboration at this point. It is interesting to
note, however, that precipitation amounts do not vary
radically throughout the year at most northeastern U.S.
stations; this contrasts especially with the western and
arid stations, whose seasonal variabilities tend to be
pronounced. It should be noted as well that actual
precipitation amounts for a given single month can vary
appreciably from the climatological averages presented
here.

Storm Tracks

Because of the difficulties noted previously with regard
to precise classification or definition of storms, a
really concise climatological summary of storm-pathway
behavior is largely impossible. Some useful information
can be generated, however, by observing the tracks of the

cyclonic (low-pressure) centers associated with major
storm systems. Klein (1958), for example, has conducted
a systematic survey of cyclonic centers in the northern
hemisphere and from this has constructed monthly climatological maps of low-pressure tracks. Figure C.3-13,
taken from the book by Haurwitz and Austin (1944),
presents the combined results of the analyses by several
previous authors. On the basis of the previous discussion it should be re-emphasized that, owing to the
complex flow processes associated with cyclonic systems,
one should not interpret the motion of these low-pressure
centers as being identical with feeder trajectories for
the storms themselves. Careful and skilled meteorological guidance is mandatory for the successful
interpretation of such information in the context of
source-receptor analyses.

Several additional points should be emphasized in the
context of Figure C.3-13. Firstly it should be noted
that this presents a long-term composite average and that
marked deviations from this pattern can be expected to
occur with season. Secondly the statistical variability
of storm tracks is such that substantial departures from
the long-term averages can be expected for any particular
year. Finally, there is substantial evidence for longer-
term shifts in average storm-track distributions (Zishka
and Smith 1980); thus presentations (such as Figure
C.3-13), which are based on historical data may vary
considerably from storm patterns to be observed over the
next 20 years. The implications of this with regard to
long-term acid-deposition forecasting are obvious.

Additional features of cyclonic storm climatology can
be found in standard climatological textbooks (e.g.,
Haurwitz and Austin 1944). Convective-storm climatology,
which tends to be much more region-specific, can be
evaluated from such references as well, although more
recent weather modification programs such as METROMEX,
NHRE, and HIPLEX have generated a considerable amount of
new information in this area.

Storm-Duration Statistics

In the preparation of regional scavenging models it often
is desirable to create some sort of statistical average
of storm characteristics so that "average" wet-removal
behavior can be defined. Although little activity has
been devoted to this area until very recently, the

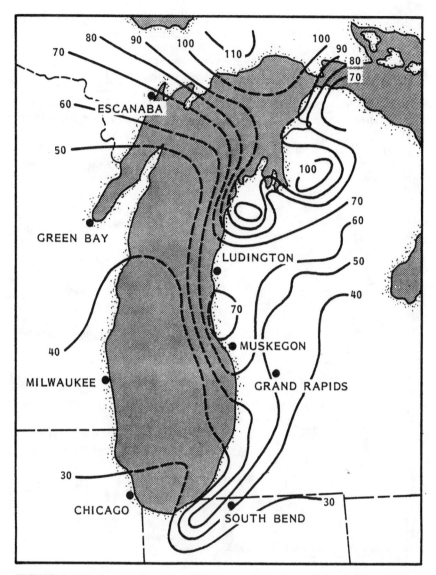

FIGURE C.3-11 Average annual snowfall pattern (inches) over Lake Michigan and environs. Adapted from Changnon (1968).

NORMAL MONTHLY TOTAL

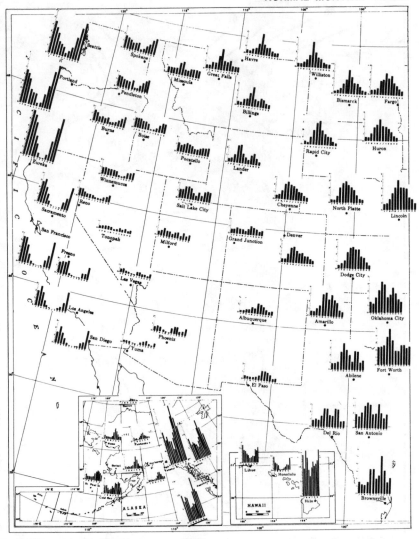

PRECIPITATION (Inches)

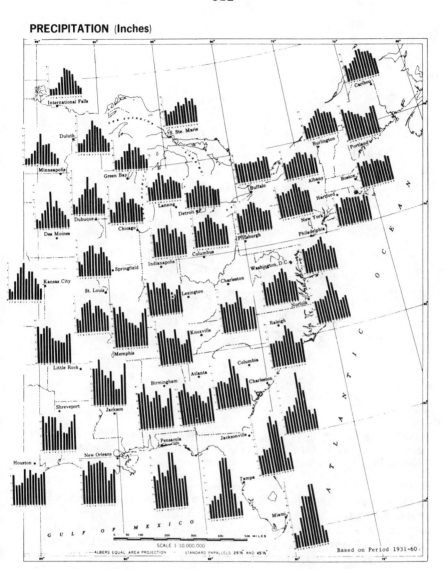

FIGURE C.3-12   Climatological summary of U.S. precipitation. From U.S. Climatological Atlas.

FIGURE C.3-13   Major climatological storm tracks for the North American Continent.
Adapted from Haurwitz and Austin (1944). Dashed line denote tropical cyclone
centers, and solid lines denote those of extratropical cyclones.

usefulness of such an approach to regional model development suggests accelerated effort during future years.

The analysis by Thorp and Scott (1982) provides an
example of one such effort. These authors compiled data
from hourly precipitation records from northeastern U.S.
stations to obtain seasonally stratified duration statistics, which were expressed in terms of probability plots
as shown in Figure C.3-14. As can be noted from these
plots, "average" storm durations during summertime are
significantly less than their wintertime counterparts,
reflecting relative influences of short-term convective
behavior. Some of the references given in Section 3.4
suggest potential modeling applications for these
statistical summaries.

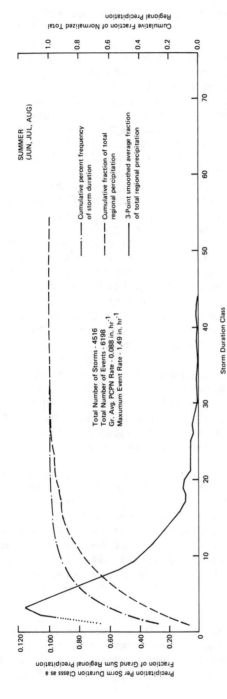

FIGURE C.3-14 Frequency statistics for storm duration, cumulative fraction of total regional precipitation, and smoothed average fraction of total regional precipitation. Summer (June, July, August) data from the MAP3S region. Frequency analysis allows up to 3 consecutive hours without precipitation to be counted as part of one storm.

3.3 SUMMARY OF PRECIPITATION-SCAVENGING FIELD INVESTIGATIONS

For the purposes of this document "field investigations" of precipitation-scavenging mechanisms will be differentiated from routine precipitation-chemistry network measurements, which are intended primarily for characterization purposes. There is of course a great deal of overlap between these two classes of measurements, and significant reciprocal benefit is generated as a consequence of each. There are some essential differences between the two, however, and it is convenient for present purposes to differentiate them accordingly.

The primary distinguishing feature of a scavenging field investigation is that the study usually is designed around the basis of some sort of conceptual or interpretive model(s) of scavenging behavior, which is tested on the basis of the field data. If the model predictions and data disagree, then some basic precepts of the model must be invalid, and additional mechanistic insights must be generated to rectify the situation. In the event that predictions and data agree, then this may be taken as evidence that the precepts may be correct. Regardless of whether positive or negative results are obtained (and assuming that the field study has been well designed and well interpreted), an advance in understanding has been achieved. The importance of such input cannot be overemphasized. Examples exist wherein field investigations have demonstrated then-accepted models to be in error by several orders of magnitude (e.g., Hales et al. 1971). Field studies have been essential in keeping the models "honest."

Field studies of precipitation scavenging began in earnest during the early 1950's to gain an understanding of radioactive fallout. Pioneering studies in this area were performed in England by Chamberlain (1953), which pertained to radioactive pollutant releases from point sources in anticipation of reactor accidents and related phenomena. These constituted the basis for the washout-coefficient approach to scavenging modeling (see Section 3.4). Other studies focused primarily on nuclear-detonation fallout and thus approached the scavenging problem from a more global point of view.

Following the English lead, nuclear-oriented studies were conducted by the United States, Canada, and the Soviet Union. These included studies of tracers as well as those of the radionuclides themselves, and although

some of this material still remains in the classified
literature, it may be stated with certainty that most of
what we know today regarding scavenging processes has
been generated as a consequence of the nuclear era. The
review "Scavenging in Perspective" by Fuquay (1970)
presents a comprehensive account of this early stage of
scavenging field studies.

During the late 1960s field-experiment emphasis
shifted to more conventional pollutants, with the general
recognition of precipitation scavenging's importance in
preserving atmospheric quality and its potential adverse
impacts of deposition on the Earth's surface ecosystem.
Since that time a variety of large and small field
studies have been conducted. These are summarized in
Table C.3-2, which provides a logical classification in
terms of source type, pollutant type, and geographical
scale.

Although field studies have been focused strongly on
quantitative aspects of precipitation scavenging, they
have provided important qualitative information regarding
acid precipitation processes as well. The ensemble of
studies listed in Table C.3-2 presents a rather cohesive
base of evidence in this regard; and although some
conflicting results and uncertainties do exist, a
generally coherent picture can be constructed in several
important areas. Although there is considerable overlap
of source-receptor distance scales among these studies,
they tend to group rather conveniently into three classes
of areal extent: 0-20 km, 0-200 km, and 0-2000 km.
These classes shall be termed loosely as "local,"
"intermediate," and "regional" scales in the following
discussion, where key qualitative features are
illustrated by considering the fate of specific acidic-
precipitation precursors (SO_x, NO_x, and HCl) as they
are transported over these increasing scales of time and
distance.

On a local scale (0-20 km) field studies have
generally demonstrated the precipitation scavenging of
sulfur and nitrogen oxides from conventional utility and
smelting sources to be minimal. The virtual absence of
excess nitrate or nitrite ion in precipitation samples
collected beneath such plumes (Dana et al. 1976) provides
strong evidence that direct uptake of primary nitric
oxide and nitrogen dioxide by precipitation and cloud
elements is a negligibly slow process.

Nonreactive scavenging of plumeborne sulfur dioxide is
solubility dependent and tends also to be a rather

TABLE C.3-2 Summaries of Some Precipitation Scavenging Field Investigations

General Source Type	Specific Source Type	Selected References
Continuous point source	Tower releases of aerosols	Chamberlain (1953), Englemann (1965), Dana (1970)
	Tower releases of radioactive gases and simulated tracers	Chamberlain (1953), Engelmann et al. (1966)
	Tower releases of SO_2	Dana et al. (1972), Hales et al. (1973)
	Tower releases of tritiated water vapor	Dana et al. (1978)
	Tower releases of organic vapors	Lee et al. (1974)
	Power-plant plumes	Barrie and Kovalick (1978), Dana et al. (1973, 1976, 1982), Enger and Hogstrom (1979), Granat and Rodhe (1973), Granat and Soderland (1975), Hales et al. (1973), Hutchenson and Hall (1974), Radke et al. (1978)
"Instantaneous" and/ or moving sources	Smelter plumes	Kramer (1973), Larson et al. (1975)
	Aircraft releases of rare-earth tracers	Changnon et al. (1981), Dingle et al. (1969), Gatz[a] (1977), Slinn (1973b), Young et al. (1976)
	Rocket releases of radioactive tracers	Burtsev et al. (1976), Shopauskas et al. (1969)
Urban sources	Uppsalla, Sweden	Hogstrom (1974)
	St. Louis, Mo.	Hales and Dana (1979a), Hales et al. (1979)
	Los Angeles, Calif.	Morgan and Liljestrand (1980)
General and regional sources	Regional pollution flowing into lake-effect storms	Scott (1981)
	Regional pollution in the eastern U.S. and Canada	Easter (1982), MAP3S/RAINE (1981)
	Regional aerosol loadings at a specific receptor point	Graedel and Franey (1977), Peters et al. (1978)
Global and stratospheric sources	Cosmogeric radionuclides	Young et al. (1973)
	Nuclear fallout	Numerous studies: see Fuquay (1970)

[a]The reference by Gatz provides a comprehensive list of past tracer studies of precipitation scavenging.

inefficient process, although it is definitely detectable in field studies conducted in relatively clean environments (Dana et al. 1973, 1976; Hales et al. 1973). This phenomenon, which is suppressed under conditions involving high rain acidity, is relatively well understood at present (Drewes and Hales, 1982).

Nonreactive scavenging of sulfate aerosol can be an efficient removal process. The preponderance of relevant field tests of Table C.3-2, however, has demonstrated that wet deposition of sulfate from local power-plant and smelter plumes occurs rather slowly. This is undoubtedly a consequence of the small amounts of primary sulfate available for scavenging under such circumstances.

Field tests conducted under situations wherein sulfur trioxide was intentionally injected into the stack of a coal-fired power plant (Dana and Glover 1975) show correspondingly high sulfate scavenging rates, and it has been suggested that under certain operating conditions some types of power plants (especially oil-fired units) will produce sufficient primary sulfate to account for appreciable local deposition. To date, however, there has been no really strong field evidence in support of this point. Hogstrom et al. (1974) reported the observation of substantial sulfate scavenging from the local plume of an oil-fired power plant in Sweden, but these results are rather dependent on the interpretation of background contributions. Granat and Soderlund (1975) performed a similar investigation in the vicinity of a second Swedish oil-fired plant and found a comparatively small scavenging rate.

Reactive scavenging of plumeborne sulfur dioxide to form rainborne sulfate is difficult to differentiate from primary sulfate removal. The previously noted findings of low excess sulfate in below-plume rain samples, however, suggests that neither process is particularly effective in near-source plume depletion.

The scavenging of hydrochloric acid to produce chloride and hydrogen ions in precipitation will most certainly be a highly effective process, depending on the quantities of hydrochloric acid available. Considerable theoretical and laboratory work has been conducted in this area for Space Shuttle impact assessment, and there are limited data suggesting that hydrogen chloride is scavenged in measurable amounts from power-plant plumes (Dana et al. 1982).

With the exception of studies conducted under rather clean ambient conditions (e.g., Dana et al. 1973, 1976) the influence of background contributions has made the interpretation of plume scavenging a difficult task. Typically the sulfate and nitrate concentrations in precipitation collected adjacent to the plume are quite variable, and subtracting this influence to determine source contributions involves substantial levels of uncertainty. This difficulty of "source attribution" at the local scale is compounded appreciably as greater scales of time and distance are considered.

On a more intermediate scale (0-200 km) an enhancement of sulfate and nitrate precipitation scavenging seems to occur, presumably because the precursors have had more opportunity to dilute and to react under these circumstances. Hogstrom (1974) reported substantial scavenging rates of sulfur compounds using an extended network of samplers in the vicinity of Uppsala, Sweden. Hales and Dana (1979a,b) observed summertime convective storms to remove appreciable fractions of urban NO_x and SO_x burdens in the vicinity of St. Louis, Missouri. Although both of these studies were subject to the usual uncertainties with regard to background contributions, there is little doubt about their general conclusions of significant scavenging under such circumstances.

On a regional scale (0-2000 km) there are relatively few data from intensive field experiments. Precipitation-chemistry network data are of some utility in this regard, however, and several analyses have applied these measurements to specific ends. One result of these analyses is the suggestion that, in the northeastern quadrant of the United States, roughly one third of the emitted NO_x and SO_x is removed by wet processes (Galloway and Whelpdale 1980). Network data for the northeast (MAP3S/RAINE, 1982) show also that the molar wet delivery rates of NO_x and SO_x are roughly equivalent. Combining this result with regional emission inventories suggests that nitrogen compounds begin to wet deposit with a significantly enhanced efficiency as distance scales become regional in extent.

The above changes in behavior with increasing scale seem to be a logical consequence of current understanding regarding the atmospheric chemistry of SO_x and NO_x. On local scales neither is scavenged very effectively owing to the chemical makeup of the primary emissions. On intermediate scales both groups have had some opportunity to react into more readily scavengable substances. Depending on ambient conditions, the nitrogen oxides will have participated to some extent in initial photolysis reactions and proceeded on to form scavengable products such as nitric acid, peroxyacetyl nitrate, and nitrate aerosol. Sulfur dioxide also will have reacted homogeneously to a limited extent; more importantly, however, this compound will have <u>diluted</u> to levels where limited reactants (and possibly catalysts) will facilitate its oxidation in the aqueous phase. On a regional scale this progression continues with the relative acceleration of NO_x scavenging.

Present field-study indications that NO_x scavenging may occur primarily through the attachment of gas-phase reaction products, while the scavenging of SO_x may depend much more heavily on aqueous-phase oxidation processes are also reflected in precipitation-chemistry data. A possible consequence of this difference in mechanisms is illustrated in Figure C.3-15, which is a time series of daily precipitation-chemistry measurements for a northeastern U.S. site. The decidedly periodic* behavior of sulfate-ion concentrations in contrast to the largely disorganized behavior of nitrate-ion concentrations has been suggested to occur as a consequence of an aqueous-phase oxidation of sulfur dioxide, which proceeds more rapidly during summer months. Whatever the cause, it is readily apparent from this figure that scavenging mechanisms for these two species differ appreciably.

---

*One should note in Figure C.3-16 that the periodic functions are fit to the total data, whereas the linear regressions are fit only for the period January 1, 1977-December 31, 1979; thus the cyclic functions are not exactly symmetric about the linear regression curves. Some idea of statistical improvement in fit may be obtained using the expression

$$\hat{r}^2 = \frac{\sigma^2_{\text{linear regression}} - \sigma^2_{\text{periodic fit}}}{\sigma^2_{\text{linear regression}}},$$

where the σ^2's pertain to variances of the data points over the three and one-half period. For sulfate in Figure C.3-16, r^2 equals 0.22, indicating a significant reduction in variance; the corresponding r^2 value for nitrate is 0.01, suggesting that no significant annual periodicity exists in this case.

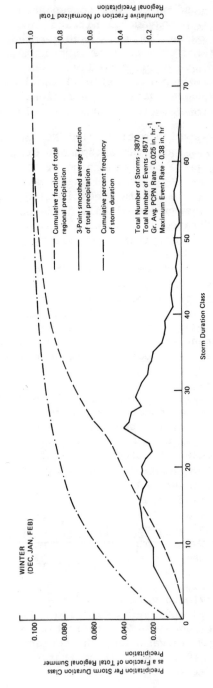

FIGURE C.3-15 Frequency statistics for storm duration, cumulative fraction of total regional precipitation, and smoothed average fraction of total regional precipitation. Winter (December, January, February) data from the MAP3S region. The frequency analysis allows up to 6 consecutive hours without precipitation to be counted as part of one storm.

As noted above, most past field experiments have experienced difficulty in resolving precisely which source(s) of pollution has been responsible for material wet-deposited at sampled receptor sites, and this problem is typically amplified as time and distance scales increase. Source attribution is particularly uncertain on a regional scale, and the basic data obtainable from standard precipitation-chemistry networks are of limited help in this regard. Combined with the lack of data from well-designed regional field studies, this aspect poses one of the most important and uncertain questions facing the acid deposition issue at present.

As a consequence of this need, a major regional field experiment has recently been designed and conducted in the northeastern United States (Easter 1982, MAP3S/RAINE 1981). Known as the Oxidation and Scavenging Characteristics of April Rains (OSCAR) study, this field experiment was based on the concept of characterizing, as completely as possible, the dynamical and chemical features of major cyclonic storm systems as they traverse the continent. Specific objectives were as follows:

1. To assess spatial and temporal variability of precipitation chemistry in cyclonic storm systems and to test the adequacy of existing networks to characterize this variability;
2. To provide a comprehensive, high-resolution data base for prognostic, regional deposition-model development;

and

3. To develop increased understanding of the transport, dynamical, and physicochemical mechanisms that combine to make up the composite wet-removal process and to identify source areas responsible for deposition at receptor sites.

The data collected and assembled by the OSCAR project are summarized in Table C.3-3. These are being made available to the general user community in a computerized data base.

A general layout of the OSCAR precipitation chemistry network is shown in Figure C.3-16. The points and triangles on this map represent locations of sequential precipitation-chemistry stations on an "intermediate-density" network, and the open square, overlapping Indiana and Ohio, depicts a concentrated network of 47 additional sites. Specific design criteria for this

TABLE C.3-3   Summary of Data Collected for the OSCAR Data Base

---

Meteorological Data

- North American standard 12-h upper-air observations (rawinsondes)
- OSCAR special rawinsonde data
- North American 3-h standard surface observations
- North American hourly precipitation amount data
- Trajectory forecast data (Limited Fine Mesh and Global Spectral Models)
- Gridded forecast data (Limited Fine Mesh Model)
- Satellite observations

Precipitation-Chemistry Data

- OSCAR network: Sequential measurements of rainfall, field pH, laboratory pH, conductivity, NO_3^-, NO_2^-, $SO_4^=$, $SO_3^=$, Cl^-, NH_4^+, Ca^{++}, Mg^{++}, K^+, Na^+, Al^{+++}, PO_4^X, total Pb
- Additional networks: Time-averaged data as available from sources such as NADP, CANSAP, CCIW, and APN
- Special rainborne H_2O_2 measurements

Aircraft Data

- Trace gases: O_3, NO/NO_x, SO_2, HNO_3, NH_3
- Aerosol parameters: scattering coefficient (b_{scat}), Aitken nuclei, aerosol sulfur, sulfate size distribution, aerosol size distribution, aerosol acidity
- Cloud water chemistry: NO_3^-, NO_2^-, $SO_4^=$, $SO_3^=$, pH, NH_4^+, conductivity, Cl^-, Ca^{++}, Mg^{++}, K^+, Na^+, total Pb
- Meteorological parameters: Temperature, humidity, liquid-water content, wind speed and direction, cloud droplet size distribution
- Position parameters: Latitude, longitude, altitude, time

Surface Air Chemistry Data

- OSCAR SAC site (Fort Wayne 40° 49.8' N, 85° 27.6' W): H_2O_2, peroxyacetyl nitrate, sulfur aerosol size distribution, NH_3, SO_2, $SO_4^=$, O_3, NO/NO_x, HNO_3, aerosol composition versus particle size, aerosol acidity
- Selected air -uality data from specific surface monitoring sites throughout eastern North America

Emissions

- MAP3S/RAINE standard inventory

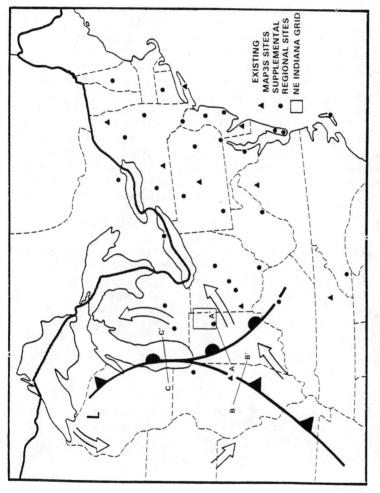

FIGURE C.3-16 General layout of OSCAR sequential precipitation chemistry network, showing hypothetical "design-basis" cyclonic system.

configuration are discussed in the supporting literature (MAP3S/RAINE 1982).

The OSCAR data set is currently under intensive analysis, and only preliminary results are available. It is of interest to consider some of these results at this point, however, to evaluate the potential future utility of this material. One early result, presented by Raynor (1981), is primarily of qualitative interest. These are the first-sample/last-sample pH data obtained by the sequential rain samplers for individual storms and are typified by the plots shown in Figures C.3-17 and C.3-18. It is interesting to note that Figure C.3-17 is strongly reminiscent of annual- or multiyear-average plots for the northeastern United States in the sense that it shows the familiar acid "core" region centered upon Pennsylvania.* The final-sample distribution in Figure C.3-18 is quite different. Besides indicating a much cleaner sample set, very little structure exists in this final distribution. This relative cleanliness of late-storm precipitation is consistent with the general OSCAR finding that most of the pollutant is scavenged comparatively early in a storm's life cycle (Easter and Hales 1983a).

Substantial source-receptor analysis is currently being conducted in conjunction with the Indiana-Ohio concentrated network. One early analysis, conducted for the April 22-24, 1981, storm, is presented in Figure C.3-19. Backtrajectories of this type are currently being combined in diagnostic scavenging models with aircraft and surface data to evaluate source-receptor relationships in greater detail (Easter and Hales 1983a,b).

---

*It should be noted in this context that field studies having higher spatial resolution (e.g., Hales and Dana, 1979b, Semonin 1976) indicate that significant fine structure typically exists in spatial pH distributions. Much of this fine structure can be expected to be hidden within the relatively coarse sampling mesh shown in Figures C.3-18 and C.3-19.

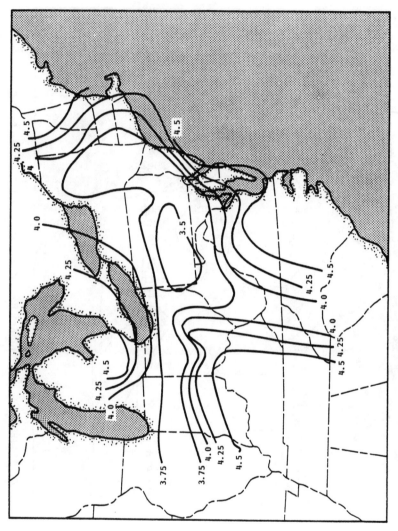

FIGURE C.3-17 Distribution of pH for initial precipitation sampled during OSCAR storm of April 22-24, 1981.

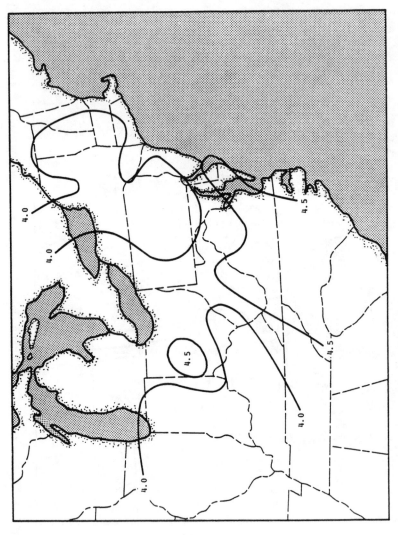

FIGURE C.3-18 Distribution of pH for final precipitation sampled during OSCAR storm of April 22-24, 1981.

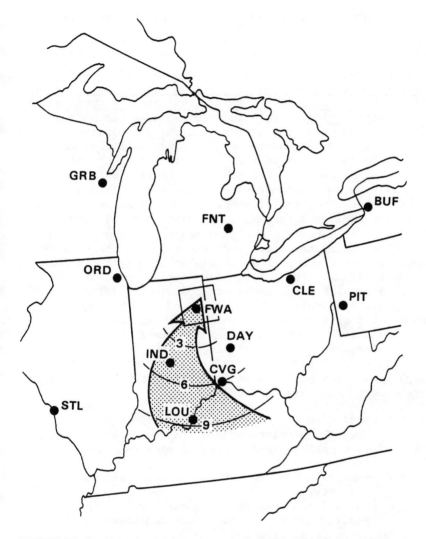

FIGURE C.3-19   Loci of points contributing pollution to the high-density network near 1400 EST on April 22, 1981. Contour intervals 3, 6, 9 represent travel times in hours from source regions. The large arrow represents the likely path of air originating from points 9 hours upwind of the receptors.

3.4 PREDICTIVE AND INTERPRETIVE MODELS OF SCAVENGING

3.4.1 Introduction

A precipitation-scavenging model can be defined as <u>any</u>
<u>conceptualization of individual or composite processes of</u>
<u>Figure C.3-2, in a manner that allows their expression in</u>
<u>mathematical form.</u> Often such models take the form of
submodels or "modules" within a larger calculational
framework, such as a composite regional pollution code.
When considered in a modular sense the lines connecting
the boxes of Figure C.3-2 can be considered as channels
for information exchange within the overall framework,
whereas the boxes (or clusters of boxes) can be identified
with the modules themselves. Scavenging models are
currently in a rapidly evolving state, and a profusion of
associated computer codes and computational formulas is
currently available. Indeed, one of the major problems
in precipitation-scavenging assessment is determining
precisely which model to select from the large number of
available candidates. A major aim of the present
subsection is to guide the reader in this pursuit.
    There are a number of potential uses for
precipitation-scavenging models, and the intended use
will to a large extent determine which model should be
employed. Some of the more important potential uses are
itemized as follows:

- Prediction of the impact on precipitation
  chemistry of proposed new sources, source
  modifications, and alternate emission-control
  strategies;
- Prediction of long-range trends in precipitation
  chemistry;
- Estimation of the relative contributions of
  specific sources to precipitation chemistry at a
  chosen receptor point;
- Estimation of transport of acidic-precipitation
  precursors across political borders;
- Estimation and prediction of air-quality
  modifications occurring as a consequence of the
  scavenging process;
- Site selection for precipitation-chemistry network
  sampling stations;
- Design of field studies of precipitation
  scavenging;

and

* Elucidation of mechanistic behavior of the
  scavenging process on the basis of field
  measurements.

In selecting an appropriate model, the user should
review his intended application carefully with regard to
the pollutant materials of interest, the time and
distance scales, the processes in Figure C.3-2 covered,
the source configuration, the precipitation type, and the
mechanistic detail required. The question of pollutant
materials is particularly important when precipitation
acidity is of interest. Acidity in precipitation is
determined by the presence of a multitude of chemical
species, and in principle one must compute (via a model)
the scavenging of each species and then estimate acidity
on the basis of an ion balance:

$$[H^+] = \Sigma \text{ Anions} - (\Sigma \text{ cations other than } H^+). \quad (C.3-1)$$

Inorganic ions usually important in precipitation
chemistry are itemized in Table C.3-4. Organic species
play a secondary role in the acidification process, which
appears to vary widely with region. Modeling of all of
these species simultaneously requires substantial effort,
and all "acid-precipitation" models up to the present
have focused on only one or just a few of the more important species, with contributions of the others estimated
on the basis of empiricism. Currently there is a tendency
for newer models to accommodate larger numbers of these
species; but complete modeling coverage will not be
achieved in the foreseeable future.
Mechanistic detail is another important feature
determining the basic composition of a scavenging model.
A comprehensive mathematical description of the scavenging
process can become rapidly overwhelming, and there is
usually a need to represent these relationships in a
comparatively simple, albeit approximate, manner. The
process of consolidating complex behavior in this fashion
is often referred to as lumping the system's parameters.
The resulting simplified expressions are termed parameterizations. Consolidating the effects of nonmodeled species
in empirical form, described in the preceding paragraph,
is one example of lumping. Numerous other examples will
arise throughout the remainder of this section.
This section will not attempt to provide the reader
with a detailed treatise on how models should be

TABLE C.3-4 Some Inorganic Ions Important in Precipitation Chemistry[a]

Cations	Anions
H^+	
NH_4^+	Cl^-
Na^+	NO_3^-
K^+	$SO_3^=$
Ca^{++}	$SO_4^=$
Mg^{++}	$PO_4^=$
	$CO_3^=$

[a]All ions are presented here in their completely dissociated states. The reader should note, however, that various states of partial dissociation are possible as well (e.g., HSO_3^-, HCO_3^-).

formulated and applied.* The approach, rather, will be to develop a basic understanding of the fundamental elements of a scavenging model and then provide a systematic procedure for choosing and locating appropriate models from the literature. The following subsection discusses the basic conservation equations, which constitute the conceptual bases for scavenging models in general. This is followed in turn by two simple applications of these relationships, which are presented to illustrate usage and to define some terms commonly

---

*For the reader interested in more detailed pursuit of this area, the works by Hales (1983) and Slinn (1983) are recommended. The Hales reference is something of a beginner's primer, while Slinn's treatment delves substantially deeper into mechanistic detail. Together they constitute a reasonable starting point for understanding and modeling basic scavenging phenomena.

used in scavenging models. The final subsection attacks
the problem of model selection, using a flow-chart
approach, which is designed to guide the user to a valid
choice in a systematic manner that avoids many of the
pitfalls normally encountered on such endeavors.

3.4.2 Elements of a Scavenging Model

3.4.2.1 Material Balances

In Figure C.3-3 the various arrows between boxes correspond physically to streams of pollutant and/or water,
and from this it is not difficult to realize that any
characterization of this system must include material
balances. Material balances thus form the underlying
structure for all scavenging models. To formulate a
material balance one simply visualizes some chosen volume
of atmosphere and sums over all inputs and outputs of the
substance in question.
  Two basic types of material balance are possible:

  1. "Microscopic" material balances, based on
      summation over a limiting small volume element of
      atmosphere;
and
  2. "Macroscopic" material balances, based on
      summation over a larger volume element of
      atmosphere (e.g., a complete storm system).

Microscopic material balances invariably lead to
differential equations, which must be integrated over
finite limits to obtain practical results. Macroscopic
balances result in mixed, integral, or algebraic
equations. Again the choice of material-balance type
depends on the specific modeling purpose at hand.
  An important general form of the differential material
balance for some chosen pollutant (denoted by subscript
A) is given by the equations* (cf. Hales 1983)

Equations (C.3-2) and (C.3-3) are quite general in the
sense that the velocity vectors denote velocity of
pollutant (rather than that of the bulk media) and thus
provide for all modes of transport (convective, diffusive,
...) without yet specifying how this transport is to
occur. These equations are not yet time-smoothed; thus
no closure assumptions have been applied at this point.

$$\frac{\partial c_{Ay}}{\partial t} = -\widetilde{\nabla} \cdot c_{Ay}\widetilde{v}_{Ay} - w_A + r_{Ay} \quad \text{(gas phase)} \qquad \text{(C.3.2)}$$

and

$$\frac{\partial c_{Ax}}{\partial t} = -\widetilde{\nabla} \cdot c_{Ax}\widetilde{v}_{Ax} + w_A + r_{Ax} \quad \text{(aqueous phase)}. \qquad \text{(C.3.3}$$

Here c_{Ay} and c_{Ax} denote concentrations of pollutant in the gaseous and condensed-water phases, respectively. The time rate of change of these concentrations within the differential volume element is related to the sum of inputs by (1) flow through the walls of the element, (2) interphase transport between the gaseous and condensed phases, and (3) chemical (and/or physical) reaction within the element. The $\widetilde{v}$ terms in Equations (C.3-2) and (C.3-3) denote velocity vectors, while $\widetilde{\nabla}$. is the standard vector divergence operator. The interphase transport term w_A accounts for all "attachment" processes (impaction, phoresis, diffusion, . . .) as well as any reverse phenomena such as pollutant-gas desorption, while the r terms denote chemical conversion rates in the usual sense. To formulate a usable model from these equations one needs to specify values for the functions v, w, and r and then solve differential equations (C.3-2) and (C.3-3) (subject to appropriate initial and boundary conditions) to obtain the desired concentration fields c_{Ay} and c_{Ax}. A simple example of this procedure is given in Section 3.4.2.

Energy Balances

Many terms in Equations (C.3-2) and (C.3-3), especially v_{Ax}, w_A, and r_{Ax}, depend strongly on the amount, state, and interconversion rates of condensed water; and it is important at this point to note that atmospheric water itself obeys material-balance expressions of this form. In selecting a scavenging model one often is confronted with the problem of deciding whether to estimate precipitation attributes and these related terms independently on the basis of assumptions or previous information or to attempt to compute the desired entities directly by solving appropriate forms of Equations (C.3-2) and (C.3-3).

If the latter of these alternatives is chosen, then the inclusion of an energy-balance equation is mandatory. This need arises because the evaporation-condensation process influences, and is influenced by, a variety of energy-related considerations. These include temperature influences on vapor pressure and latent-heat effects and can be incorporated in the model via an energy balance performed over the same element of atmosphere as that of the associated material balances. In microscopic form, a general expression of the energy balance (cf. Bird et al. 1960), is

$$\rho C_v \frac{\partial T}{\partial t} = - \nabla \cdot \widetilde{h} - p\nabla \cdot \widetilde{v} + \Gamma - D . \qquad (C.3-4)$$

Here the time rate of change of temperature is related to the sum of inputs by (1) flow through the walls of the element and (2) generation via (a) compression work, (b) latent heat effects, and (c) frictional dissipation. The vector terms h and v denote sensible heat flux and fluid velocity, respectively, while Γ and D pertain to latent heat and dissipation. ρ and C_v denote fluid density and specific heat in the usual sense. A straightforward example of the incorporation of Equation (C.3-4) for scavenging modeling purposes is given by Hales (1983).

Momentum Balances

Solutions to Equations (C.3-2)-(C.3-4) depend on the existence of some previous description of fluid velocity v [or v_{Ay} in the case of Equation (C.3-2)]. As was the case for the preceding parameters associated with the energy balance, velocity may be specified for the model on the basis of previous measurements or assumptions. Flow patterns in storm systems may be sufficiently complex to defy empirical specification, however, and the modeler may wish to compute the associated fields on the basis of a modeling approach. If this is to be done, a momentum-balance equation must be employed. In microscopic form the general momentum balance may be expressed (cf. Bird et al. 1960) as

$$\frac{\partial}{\partial t} \rho v = -\nabla \cdot \rho \widetilde{vv} - \nabla p - \widetilde{F}_v + \rho g. \qquad (C.3-5)$$

Here the time rate of change of momentum (ρv) is expressed as the sum of inputs by (1) flow through the walls of the element, (2) pressure forces, (3) viscous drag forces, and (4) gravitational forces. To apply Equation (C.3-5) for modeling purposes one specifies frictional, pressure, and gravitational terms and solves the differential equation subject to appropriate initial and boundary conditions to obtain fields of the velocity vector $\tilde{v}$. An example of application of Equation (C.3-5) for scavenging modeling purposes is given by Hane (1978).

Incorporation of energy and momentum balances Equations (C.3-4) and (C.3-5) into a scavenging model is a rather challenging exercise, and a relatively small number of models exist that apply these equations for this purpose. The usual tack is simply to "prespecify" the required parameters and proceed with material-balance calculations alone. Numerous examples of both types of models will be presented in Section 3.4.5.

3.4.3 Definitions of Scavenging Parameters

Four key parameters often arise in the context of scavenging models, and it is appropriate at this point to define these terms and indicate their general application. Reference to these entities as "parameters" is consistent with the usage applied in the previous section, in that they serve to "lump" the effects of a number of mechanistic processes in a simple formulation. These will be discussed sequentially in the following paragraphs.

The first parameter to be defined is the <u>attachment efficiency</u>. Also known as the <u>capture efficiency</u>, this term can be visualized most easily by considering a hydrometeor falling through a volume of polluted air space, as shown in Figure C.3-20. This hydrometeor sweeps out a volume of air during its passage; and attachment efficiency is defined as the amount of collected pollutant divided by the amount that was initially in this volume. The efficiency can exceed 1.0 if pollutant from outside the swept volume becomes attached to the drop.

From the discussion of attachment mechanisms in Section 3.2 it is seen that the attachment efficiency accounts for a multitude of processes. Usually the efficiency is less than 1; but mechanisms such as diffusion, electrical effects, and interception can give rise to larger values, especially when the collecting

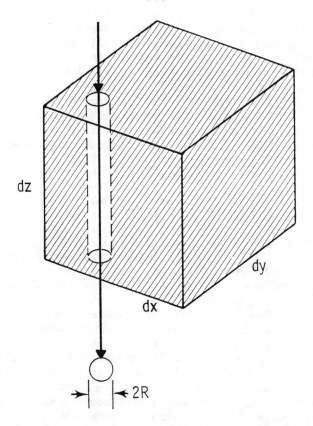

FIGURE C.3-20 Schematic of a scavenging Hydrometeor falling through a volume element.

element's fall velocity is small. Efficiencies can be negative if the element is releasing pollutant to the surrounding atmosphere, such as in the case of pollutant-gas desorption. Typical efficiencies for aerosol particles collected by raindrops are shown in Figure C.3.4.

Another important parameter is the <u>scavenging coefficient</u>. This entity is basically an expression of the law of mass action, and is defined by the form

$$\Lambda = \frac{w_A}{c_{AY}} , \qquad (C.3-6)$$

where [in a manner consistent with Equations (C.3-2) and C.3-3)] w_A is the rate of depletion of pollutant A from the gaseous phase by attachment to the aqueous phase in a differential volume element. This is similar to a rate expression for a first-order, irreversible chemical reaction, and as such it applies strictly only to irreversible attachment processes (e.g., aerosols or highly soluble gases). Λ can be related to the attachment efficiency E by the form (which assumes spherical hydrometeors)

$$\Lambda(a) = -\pi N_T \int_0^\infty R^2 v_z(R) E(R,a) f_R(R) dR, \qquad (C.3-7)$$

where a and R denote aerosol and hydrometeor radii, respectively, v_z is the hydrometeor fall velocity, and N_T and f_R are the total number and probability-density functions for the size-distributed hydrometeors residing in the volume element of Figure C.3-20 at any instant in time. From this one can note that Λ essentially extends the paramaterization over the total spectrum of hydrometeor sizes.

Atmospheric aerosol particles are typically distributed over extensive size ranges, and because of this it is often desirable to possess some sort of an effective scavenging coefficient, which represents a weighted average over the aerosol size spectrum. Figure C.3-21 presents a family of curves corresponding to such averages, which are based on assumed log-normal particle-size spectra, with different geometric standard deviations. From these curves one can observe that for the same geometric mean particle size, changes in spread of the size distribution can result in dramatic changes in the effective scavenging coefficient.

Inclusion of reversible attachment processes in a scavenging model usually involves utilization of the mass-transfer coefficient. This parameter can be defined in terms of the flux of pollutant moving from the scavenging element as

$$\text{Flux} = -\frac{K_y}{c}(c_{Ay} - h'\hat{c}_A). \qquad (C.3-8)$$

Here K_y is the mass-transfer coefficient and c_A is the concentration, within the scavenging element, of

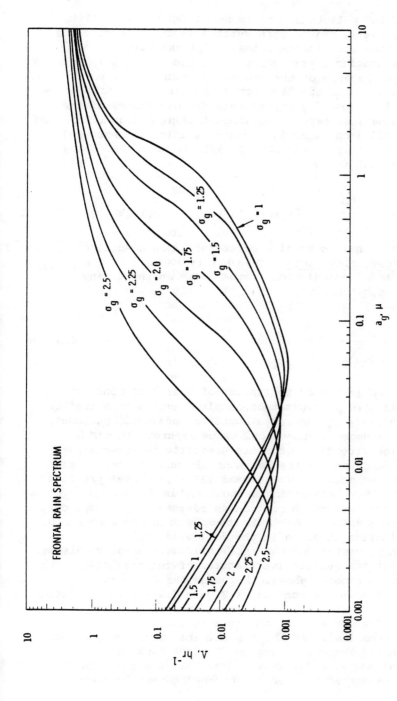

FIGURE C.3-21 Computed effective scavenging coefficients for size-distributed aerosols. Based on a log-normal aerosol radius distribution with geometric means and standard deviations a_g and σ_g. A typical frontal-rain drop-size spectrum is assumed. From Dana and Hales (1976).

collected pollutant. h' is essentially a solubility
coefficient, which, when multiplied by c_A, produces a
gas-phase equilibrium value. c is the molar
concentration of air molecules, which appears in Equation
(C.3-8) because of the manner in which K_y has been
defined. Thus the flux can be either to the drop or away
from it, depending on the relative magnitudes of the
parenthetical terms. Equation (C.3-8) can be integrated
over all drop sizes in a manner similar to that used in
Equation (C.3-7) (cf. Hales 1972) to form the following
expression for w_A:

$$w = \frac{4\pi N_T}{c} \int_0^\infty R^2 f_R(R) K_y(R) (c_{Ay} - h' \hat{c}_A) \, dR. \qquad (C.3-9)$$

The final scavenging parameter to be described here is
the scavenging ratio. This entity is usually the result
of a model calculation, rather than an input, and is
defined by the form

$$\xi = \frac{\hat{c}_A}{c_{Ay}}, \qquad (C.3-10)$$

where C_A is the concentration of pollutant contained in
a collected precipitation sample. ξ is a term that is
immediately usable for a number of pragmatic purposes,
because once its numerical value is known it can be
applied directly to compute precipitation-chemistry
concentrations on the basis of air-quality measurements.
Tables of measured (Engelmann 1971) and model-predicted
(Scott 1978) scavenging washout ratios have been
published, although caution is advised in the application
of these values. A simple example of scavenging-ratio
application is given in the following section.

It is useful for the sake of visualization to discuss
briefly the qualitative features of the scavenging
parameters noted above. The parameter E is easy to
visualize in the context of Figure C.3-20; it is, simply,
the collection efficiency of an individual cloud or
precipitation element and as such should be expected to
fall numerically in the approximate range between zero
and one. The scavenging coefficient Λ can be
visualized as a first-order removal rate, in much the
same manner as that of a first-order reaction-rate

coefficient. As such it may be utilized roughly as a characteristic time scale for wet removal. $\Lambda = 1$ h^{-1}, for example, would imply that the scavenging process will cleanse $100(1 - 1/e)$ percent of the pollutant in 1 h if conditions remain constant and competitive processes do not occur. From this one can note that 1 h^{-1} is a moderately large scavenging coefficient. Λs ranging from zero to 1 h^{-1} and beyond have been reported in the literature (cf. Figure C.3-21).

The mass-transfer coefficient K_y is essentially a normalized interfacial flux of pollutant between the atmosphere and an individual droplet. Little needs to be said here regarding magnitudes of K_y, except to note that a variety of different <u>definitions</u> of K_y exist, and one must be cognizant of these definitions when employing values obtained from outside sources. The washout ratio, ξ, is essentially a measure of the <u>concentrating power</u> of precipitation in its extraction of pollutant from the atmosphere. As will be noted in the next section, precipitation often has the ability to concentrate airborne pollution by a factor of a million or more. ξs ranging from below 100 up through 10^8 and higher have been reported in the literature.

The expected magnitudes and uncertainty levels associated with the scavenging parameters listed in this section depend strongly on the substance being scavenged and the environment in which the scavenging takes place. Large aerosol particles in below-cloud environments, for example, are characterized by scavenging efficiencies in the range of 1.0 (cf. Figure C.3-4), which can be estimated with relatively high precision. Smaller particles, especially those in the "Greenfield-gap" region, are much more difficult to simulate, and associated errors in estimated efficiencies may approach an order of magnitude or more. Errors in these efficiency estimates will of course be compounded by uncertainties in raindrop size spectra, if extended to scavenging coefficients via Equation (C.3-7). In the case of gases, the mass-transfer coefficient usually can be estimated to within a factor of 2 or less; again this error can be expected to compound when integrated over assumed raindrop size-spectra.

In the case of in-cloud scavenging of aerosols our capability for estimating transport parameters is seriously impeded, owing to the profusion of mechanisms and the complex environments involved. Typical uncertainties in both Λ and ζ can be expected to approach an order of magnitude in some cases. Some appreciation for the factors influencing in-cloud scavenging coefficients can be obtained from the work of Slinn (1977), who attempts to evaluate theoretical, "storm-averaged" values for Λ. An idea of the magnitudes and uncertainties of ξ is given in Figure C.3-23.

In all cases involving reactive gases the values of E, Λ, and ξ are heavily contingent on the aqueous-phase chemical processes involved. Much remains to be accomplished in our understanding of aqueous-phase chemistry before a meaningful assessment of associated uncertainties is possible.

As a final note in this context it should be emphasized that uncertainties in scavenging <u>parameters</u> dictate uncertainties in scavenging <u>calculations</u> in a complex fashion and that errors associated with the microscopic phenomena can be either amplified or attenuated by their applications in macroscopic models to produce practical results. Uncertainties associated with macroscopic modeling applications will be discussed at some length in a later section.

3.4.4 Formulation of Scavenging Models: Simple Examples of Microscopic and Macroscopic Approaches

As noted previously, the description given in this document will refrain in general from deriving and applying scavenging models explicitly. This is too broad and complex a topic to be discussed in detail here, and the reader is referred to the previously cited literature for more detailed pursuit of this subject. For purposes of illustration, however, it is worthwhile to consider two simple examples of scavenging-model formulation, which demonstrate microscopic and macroscopic approaches to the problem. The present subsection addresses this task.

The microscopic material balance approach will be considered first. For this example it is useful to visualize an idealized situation where rain of known characteristics is falling through a stagnant volume of atmosphere, which contains a well-mixed, nonreactive pollutant with concentration c_{Ay}. The air velocity is known ($v = 0$) so solution of the momentum equation (C.3-5) is not required. The raindrop size distribution is presumed to remain constant; thus evaporation-condensation and other energy-related effects are immaterial, and the energy equation (C.3-4) may be disregarded.

Since the pollutant is well mixed, no concentration gradients occur; thus the divergence term in Equation (C.3-2) is zero. Because of nonreactivity the reaction term is zero as well.

Now presume that the pollutant is an aerosol, whose attachment can be characterized in terms of the known scavenging coefficient Λ, using Equation (C.3-6). The corresponding reduced form of Equation (C.3-2) is, then,

$$\frac{\partial c_{Ay}}{\partial t} = -\Lambda c_{Ay}. \qquad\qquad (C.3-2a)$$

Given some initial pollutant concentration c_{Ayo}, Equation (C.3-2a) can be integrated to obtain the form

$$c_{Ay}(t) = c_{Ayo} \exp(-\Lambda t), \qquad\qquad (C.3-11)$$

which expresses the decrease of the gas-phase pollutant concentration with time. Counterpart expressions for rainborne concentrations may be derived by subjecting Equation (C.3-3) to a similar treatment.

The reader is cautioned to consider this treatment as an example only and to recognize that actual atmospheric conditions seldom conform to the idealizations invoked above. Gas-phase concentrations are usually not uniformly distributed in space, raindrop characteristics are usually not invariant with time, wind fields are usually not well characterized by $v = 0$. Λ is usually not a time-independent constant, and many pollutants are usually not well characterized by the washout coefficient approximation, anyway. The pollutant often is not unreactive. Examples of existing models where these constraints are relaxed in various ways are presented in the following subsection.

Figure C.3-22 illustrates the formulation of a macroscopic type of scavenging model. Here, in contrast to the differential-element approach, the material balances are formulated around a large volume element, in this case a total storm. If one denotes concentrations and flow rates of water and pollutant as follows:

c_{Ay} = airborne concentration of pollutant,
H = airborne concentration of water vapor into cloud,

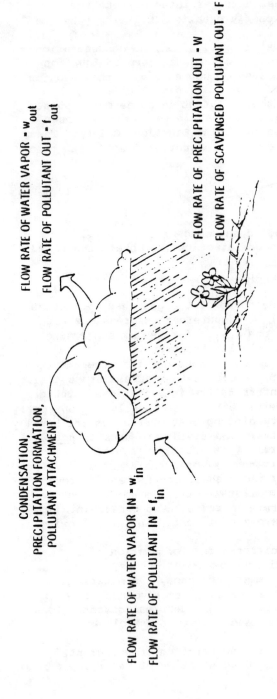

CONDENSATION,
PRECIPITATION FORMATION,
POLLUTANT ATTACHMENT

FLOW RATE OF WATER VAPOR - w_{out}
FLOW RATE OF POLLUTANT OUT - f_{out}

FLOW RATE OF WATER VAPOR IN - w_{in}
FLOW RATE OF POLLUTANT IN - f_{in}

FLOW RATE OF PRECIPITATION OUT - W
FLOW RATE OF SCAVENGED POLLUTANT OUT - F

DEFINITIONS OF EFFICIENCIES:

WATER REMOVAL   POLLUTANT REMOVAL

$$\epsilon_p = \frac{W}{w_{in}} \qquad \epsilon = \frac{F}{f_{in}}$$

FIGURE C.3-22   Schematic of a typical macroscopic material balance.

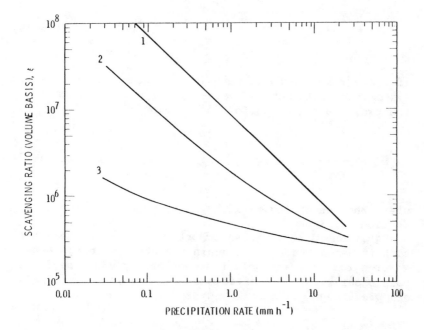

FIGURE C.3-23 Scott's scavenging ratio curves: 1, convective storms; 2, warm, non-convective storms; 3, cold storms, where Bergeron-Findeisen process is active (Scott 1978).

$\hat{c}_A$	=	concentration of scavenged pollutant in rainwater,
ρ_w	=	density of condensed water,
w_{in}	=	flow rate of water vapor into the storm,
w_{out}	=	flow rate of water vapor out of the storm,
f_{in}	=	flow rate of pollutant into the storm,
f_{out}	=	flow rate of pollutant out of the storm,
W	=	low rate of precipitation out of the storm,
F	=	flow rate of scavenged pollutant out of the storm,

then extraction efficiencies for water vapor and pollutant can be defined, respectively, as

$$\epsilon_p = \frac{W}{w_{in}} \qquad\qquad (C.\ 3\text{-}11)$$

and

$$\epsilon = \frac{F}{f_{in}}.$$ (C.3-12)

If one further performs material balances over this storm system for pollutant and water vapor, and then combines the two, the following form is obtained:

$$\xi = \frac{\hat{c}_A}{c_{AY}} = \frac{\epsilon_p \rho_w}{\epsilon H},$$ (C.3-13)

where the scavenging ratio, ξ, was defined earlier in Section 3.4.3.

Equation (C.3-13) is an important result in the sense that it demonstrates once again the strong linkage between water-extraction and pollutant-scavenging processes. If both occur with equal efficiency ($\epsilon_p = \epsilon$),* for example, then

$$\xi = \frac{\rho_w}{H} \approx 10^5 - 10^6.$$ (C.3-14)

Experimentally measured scavenging ratios often fall in this range, although wide variability often may be observed.

Utilizing a rather involved series of arguments pertaining to cloud-physics processes and attachment mechanisms, Scott (1978) has created a family of curves expressing aerosol scavenging ratio as a function of precipitation rate. Shown in Figure C.3-23, curves 1, 2,

---

*There is no direct reason to expect that ϵ_p should be similar to ϵ in magnitude. In the absurd circumstance where all the pollutant were concentrated into one particle, for example, then scavenging of that pollutant by a very light rainfall would yield $\epsilon \simeq 1.0 >> \epsilon_p$. Conversely, a large storm processing an insoluble gaseous pollutant (SF_6^p, say) would provide $\epsilon \approx 0 << \epsilon_p$. For practical conditions involving acid-forming aerosols, however, the scavenging of water vapor and pollutant appear to be sufficiently related to allow $\epsilon_p \approx \epsilon$ to be employed as an approximate rule of thumb.

and 3 pertain, respectively, to convective storms, nonconvective warm-rain process storms, and cold storms where the Bergeron-Findeisen process is active.

A major assumption of Scott's analysis is that the pollutant is ingested by the storm in the form of aerosol particles that are active as cloud condensation nuclei. The analysis also assumes a steady-state storm system and complete vertical mixing of pollutant between the storm height and the surface. Under such conditions Scott's curves can be considered as reasonably good estimators of actual scavenging behavior. More elaborate systems, involving reactive pollutants, gases, and nonhomogeneous systems are discussed in references given in the following section.

3.4.5 Systematic Selection of Scavenging Models: A Flow-Chart Approach

Hales (1983) has suggested a flow-chart approach to aid in the process of scavenging-model selection. Presented as a decision tree in Figure C.3-24, the user proceeds by answering a series of questions that relate to the model's intended use, the temporal and geographical scales, the pollutant characteristics, the choice between macroscopic and microscopic material balances, and the type of conservation (i.e., material, energy, momentum) equations involved. Various pathways through this decision tree are discussed in the original reference.

Proceeding through Figure C.3-24 in this manner the user can arrive at simple or complex end points, depending on the nature of his particular application. A trivial example is pathway 1-5-6, which instructs the user to disregard modeling totally and rely solely on past measurements. The simple microscopic-balance example of Section 3.4.4 can be traced through pathway 1-2-7-8-21-23-15-16.

Table C.3-5 presents an itemization of some currently available models, which can be related directly to the pathways of Figure C.3-24. This provides the reader with a rapid and efficient means of access to current modeling literature, while minimizing the chance of pitfall encounters that can arise from the inadvertent invocation of inappropriate physical constraints. For a more definitive description of this model selection process, the reader is referred to Hales's original reference.

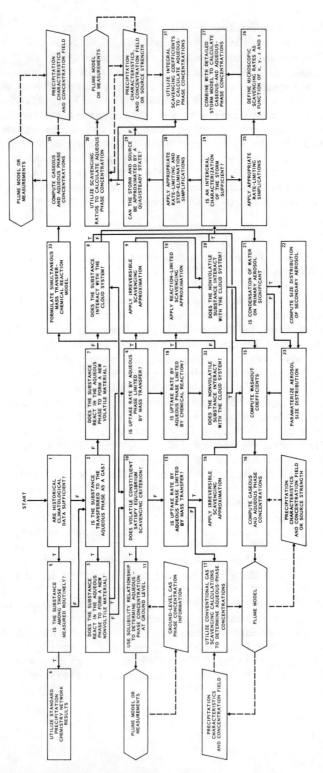

FIGURE C.3-24 Flow chart for scavenging calculations.

3.5 PRACTICAL ASPECTS OF SCAVENGING MODELS: UNCERTAINTY LEVELS AND SOURCES OF ERROR

Quantitative assessment of the predictive capability of present wet-removal models is a complex task and is well beyond the scope of this document. There are, however, a number of general statements that are highly useful in focusing in on this question and in providing insights pertaining to model reliability. These are itemized sequentially below.

• The predictive capability of a scavenging model is strongly contingent on its desired application.

As noted in Section 3.4.1, there exists a variety of different applications of scavenging models, and some are much more difficult to fulfill than others. One can, for example, employ existing regional models to reproduce distributions of annually averaged, wet-deposited, sulfate ion in eastern North America with moderate success. If one is charged with the task of relating specific sources to deposition at a chosen receptor site, however, our predictive capability can be expected to be relatively imprecise. Similarly, if one is expected to forecast the change in deposition that would occur in response to some future change in emissions, then the associated uncertainty level would be very high indeed. The question of nonlinear response is of paramount importance in this last application.

A large component of our uncertainty in predicting source attribution and transient response is based simply on the fact that we do not have adequate data bases for testing model performance for these applications. Our present models may in actuality be better predictors in this respect than anticipated; but because we have no immediate way of confirming this our uncertainty level remains high.

Regardless of the above considerations it should be emphasized strongly that the first step in scavenging model evaluation must be the precise definition of the intended uses of the model. All subsequent efforts will be confounded in the absence of this focal point.

• The predictive capability of a scavenging model is dependent on the choice of model.

At first sight this appears to be a self-evident and trivial statement. A profusion of scavenging models exists, however, and it is not at all difficult to choose

TABLE C.3-5 Pertinent Literature References for
Wet-Removal Models

Model	Type of Balance Equations	Mechanisms
1. Classical washout coefficient	Material (differential)	Irreversible attachment
2. Distributed washout coefficient	Material (differential)	Irreversible attachment
3. "Two-stage" nucleation-accretion	Material (differential)	Irreversible attachment
4. Nonreactive gas scavenging	Material (differential)	Reversible attachment
5. Reactive gas scavenging	Material (differential)	Reversible attachment with aqueous-phase reaction
6. In-cloud aerosol scavenging	Material (differential)	Irreversible attachment
7. In-cloud aerosol scavenging	Material (integral)	Irreversible or reversible attachment
8. In-cloud reactive gas and aerosol scavenging	Material (differential)	Transport, reaction, and deposition
9. In-cloud reactive gas and aerosol scavenging	Material (integral)	Irreversible or reversible attachment with chemical reaction
10. Composite analytical	Material (differential)	Transport, reaction, and deposition
11. Composite trajectory	Material (differential)	Transport, reaction, and deposition
12. Composite grid	Material (differential)	Transport, reaction, and deposition
13. Composite statistical	Material	Transport, reaction, and deposition
14. Nonreactive	Material energy and momentum (differential)	Irreversible attachment, nonreactive
15. Reactive	Material and energy (differential)	All modes of scavenging including chemical reaction

Typical Application	Pertinent References
Below-cloud scavenging of aerosols and reactive gases	Chamberlain (1953), Engelmann (1968), Fisher (1975), Scriven and Fisher (1975), Wangen and Williams (1978)
Below-cloud scavenging of size-distributed aerosols	Dana and Hales (1976), Slinn (1982)
Condensation-enhanced below-cloud scavenging of aerosols	Radke et al. (1978), Slinn (1982)
Below-cloud scavenging of nonreactive gases	Barrie (1978), Hales et al. (1973), Slinn (1974b)
Below-cloud scavenging of reactive gases	Adamowitz (1979), Drewes and Hales (1982), Durham et al. (1981), Hill and Adamowitz (1977), Overton et al. (1979)
Scavenging in storm systems (nonreactive)	Dingle and Lee (1973), Junge (1963), Klett (1977), Lange and Knox (1977), Slinn (1982), Storebo and Dingle (1974)
Scavenging in storm systems	Engelmann (1971), Gatz (1972), Hales and Dana (1979), Scott (1978), Slinn (1982)
Scoping studies	Gravenhorst et al. (1975), Omstedt and Rodhe (1978)
Interpretation of study data	Scott (1982)
Regional-scale deposition	Astarita et al. (1979), Fay and Rosenzweig (1980)
Regional-scale deposition	Bass (1980), Bhumralker et al. (1980), Bolin and Persson (1975), Eliassen (1978), Fisher (1978), Hales (1977), Heffter (1980), Henmi (1980), Kleinman et al. (1980), McNaughton et al. (1981), Patterson et al. (1981), Sampson (1980), Shannon (1981), Voldner (1982)
Regional-scale deposition	Carmichael and Peters (1981), Lamb (1981), Lavery et al. (1980), Lee (1981), Liu and Durran (1977), Prahm and Christensen (1977), Wilkening and Ragland (1978)
Scoping studies and life-time assessment	Rodhe and Grandell (1972)
In-cloud scavenging analysis	Hane (1978), Kreitzburg and Leach (1978), Molenkamp (1974)
In-cloud scavenging analysis	Hales (1982a)

an inappropriate candidate inadvertently. Such
inappropriate selections have on occasion resulted in
reported calculations that have been in error by several
orders of magnitude.

This component of error may of course be totally
eliminated by selection of the most appropriate model for
the intended application. The flow chart presented in
Figure C.3-24 is a useful guide for this purpose,
especially for those only casually familiar with the
field.

• The predictive capability of a scavenging model
depends strongly on the processes modeled.

As noted in the context of Figure C.3-2 a scavenging
model may encompass one, several, or all of the steps in
the composite wet-removal sequence. If only a small
portion of this sequence is being considered, the model
depends heavily on information supplied from the
remaining components. This information may originate
from assumptions, from empirical measurements, or from
the output of other models.

Assuming that all input information is error-free,
then it may be stated generally that, the more steps in
Figure C.3-2 encompassed by a given model, the greater
will be its predictive uncertainty. This is simply a
consequence of propagating errors and must be considered
as a primary factor when addressing the validation of
wet-removal calculations.

• The predictive capability of a scavenging model
is dependent on its areal range.

This statement is largely a corollary of the one
immediately above. As a scavenging model is extended to,
say, a regional scale it is forced to include essentially
all the components of Figure C.3-2. As noted previously,
this is likely to increase uncertainty levels
appreciably.

• The predictive capability of a scavenging model
is contingent on its temporal averaging time.

Owing to the propensity of stochastic phenomena to
average out to mean values, the predictive capabilities
of (especially regional) scavenging models can be
expected to improve somewhat as averaging times
increase. This improvement is, of course, gained at the
expense of sacrificing temporal resolution, and a value
judgment is necessary (again requiring a precise

definition of intended model application) at this
juncture.*

This observation should be tempered by the fact that,
in addition to random errors, scavenging models can be
expected to possess substantial systematic biases. In
general these biases do not decrease with averaging time
and in fact many lead to cumulative discrepancies on
occasion. Examples of systematic errors are biases in
trajectory calculations and artificial offsets induced by
the superimposition of random events on nonlinear
processes. Again the seriousness of such factors is
heavily contingent on the intended model application.

In general summary it may be stated that several
important factors lead to widely varying levels of
uncertainty in scavenging-model predictions. One may
predict, for example, the scavenging of SO_2 from a
local power-plant plume using existing models and expect
to match measured results within a factor of 2. On the
other hand, similar predictions of, say, the fraction of
sulfate at a given receptor that originated from some
particular source can be expected to have orders-of-
magnitude associated uncertainty. Both a comprehensive
model-evaluation effort and a substantially improved data
base will be required before this situation can be
remedied to any appreciable extent.

3.5 CONCLUSIONS TO SECTION 3

This section has provided an overview of meteorological
processes contributing to wet removal of pollutants and
has summarized the current state of our capability to
describe these complex phenomena in mathematical form.
Because of the magnitude of this problem it has been
necessary to refrain from detailed descriptions of models
and modeling techniques; rather, we have chosen to
describe the general mathematical basis for wet-removal
modeling, to give two simple examples of direct
application, and then to supply the reader with a means

---

*This issue is especially pertinent in view of the
contention, often voiced by some scientists within the
acid-precipitation effects community, that temporally
averaged results (averaging times of a few months or
more) are totally adequate for assessment purposes.

for efficiently pursuing the available literature for
specific applications of interest.

In conclusion to this discussion it is appropriate to
summarize the state of these calculational techniques by
asking the following questions:

- Just how accurate and valid are current
wet-removal modeling techniques as predictions of
precipitation chemistry and wet deposition; that is, how
well do they fulfill the needs itemized in Section 3.4.1?
- What must be accomplished before the present
capabilities can be improved?

The answers to these questions are somewhat mixed.
Certainly the techniques discussed in this section, if
used appropriately, are capable of order-of-magnitude
determinations in many circumstances; and under
restricted conditions they can even generate predictions
having factor-of-2 accuracy or better. Moreover, there
is ample explanation in existing theories of wet removal
to account easily for the spatial and temporal
variabilities observed in nature.

These capabilities, however, cannot be considered to
be satisfactory in the context of current needs. The
noted ability to _explain_ spatial and temporal variability
on a semiquantitative basis has not resulted in a large
competence in _predicting_ such variability in specific
instances. Moreover, we possess little competence in
identifying specific sources responsible for wet
deposition at a given receptor site. Finally, the
order-of-magnitude predictive capability noted above can
hardly be judged satisfactory for most assessment
purposes. In reviewing the discussions of this section
against the backdrop of these deficits, several research
needs become apparent. The most important of these are
itemized in the following paragraphs.

- Much more definitive information is needed with
regard to the scavenging efficiencies of submicrometer
aerosols, for both rain and snow. Especially important
in this regard is the effect of condensational growth of
such aerosols in below-cloud environments.
- We need to know much more about aqueous-phase
conversion processes, which are potentially important as
alternate mechanisms resulting in the presence of species
such as sulfate and nitrate in precipitation. Since
virtually nothing is known at present regarding the

chemical formation of such species in clouds and precipitation, there is a tendency to lump these effects with <u>physical</u> removal processes in most modeling efforts, expressing them in terms of pseudo scavenging coefficients or collection efficiencies. Such phenomena must be resolved in finer mechanistic detail than this before a satisfactory treatment is possible, and this requires a knowledge of chemical transformation processes that is much more advanced than existing at present.

• Much more extensive understanding of the competitive nucleation capability of aerosols in in-cloud environments is needed, especially for those substances that do not compete particularly well in the nucleation process. The influence of aerosol-particle composition--especially for "internally mixed aerosols"*--is particularly important in this regard.

• The identification of specific sources responsible for chemical deposition at a given receptor location requires that we possess a much more accomplished capability to describe long-range pollution transport. Progress in this area during recent years has been encouraging, but much more remains to be achieved before we have a proficiency that is really satisfactory for reliable source-receptor analysis.

• We still need to enhance our understanding of the detailed microphysical and dynamical processes that occur in storm systems. Besides providing required knowledge of basic physical phenomena, such research is important in providing valid parameterizations of wet removal for subsequent use in composite regional models.

As a final note, it is useful to reflect once again on the fact that scavenging modeling research--as treated in this section--has been in a rather continuous state of development over the past 30 years. While progress has been indeed significant during this period, a number of important and unsolved problems still exist. Accordingly, one must use this perspective in assessing our rate of advancement during future years. Reasonable progress in resolving the above items can be expected over the next decade; but the complexity of these problems demands that a serious and sustained effort be applied for this purpose.

---

*Those containing individual particles composed of a mixture of chemical species.

364

3.7 REFERENCES

Adamowitz, R.F. 1979. A model for the reversible washout of sulfur dioxide, ammonia, and carbon dioxide from a polluted atmosphere and the production of sulfate in raindrops. Atmos. Environ. 13:105-122.

Astarita, G., J. Wei, and G. Iorio. 1979. Theory of dispersion transformation and deposition of atmospheric pollution using modified Green's functions. Atmos. Environ. 13:239-246.

Baker, M.G., H. Harrison, J. Vinelli, and K.B. Erickson. 1969. Simple stochastic models for the sources and sinks of two aerosol types. Tellus 31:1-39.

Barrie, L.A. 1978. An improved model of reversible SO_2 washout by rain. Atmos. Environ. 12:402-412.

Barrie, L.A., and J. Kovalick. 1978. A wintertime investigation of the deposition of pollutants around an isolated power plant in northern Alberta. Atmospheric Environment Service, Environment Canada, REP ARQT-4-78.

Bass, A. 1980. Modeling long range transport and diffusion. In Proceedings Second Conference on Applied Air Pollution Meteorology. AMS/APCA, New Orleans.

Berry, E.X., and R.L. Reinhardt. 1974. An analysis of cloud drop growth by collection. Part IV. A new parameterization. J. Atmos. Sci. 31:2127-2135.

Bhumralkar, C.M., W.B. Johnson, R.H. Mancusco, R.H. Thuillier, and D.E. Wolf. 1980. Interregional exchanges of airborne sulfur pollution and deposition in eastern North America. In Proceedings Second Conference on Applied Air Pollution Meteorology. MS/APCA, New Orleans.

Bird, R.B., W.E. Stewart, and E.N. Lightfoot. 1960. Transport Phenomena. New York: John Wiley and Sons.

Bolin, B., and C. Persson. 1975. Regional dispersion and deposition of atmospheric pollutants with particular application to sulphur pollution over western Europe. Tellus 27:281-309.

Browning, K.A., M.E. Hardman, T.W. Harrold, and C.W. Pardoe. 1973. The structure of rainbands within a midlatitude cyclonic depression. Q. J. Roy. Meteorol. Soc. 99:215-231.

Burtsev, I.E., L.V. Burtsevva, and S.G. Malakhov. 1976. Washout characteristics of a 32P aerosol injected into a cloud. Atmospheric Scavenging of Radioisotopes. Symp. Proc. Palanga, USSR.

Cadle, R.D. 1965. Particle Size. New York: Reinhold
   Publishing Company. 390 pp.
Carmichael, G.R., and L.K. Peters. 1981. Application of
   the sulfur transport Eulerian model (STEM) to a SURE
   data set. 12th International Technical Meeting on Air
   Pollution. Modelling and Its Applications. NATO. Palo
   Alto, Calif.
Chamberlain, A.C. 1953. Aspects of travel and deposition
   of aerosols and vapor clouds. AERE Harwell Report
   R1261. London: HMSO.
Changnon, S.A. 1968. Precipitation scavenging of Lake
   Michigan Basin. Illinois State Water Survey Report,
   Bull. 52, Urbana, Ill.
Changnon, S.A., A. Auer, R. Brahm, J. Hales, and R.
   Semonin. 1981. METROMEX-A Review and Summary.
   Meteorological Monograph, Vol. 18. Boston, Mass.:
   American Meteorological Society.
Climatic Atlas of the United States. 1968. Washington,
   D.C.: U.S. Dept. of Commerce.
Court, A. 1966. Fog frequency in the United States. Geog.
   Rev. N.Y. 56:543-550.
Dana, M.T. 1970. Scavenging of soluble dye particles by
   rain. In Precipitation Scavenging 1970. R.J. Engelmann
   and W.G.N. Slinn, eds. AEC Symposium Series.
Dana, M.T., and D.W. Glover. 1975. Precipitation
   scavenging of power plant effluents: rainwater
   concentrations of sulfur and nitrogen compounds and
   evaluation of rain samples desorption of SO_2. PNL
   Annual Report to U.S. AEC, BNWL-1950.
Dana, M.T., and J.M. Hales. 1976. Statistical aspects of
   the washout of polydisperse aerosols. Atmos. Environ.
   10:45-50.
Dana, M.T., J.M. Hales, and M.A. Wolf. 1972. Natural
   precipitation washout of sulfur dioxide. Battelle-
   Northwest Report to EPA. BNW-389.
Dana, M.T., J.M. Hales, W.G.N. Slinn, and M.A. Wolf.
   1973. Natural precipitation washout of sulfur
   compounds from plumes. Battelle-Northwest Report to
   EPA. EPA-R3-73-047.
Dana, M.T., D.R. Drewes, D.W. Glover, and J.M. Hales.
   1976. Precipitation scavenging of fossil fuel
   effluents. Battelle-Northwest Report to EPA.
   EPA-600/4-76-031.
Dana, M.T., N.A. Wogman, and M.A. Wolf. 1978. Rain
   scavenging of tritiated water (HTO): a field
   experiment and theoretical considerations. Atmos.
   Environ. 12:1523-1529.

Dana, M.T., A.A.N. Patrinos, E.G. Chapman, and J.M. Thorp. 1982. Wintertime precipitation chemistry in North Georgia. In Proceedings ACS Symposium on Acid Rain, Las Vegas, Nev.

Davenport, H.M., and L.K. Peters. 1978. Field studies of atmospheric particulate concentration changes during precipitation. Atmos. Environ. 12:997-1008.

Davies, C.N. 1966. Aerosol Science. New York: Academic Press.

Dingle, A.N., and Y. Lee. 1973. An analysis of in-cloud scavenging. J. Appl. Meteorol. 12:1295-1302.

Dingle, A.N., D.F. Gatz, and J.W. Winchester. 1969. A pilot experiment using indium as tracer in a convective storm. J. Appl. Meteorol. 8:236-240.

Drewes, D.R., and J.M. Hales. 1982. SMICK: a scavenging model incorporating chemical kinetics. Atmos. Environ. 16:1717-1724.

Durham, J.L., J.H. Overton, and V.P. Aneja. 1981. Influence of gaseous nitric acid on sulfate production and acidity in rain. Atmos. Environ. 15:1059-1068.

Easter, R. C. 1982. The OSCAR Experiment. In Proceedings ACS Symposium on Acid Rain, Las Vegas, Nev.

Easter, R.C., and J.M. Hales. 1983a. Interpretations of the OSCAR data for reactive gas scavenging. Proceedings Fourth International Conference on Precipitation Scavenging, Dry Deposition and Resuspension, Santa Monica, Calif.

Easter, R.C., and J.M. Hales. 1983b. Mechanistic evaluation of precipitation-scavenging data using a one-dimensional reactive storm model. Battelle-Northwest Report to EPRI. EPRI RP-2022-1.

Eliassen, A. 1978. The OECD study of long-range transport of air pollutants. Atmos. Environ. 12:479-487.

Engelmann, R.J. 1965. Rain scavenging of zinc sulphide particles. J. Atmos. Sci. 22:719-724.

Engelmann, R.J. 1968. The calculation of precipitation scavenging. In Meteorology and Atomic Energy 1968. D. Slade, ed. U.S. AEC.

Engelmann, R.J. 1971. Scavenging prediction using ratios of air and precipitation. J. Appl. Meteorol. 10:493-497.

Engelmann, R.J., R.W. Perkins, D.I. Hagen, and W.A. Haller. 1966. Washout coefficients for selected gases and particles. U.S. AEC Report. BNWL-SA-657.

Enger, L., and U. Hogstrom. 1979. Dispersion and wet deposition of sulfur from a power-plant plume. Atmos. Environ. 13:789-810.

Falconer, R.E., and P.D. Falconer. 1980. Determination of cloud water acidity at a mountain observatory in the Adirondack Mountains of New York State. J. Geophys. Res. 85:7465-7470.

Fay, J.A., and J.J. Rosenzweig. 1980. An analytical diffusion model for long-distance transport of air pollutants. Atmos. Environ. 14:355-365.

Fisher, B.E.A. 1975. The long range transport of sulfur dioxide. Atmos. Environ. 9:1063-1070.

Fisher, B.E.A. 1978. The calculation of long term sulphur deposition in Europe. Atmos. Environ. 12:489-501.

Fitzgerald, J.W. 1974. Effect of aerosol composition of cloud-droplet size distribution: a numerical study. J. Atmos. Sci. 31:1358-1367.

Fuchs, N.A. 1964. The Mechanics of Aerosols. Oxford: Pergamon Press. 407 pp.

Fuquay, J.J. 1970. Scavenging in perspective. In Precipitation Scavenging 1970. R.J. Engelmann and W.G.N. Slinn, eds. AEC Symposium Series 22.

Galloway, J.N., and D.M. Whelpdale. 1980. An atmospheric sulfur budget for eastern North America. Atmos. Environ. 14:409-417.

Gatz, D.F. 1972. Washout ratios in urban and non-urban areas. In Proceedings AMS Conference on Urban Environment, Philadelphia, Pa.

Gatz, D.F. 1977. A review of chemical tracer experiments onprecipitation systems. Atmos. Environ. 11:945-953.

Godske, C.L., T. Bergeron, J. Bjerkness, and R.E. Bundgaard. 1957. Dynamic Meteorology and Weather Forecasting. Boston, Mass.: American Meteorological Society.

Graedel, T.E., and J.P. Franey. 1977. Field measurements of submicronaerosol washout by rain. In Precipitation Scavenging 1974. ERDA Symposium Series 41.

Granat, L., and H. Rodhe. 1973. A study of fallout by precipitation around an oil-fired power plant. Atmos. Environ. 7:781-792.

Granat, L., and R. Soderlund. 1975. Atmospheric deposition due to long and short distance sources with special reference to wet and dry deposition of sulphur compounds around an oil-fired power plant. MISU Report A-32, Stockholm University, Sweden.

Gravenhorst, G., T. Janssen-Schmidt, D.H. Ehhalt, and E.P. Roth. 1975. Long-range transport of airborne material and its removal by deposition and washout. Atmos. Environ. 9:49-68.

Greenfield, S.M. 1957. Rain scavenging of radioactive particulate matter from the atmosphere. J. Meteorol. 14:115-123.

Hales, J.M. 1972. Fundamentals of the theory of gas scavenging by rain. Atmos. Environ. 6:635-659.

Hales, J.M. 1977. An air pollution model incorporating nonlinear chemistry, variable trajectories, and plume-segment diffusion. Battelle-Northwest Report to EPA. EPA-450/3-77-012.

Hales, J.M. 1983. Precipitation chemistry: its behavior and its calculation. In Air Pollutants and Their Effects on the Terrestrial Ecosystem. S.V. Krupa and A.H. Legge, eds. New York: John Wiley and Sons.

Hales, J.M., and M.T. Dana. 1979a. Precipitation scavenging of urban pollutants by convective storm systems. J. Appl. Meteorol. 18:294-316.

Hales, J.M., and M.T. Dana. 1979b. Regional scale deposition of sulfur dioxide by precipitation scavenging. Atmos. Environ. 13:1121-1132.

Hales, J.M., J.M. Thorp, and M.A. Wolf. 1971. Field investigation of sulfur dioxide washout from the plume of a large coal-fired power plant by natural precipitation. Battelle-Northwest Final Report to Environmental Protection Agency. NTIS PB 203-129.

Hales, J.M., M.A. Wolf, and M.T. Dana. 1973. A linear model for predicting the washout of pollutant gases from industrial plumes. AICHE J. 19:292-297.

Hane, C.E. 1978. Scavenging of urban pollutants by thunderstorm rainfall: numerical experimentation. J. Appl. Meteorol. 17:699-710.

Haurwitz, B., and J.M. Austin. 1944. Climatology. New York: McGraw-Hill Book Company.

Heffter, J.L. 1980. Air resources laboratories atmospheric transport and dispersion model (ARL-ATAD). NOAA Tech. Memo. ERL-81.

Henmi, J. 1980. Long-range transport model of SO_2 and sulfate and its application to the eastern United States. J. Geophys. Res. 85:4436-4442.

Hill, F.B., and R.F. Adamowitz. 1977. A model for reversible washout of sulfur dioxide, ammonia, and carbon dioxide from a polluted atmosphere, and the production of sulfates in raindrops. Atmos. Environ. 11:912-927.

Hobbs, P.V. 1978. Organization and structure of clouds and precipitation on the mesoscale and microscale in cyclonic storms. Rev. Geophys. Space Sci. 16:741-755.

Hobbs, P.V. 1979. A reassessment of the mechanism responsible for the sulfur content of acid rain. In Proceedings of Advisory Workshop to Identify Research Needs on Formation of Acid Precipitation. EPRI Report, EA-10 74, WS-78-98.

Hogstrom, U. 1974. Wet fallout of sulfurous pollutants emitted from a city during rain or snow. Atmos. Environ. 8:1291-1303.

Hutchenson, M.R., and F.P. Hall. 1974. Sulfate washout from a coal-fired power plant plume. Atmos. Environ. 8:23-28.

Junge, C.E. 1963. Air Chemistry and Radioactivity. New York: Academic Press.

Junge, C.E. 1964. The modification of aerosol size distribution in the atmosphere. Final Tech. Report, Meteor. Geophys. Inst., Johannes Gutenberg Universitat. U.S. Army Contract DA-91-591-EVC2979.

Junge, C.E. 1974. Residence time and variability of tropospheric trace gases. Tellus 26:477-488.

Klein, W.H. 1958. The frequency of cyclones and anticyclones in relation to the mean circulation. J. Meteorology 15:98-102.

Kleinman, L.J., J.G. Carney, and R.E. Meyers. 1980. Time dependence on average regional sulfur oxide concentrations. Proc. Second Conf. on Applied Air Pollution Meteorology. AMS/APCA, New Orleans.

Klett, J. 1977. Precipitation scavenging in rainout assessment: the ACRA system and summaries of simulation results. LASL Report to ERDA, LA6763.

Kramer, J.R. 1973. Atmospheric composition and precipitation of the Sudbury Region. Alternatives 2:18-25.

Kreitzberg, C.W., and M.J. Leach. 1978. Diagnosis and prediction of troposhperic trajectories and cleansing. Proc. 85th National Meeting AIChE, Philadelphia, Pa.

Lamb, R.G. 1981. A regional scale model of photochemical air pollution. Draft Report, Meteorology and Assessment Division, EPA/ESRL, Research Triangle Park, N.C.

Lange, R., and J.B. Knox. 1977. Adaptation of a three-dimensional atmospheric transport-diffusion model to rainout assessments. In Precipitation Scavenging 1974. R.S. Semonin and R.W. Beadle, eds. ERDA Symposium Series 41, CONF 741003.

Larson, T.V., R.J. Charlson, E.J. Knudson, G.D. Shristian, and H. Harrison. 1975. The influence of a sulfur dioxide point source on the rain chemistry of a single storm in the Puget Sound region. Water Air Soil Pollut. 4:319-328.

Lavery, T.L., et al. 1980. Development and validation of a regional model to simulate atmospheric concentrations of SO_2 and sulfate. In Proceedings Second Joint Conference on Applied Air Pollution Meteorology, New Orleans, La.

Lee, H.N. 1981. An alternate pseudospectral model for pollutant transport. Diffusion and deposition in the atmosphere. Atmos. Environ. 15:1017-1024.

Levich, V.G. 1962. Physicochemical Hydrodynamics. Englewood Cliffs, N.J.: Prentice-Hall. 700 pp.

Liu, M.K., and D. Durran. 1977. The development of a regional air pollution model and its application to the northern Great Plains. EPA Report EPA-908/1-77-001.

Lovett, G.M., W.A. Reiners, and R.K. Olson. 1982. Cloud droplet deposition in subalpine balsam fir forests: hydrological and chemical inputs. Science 218:1303-1304.

MAP3S/RAINE. 1981. Biennial Progress Report. NTIS PNL-4096, U.S. EPA/DOE.

MAP3S/RAINE. 1982. The MAP3S/RAINE precipitation chemistry network: statistical overview for the periods 1976-1980. Atmos. Environ. 16:1603-1631.

Mason, B.J. 1971. The Physics of Clouds. Oxford: Clarendon Press, p. 579.

McNaughton, D., D. Powell, and C. Berkowitz. 1981. A User's Guide to RAPT. MAP3S/RAINE Report, PNL-3390.

Millan, M.M., S.C. Barton, N.D. Johnson, B. Weisman, M. Lusis, W. Chan, and R. Vet. 1982. Rain scavenging from tall stacks: a new experimental approach. Atmos. Environ. 16:2709-2714.

Molenkamp, C.R. 1974. A one-dimensional numerical model of precipitation scavenging with application to rainout of radioactive debris. Lawrence Livermore Laboratory Report to U.S. AEC. UCRL-51627.

Morgan, J.J., and H.M. Liljestrand. 1980. Measurements and interpretation of acid rainfall in the Los Angeles Basin. Cal Tech Final Report AC-2-80, Pasadena, Calif.

Mosiac. 1979. Acid from the sky. Mosiac (National Science Foundation) 10:35-40.

Newell, R.E., J.W. Kidson, D.G. Vincent, and G.J. Baer. 1972. The General Circulation of the Tropical Atmosphere. Vols. 1 and 2. Cambridge, Mass.: MIT Press.

Omstedt, G., and H. Rodhe. 1978. Transformation and removal processes for sulfur compounds as described by a one-dimensional time-dependent diffusion model. Atmos. Environ. 12:503-509.

OSCAR. 1981. Chapter 4 of MAP3S/RAINE Biennial Progress Report. EPA Report PNL-4096.

Overton, J.H., V.P. Aneja, and J.L. Durham. 1979. Production of sulfate in rain and raindrops in polluted atmospheres. Atmos. Environ. 13:355-367.

Patterson, D.E., R.B. Husar, W.E. Wilson, and L.F. Smith. 1981. Monte Carlo simulation of daily regional sulfur distribution. J. Appl. Meteorol. 20:404-420.

Prahm, L.V., and O. Christensen. 1977. Long-range transmission of pollutants simulated by a two-dimensional pseudospectral dispersion model. J. Appl. Meteorol. 16:896-910.

Pruppacher, H.R., and J.D. Klett. 1978. Microphysics of Clouds and Precipitation. Dordrecht, Netherlands: D. Reidel Publishing Company.

Radke, L.F., M.W. Eltgroth, and P.V. Hobbs. 1978. Precipitation scavenging of aerosol particles. In Proceedings Cloud Physics and Atmospheric Electricity. Boston, Mass.: American Meteorological Society.

Raynor, G.S. 1981. Design and preliminary results of the intermediate density precipitation-chemistry experiment. Report BNL 29992. For presentation at Third Joint AMS/APCA Conference on Applications of Air Pollution Meteorology, January. San Antonio, Tex.

Rodhe, H., and J. Grandell. 1972. On the removal time of aerosol particles from the atmosphere by precipitation scavenging. Tellus 24:442-454.

Rodhe, H., and J. Grandell. 1981. Estimates of characteristic times for precipitation scavenging. J. Atmos. Sci. 38:370-386.

Saffman, P.G., and J.S. Turner. 1955. On the collision of drops in turbulent clouds. J. Fluid Mech. 1:16-30.

Sampson, P.J. 1980. Trajectory analysis of summertime sulfate concentrations in the northeastern United States. J. Appl. Meteorol. 19:1382-1394.

Scott, B.C. 1978. Parameterization of sulfate removal by precipitation. J. Appl. Meteorol. 1375-1389.

Scott, B.C. 1981. Sulfate washout ratios in winter storms. J. Appl. Meteorol. 20:619-625.

Scott, B.C. 1982. Predictions of in-cloud conversion rates of SO_2 to SO_4 based upon a simple chemical and kinematic storm model. Atmos. Environ. 16:1735-1752.

Scott, B.C., and N.S. Laulainen. 1979. On the concentration of sulfate in precipitation. J. Appl. Meteorol. 18:138-147.

Scriven, R.A., and B.E.A. Fisher. 1975. The long range transport of airborne material and its removal by deposition and washout. Atmos. Environ. 9:49-68.

Semonin, R.G. 1976. The variability of pH in convective storms. In Proceedings First International Symposium on Acid Precipitation and the Forest Ecosystem. USDA Tech. Rept. NE-23, pp. 349-361.

Shannon, J. 1981. A regional model of long-term average sulfur atmospheric pollution, surface removal, and net horizontal flux. Atmos. Environ. 5:689-701.

Shopauskas, K., B. Styra, and E. Verba. 1969. Spreading and rainout of passive admixture injected into a cloud. In Seventh International Conference on Condensation and Ice Nuclei, Vienna, Austria.

Slinn, W.G.N. 1973a. Fluctuations in trace gas concentrations in the troposphere. J. Geophys. Res. 78:574-576.

Slinn, W.G.N. 1973b. In-cloud scavenging studies. Annual Report to US AEC/DBER. Battelle-Northwest, BNWL-1751 pt. 1.

Slinn, W.G.N. 1974a. Rate limiting aspects of in-cloud scavenging. J. Atmos. Sci. 31:1172-1173.

Slinn, W.G.N. 1974b. The redistribution of a gas plume caused by reversible washout. Atmos. Environ. 8:233-239.

Slinn, W.G.N. 1977. Some approximations for the wet and dry removal of particles and gases from the atmosphere. J. Water Air Soil Pollut. 7:513-543.

Slinn, W.G.N. 1983. Precipitation scavenging. In Atmospheric Sciences and Power Production. D. Randerson, ed. U.S. DOE.

Slinn, W.G.N., and J.M. Hales. 1971. A reevaluation of the role of thermophoresis as a mechanism of in- and below-cloud scavenging. J. Atmos. Sci. 28:1465-1471.

Slinn, W.G.N., and J.M. Hales. 1983. Wet removal of atmospheric particles. EPA Monograph Series.

Squires, P., and S. Twomey. 1960. The relation between cloud droplet spectra and the spectrum of cloud nuclei. In Physics of Precipitation. NAS/NRC Monograph No. 5. Washington, D.C.: American Geophysical Union.

Storebo, P.B., and A.N. Dingle. 1974. Removal of pollution by rain in a shallow air flow. J. Atm. Sci. 31:533-542.

Summers, P.W., and B. Hitchon. 1973. Source and budget of sulfate in precipitation from Central Alberta, Canada. JAPCA 23:194-199.

Thorp, J.M., and B.C. Scott. 1982. Preliminary calculations of average storm duration and seasonal precipitation rates for the northeast sector of the United States. Atmos. Environ. 16:1763-1774.

Voldner, E.C., K. Olson, K. Oikawa, and M. Loiselle. 1982. Comparison between measured and computed concentrations of sulfur compounds in eastern North America. J. Geophys. Res.

Waldman, J.M., J.W. Munger, D.J. Jacob, R.C. Flagan, J.J. Morgan, and M.R. Hoffman. 1982. Chemical composition of acid fog. Science 218:677-679.

Wang, P.K., and H.R. Pruppacher. 1977. An experimental determination of the efficiency which aerosol particles are collected by water drops in sub-saturated air. J. Atmos. Sci. 34:1664-1669.

Wangen, L.E., and M.D. Williams. 1978. Elemental deposition downwind of a coal-fired power plant. Water Air Soil Pollut. 10:33-44.

Wilkening, K.E., and K.W. Ragland. 1980. Users Guide for the University of Wisconsin Atmospheric Sulfur Computer Model (UWATM-SOX). Report to EPA/Duluth Research Laboratory.

Young, J.A., C.W. Thomas, and N.A. Wogman. 1973. The use of natural and man-made radionuclides to study in-cloud scavenging processes. PNL Annual Report for 1972 to U.S. AEC/DBER. BNWL-1751, pt. 1.

Young, J.A., T.M. Tanner, C.W. Thomas, and N.A. Wogman. 1976. The entrainment of tracers near the sides of convective clouds. Annual Report to ERDA/DBER. Battelle-Northwest, BNWL-2000, pt. 3.

Zishka, K.M., and P.J. Smith. 1980. The climatology of cyclones and anticyclones over North America and surrounding ocean environs for January and July, 1950-77. Mon. Weather Rev. 108:387-401.

Appendix **D** Biographical Sketches of Committee Members

JACK G. CALVERT has been a Senior Scientist in the Atmospheric Chemistry and Aeronomy Division of the National Center for Atmospheric Research since 1982. He received his Ph.D. in physical chemistry from U.C.L.A. in 1949 and served one year as National Research Fellow (Ottawa, Canada) in 1950. He joined the faculty of The Ohio State University at that time and was Kimberly Professor of Chemistry from 1974 to 1981. His major research interests are in photochemistry and tropospheric chemistry.

JAMES N. GALLOWAY is an Associate Professor in the Department of Environmental Sciences, University of Virginia, Charlottesville, Virginia. He received his Ph.D. in chemistry from the University of California, San Diego, in 1972. His research interests are in aquatic and atmospheric chemistry.

JEREMY M. HALES is Associate Department Manager for the Geosciences Research and Engineering Department of Battelle-Pacific Northwest Laboratories. He currently is Guest Researcher at the Meteorological Institute of Stockholm University. He received his Ph.D. in chemical engineering from the University of Michigan in 1968. His research interests are in simultaneous mass transfer and chemical reaction in polluted atmospheric environments, with special applications to reactive storm-model development.

GEORGE M. HIDY is Vice President and Chief Scientist for Environmental Research & Technology, Inc. He received his D.Eng. in chemical engineering from The Johns Hopkins University in 1962. Since then he has devoted his research primarily to aerosol science, with particular concern for atmospheric chemical processes. Dr. Hidy has pioneered in the study of regional-scale atmospheric chemistry of sulfates; he was one of the designers and principal investigators of the Sulfate Regional Experiment (SURE).

JAY S. JACOBSON is a plant physiologist in the Environmental Biology Program of the Boyce Thompson Institute in Ithaca, New York. He received his Ph.D. in botany from Columbia University in 1960. He has adjunct appointments in the Department of Natural Resources and Center for Environmental Research at Cornell University. His research interests are in plant physiology, agriculture, and analytical chemistry relating especially to the air pollutants hydrogen fluoride, ozone, sulfur dioxide, and acid rain.

ALLAN LAZRUS is a Senior Scientist and Project Leader of the Reactive Gases and Particles Project at the National Center for Atmospheric Research in Boulder, Colorado. He completed three years of graduate work toward the Ph.D. in organic chemistry at the University of Colorado. His research interests include oxidation processes occurring in clouds leading to the formation of sulfate and nitrate and measuring sulfur and halogen compounds in volcanic eruptions and trace chemistry of the stratosphere.

JOHN M. MILLER has been Coordinator of all NOAA activities in precipitation chemistry and acid rain since 1978. He received his Ph.D. in meteorology from Pennsylvania State University in 1972 and completed postdoctoral studies in atmospheric chemistry at the University of Frankfurt. He is a member of the Commission on Atmospheric Chemistry and Global Pollution (IUGG). His research interests are precipitation chemistry and the use of meteorological parameters to evaluate long-range transport of acidic materials.

VOLKER MOHNEN is Director of the Atmospheric Sciences Research Center and Research Professor at the State University of New York at Albany. He received his Ph.D. in physics, with a minor in astrophysics and meteorology, from the University of Munich, Germany, in 1966. He joined the State University of New York system in 1967. His major research interests are in aerosol physics and heterogeneous atmospheric chemistry. Currently he also serves on the National Research Council's Panel on Global Tropospheric Chemistry and is a member of the Advisory Committee for Atmospheric Sciences or the National Science Foundation.